T0332893

Gravity and the Behavior of Unicellular Organisms

Unicellular organisms use gravity as an environmental guide to reach and stay in regions optimal for their growth and reproduction. These single cells play a significant role in food webs, and these factors together make the effects of gravity on unicellular organisms a fascinating and important subject for scientific study. In addition, they present valuable model systems for studying the mechanisms of gravity perception – a topic of increasing interest in these days of experimentation in space. This book reveals how single cells achieve the same sensoric capacity as multicellular organisms, such as plants or animals. It reviews the field, discussing the historical background, ecological significance, and related physiology of unicellular organisms, as well as various experimental techniques and models with which to study them. Those working on the biology of unicellular organisms – as well as in related areas of gravitational and space science – will find this book of value.

Professor Donat-Peter Häder, PhD, holds the Chair in Botany at the Friedrich-Alexander Universität, in Erlangen, Germany. He has worked in gravitational biology and space research for more than 20 years and has been involved in numerous space shuttle, sounding rocket, satellite, and parabolic flight experiments. He is the author and editor of more than a dozen books and has published more than 480 papers in scientific journals.

Ruth Hemmersbach, PhD, is a zoologist and cell biologist in the German Aerospace Center (DLR) and the Rheinische Friedrich–Wilhelms Universität in Bonn, Germany and she has been active in gravity-related research for more than 20 years. She has been principal investigator for several biological experiments under varied gravitational stimulation in microgravity (parabolic flights and space shuttle), functional weightlessness (clinostats), and hypergravity (centrifuges).

Michael Lebert, PhD, is a botanist in the Friedrich–Alexander Universität in Erlangen, Germany, who has been active in gravity-related research and computer science for 10 years. He has been involved as co-investigator in numerous biological experiments in microgravity on airplanes and sounding rockets.

Developmental and Cell Biology Series

SERIES EDITORS

The aim of the series is to present relatively short critical accounts of areas of developmental and cell biology, where sufficient information has accumulated to allow a considered distillation of the subject. The fine structure of cells, embryology, morphology, physiology, genetics, biochemistry, and biophysics are subjects within the scope of the series. The books are intended to interest and instruct advanced undergraduates and graduate students, and to make an important contribution to teaching developmental and cell biology. At the same time, they should be of value to biologists who, while not working directly in the area of a particular volume's subject matter, wish to keep abreast of developments relevant to their particular interests.

RECENT BOOKS IN THE SERIES

18. C. J. Epstein *The consequences of chromosome imbalance: principles, mechanisms and models* 0521 25464 7
19. L. Saxén *Organogenesis of the kidney* 0521 30152 1
20. V. Raghavan *Developmental biology of the fern gametophytes* 0521 33022 X
21. R. Maksymowych *Analysis of growth and development in Xanthium* 0521 35327 0
22. B. John *Meiosis* 0521 35053 0
23. J. Bard *Morphogenesis: the cellular and molecular processes of developmental anatomy* 0521 43612 5
24. R. Wall *This side up: spatial determination in the early development of animals* 0521 36115 X
25. T. Sachs *Pattern formation in plant tissues* 0521 24865 5
26. J. M. W. Slack *From egg to embryo: regional specification in early development* 0521 40943 8
27. A. L. Farbman *Cell biology olfaction* 0521 36438 8
28. L. G. Harrison *Kinetic theory of living pattern* 0521 30691 4
29. N. Satoh *Developmental biology of Ascidians* 0521 35221 5
30. R. Holliday *Understanding ageing* 0521 47802 2
31. P. Tsonis *Limb regeneration* 0521 44149 8
32. R. Rappaport *Cytokinesis in animal cells* 0521 40173 9
33. D. L. Kirk *Volvox: molecular genetic origins of multicellularity and cellular differentiation* 0521 45207 4
34. R. L. Lyndon *The shoot apical meristem: its growth and development* 0521 40457 6
35. D. Moore *Fungal morphogenesis* 0521 55295 8
36. N. Le Douarin & C. Kalcheim *The neural crest*, second edition 0521 62010 4
37. P. R. Gordon-Weeks *Neuronal growth cones* 0521 44491 8
38. R. H. Kessin *Dictyostelium: evolution, cell biology, and the development of multicellularity* 0521 58364 0
39. L. I. Held *Imaginal discs: the genetic and cellular logic of pattern formation* 0521 58445 0
40. D.-P. Häder, R. Hemmersbach & M. Lebert *Gravity and the behavior of unicellular organisms* 0521 82052 9

Gravity and the Behavior of Unicellular Organisms

DONAT-PETER HÄDER
Friedrich-Alexander Universität, Erlangen, Germany

RUTH HEMMERSBACH
German Aerospace Center (DLR), Köln, Germany
Rheinische Friedrich-Wilhelms Universität, Bonn, Germany

MICHAEL LEBERT
Friedrich-Alexander Universität, Erlangen, Germany

Shaftesbury Road, Cambridge CB2 8EA, United Kingdom

One Liberty Plaza, 20th Floor, New York, NY 10006, USA

477 Williamstown Road, Port Melbourne, VIC 3207, Australia

314–321, 3rd Floor, Plot 3, Splendor Forum, Jasola District Centre, New Delhi – 110025, India

103 Penang Road, #05–06/07, Visioncrest Commercial, Singapore 238467

Cambridge University Press is part of Cambridge University Press & Assessment, a department of the University of Cambridge.

We share the University's mission to contribute to society through the pursuit of education, learning and research at the highest international levels of excellence.

www.cambridge.org
Information on this title: www.cambridge.org/9780521820523

First published 2005

A catalogue record for this publication is available from the British Library

Library of Congress Cataloging-in-Publication data
Häder, Donat-Peter
Gravity and the behavior of unicellular organisms / Donat-Peter Häder, Ruth Hemmersbach, Michael Lebert
p. cm. – (Developmental and cell biology series; 40)
Includes bibliographical references (p.) and index.
ISBN 0-521-82052-9 (alk. paper)
1. Gravity – Physiological effect. 2. Unicellular organisms.
I. Hemmersbach, Ruth, 1958– II. Lebert, Michael, 1959– III. Title. IV. Series

QH657.H33 2004
571.4′3529 – dc22 2004055079

ISBN 978-0-521-82052-3 Hardback

Contents

List of Abbreviations

AAEU	aquatic animal experiment unit
A/D	analog to digital
AGC	automatic gain control
AM	acetoxy methyl ester
AOTF	acousto-optical tunable filters
BAPTA	1,2-bis(*o*-aminophenoxy)ethane-*N*,*N*,*N*′,*N*′-tetraacetic acid
CCD	charge-coupled device
CCIR	Commission Consultative Internationale de Radiodiffusion (video format)
CEBAS	Closed Equilibrated Biological Aquatic System
CSK	cytoskeleton network
DHP	dihydropyridine
DLR	Deutsches Zentrum für Luft- und Raumfahrt (German Aerospace Center)
EGF	epidermal growth factor
EGTA	ethyleneglycol-bis(β-aminoethyl ether)-*N*,*N*,*N*′,*N*′-tetraacetic acid
ESA	European Space Agency
FC	flagellar current
FFM	free-fall machine
FLM	fluorescence lifetime measurement
IBMX	3-isobutyl-1-methylxanthine
ISS	International Space Station
JAMIC	Japan Microgravity Center
LED	light-emitting diode
LUT	look-up table
MAXUS	"Super" TEXUS
MASER	Materials Science Experiment Rocket

MELiSSA	Microecological Life Support System Alternative
MIR	Russian space station
MscL	mechanosensitive channel large
MTR	microtubular rootlet
NIZEMI	Niedergeschwindigkeits-Zentrifugenmikroskop (slow rotating centrifuge microscope)
NP-EGTA	nitrophenyl-ethyleneglycol-bis(ß-aminoethyl ether)-*N,N,N′,N′*-tetraacetic acid
PAB	paraxonemal body
PAC	photoactivated adenylyl cyclase
PAR	paraxonemal rod
PC	photoreceptor current
PCR	polymerase chain reaction
PFB	paraflagellar body
PKC	protein kinase C
PYP	photoactive yellow protein
SAC	stretch-activated channel
STATEX	Statolithen-Experiment (statolith experiment)
TEXUS	Technologische Experimente unter Schwerelosigkeit (technological experiments under microgravity)
TPMP	triphenyl methyl phosphonium
2D	two-dimensional
3D	three-dimensional
UV	ultraviolet radiation
VCR	videocassette recorder
ZARM	Zentrum für Angewandte Raumfahrttechnologie und Mikrogravitation (Center of Applied Space Technology and Microgravity)

Preface

There comes a point in the career of a scientist when he or she should write a book about his or her subject of interest. Two of us always wondered when and how this was going to happen. Now we know: by pure accident. And, here is one word of advice: You are often warned not to get involved in the book business. Please consider those who are warning you as your best friends; they know what they are talking about. However, one day, we received an e-mail (actually much longer ago than we would have anticipated) asking whether we would be willing to write a book about the effects of gravity on single cells. One of us knew what that meant; he warned us, but we agreed anyway. Finally, all three of us completed the project, and we learned a lot in the process. So, thank you, Peter Barlow and Cambridge University Press for keeping your faith in us.

Those who teach about gravity effects on living systems, including single cells, quickly realize that this weak force seems to have escaped human attention. Although we all had strong fights with gravity, especially during the early phase of our lives, it seems that afterward, we have almost completely forgotten about it. However, for all living organisms in our world, it is the one parameter most steadily encountered. Gravity is so basic for all of us that it is almost hardwired into our interpretation of reality. Gravity is not only related to living organisms; convection and the weather are two other subjects that come to mind when thinking about gravity.

For more than 100 years, scientists have been fascinated to observe the effects of gravity on single, free-swimming cells. The reason is that these little cells have the same capability as humans to tell up from down, but they do it in a single cell. And, even though it may seem to be an eccentric subject to study, this swimming behavior bears a much closer relation to daily life than one might expect. First, it becomes more and more clear that, in terms of biochemistry, single cells detect gravity in much the same way as do higher, more organized, multicellular organisms – and that is one of the things we want to show in this book. In addition, single cells are heavily involved in assembly and disassembly (either as consumer or as producer) of organic matter, and by this means are essential for food webs. Finally, photosynthetic cells are important oxygen sources and carbon dioxide sinks – topics coming more strongly to public attention in these times of global warming and climate change.

Lastly, we would like to thank all the people who supported us, including our families, for bearing with us during the process of writing. We would also like to thank Peter Barlow for bringing up the idea of this book. Critical discussions were the source of many new fruitful insights – thanks to I. Block, M. Braun, R. Bräucker, E. Brinckmann, K. Slenzka, and D. Volkmann. Thanks are due to U. Trenz and M. Schuster for helping to prepare the manuscript, M. Häder for the drawing of *Euglena*, E. Ariskina and M. Vainshtein for supplying the image of magnetotactic bacteria, D. Volkmann for supplying the *Lepidium* images, M. Braun for supplying the *Chara* electron microscopic images, I. Block for supplying diagrams of *Physarum*, A. Schatz for supplying the scheme of the clinostat principle, K. Slenzka for supplying the CEBAS diagram, W. Engler for producing the TEXUS image, and W. Foissner for supplying the scanning electron micrograph of *Paramecium*. Finally, we thank the national and international agencies for financial support of the research: German Space Agency (DLR), European Space Agency (ESA), National Agency of Space and Aeronautics (NASA), and the German Ministry of Research and Technology (BMBF).

Donat-Peter Häder — *Spring 2003*
Ruth Hemmersbach
Michael Lebert

1

Introduction

1.1 Historical background

The phenomenon that some free-swimming unicellular organisms tend to swim to the top of a tube and gather there – independent of whether the tube is open or closed – has been observed more than 100 years ago. This behavior was termed **geotaxis** (orientation with respect to the gravity vector of the Earth) – negative geotaxis if the organisms orient upward and positive geotaxis if they swim downward (cf. Section 1.2). Nowadays, this term has been replaced by **gravitaxis**. Many early and detailed studies between 1880–1920 provided descriptive observations limited by optical and analytical means. This led to the establishment of various hypotheses that have been reviewed by different authors (Bean, 1984; Davenport, 1908; Dryl, 1974; Haupt, 1962b; Hemmersbach et al., 1999b; Jennings, 1906; Kuznicki, 1968; Machemer & Bräucker, 1992).

The results were rather conflicting and led to controversial interpretations. While Stahl (1880) stated that *Euglena* and *Chlamydomonas* do not orient with respect to gravity, Schwarz (1884) concluded from his observations that *Euglena* moves upward by an active orientational movement and is not passively driven (e.g., by currents in the water or attracted by oxygen at the surface). He found that the force of gravity could be replaced by centrifugal force and that *Euglena* could move upward against forces of up to $8.5 \times g$. The author also concluded that *Euglena* belongs to the negative geotactic organisms. Aderhold (1888) stressed that positive aerotaxis is the major reason for the upward movement of *Euglena*. The interaction of light and gravity on the orientation of *Euglena* was investigated by Wager (1911): "If light is strong, gravity may play little part in controlling the movements; if the light is weak or absent, gravity appears to be the sole determining factor." He determined the mean specific weight of a *Euglena* cell with 1.016 g cm^{-3}. Massart (1891) investigated different species (bacteria and ciliates) and found that they exhibited gravitaxis; the direction of gravitaxis appeared species-specific and, in the case of the flagellate *Chromulina*, temperature-dependent.

Gerhardt (1913) investigated the gravitactic behavior of *Closterium* (Desmids). Distribution and movement direction of these algae on a glass plate within a closed and completely water-filled tube revealed that, after one night in darkness, most of the organisms had moved upward on vertical (slightly meandering) paths, noted by the colored slime threads left behind. To prove whether crystals within the cells were involved in the perception of gravity, he proposed separating these crystals from the cells. Although this was not possible mechanically, he proposed cultivating them in a calcium-free medium, which should have at least reduced the number of crystals. Gerhardt also described the ecological suitability of gravitaxis, because it helps an organism to orient in darkness (e.g., after being buried in the mud). There are methods now available in gravitational biology, such as destruction of the gravisensor by means of a laser beam [e.g., in the case of the ciliate *Loxodes* (cf. Chapters 2 and 4)] or reducing the number of statoliths by, for example, cultivation of *Chara* rhizoids in an artificial barium-free medium (Kiss, 1994).

Moore (1903) stated that the sign of geotaxis in *Paramecium* depended on feeding status and temperature. Sufficient food and elevated temperature favor negative gravitaxis, whereas the cells show positive gravitaxis under "unfavorable conditions, such as lack of food or ice formation on the surface of the water column."

Although gravitaxis is rather easy to observe, an additional gravity-induced response of swimming microorganisms remained undiscovered for a long time. Dembowski (1929b) stated no difference between upward and downward swimming velocities of *Paramecium* cells. This phenomenon could be analyzed in detail by means of computer-controlled image analysis and automatic measurement of high numbers of cell tracks (cf. Chapter 3). This led to the identification of the phenomenon **gravikinesis** – a direction-dependent kinetic response to compensate at least part of the cell's sedimentation rate (cf. Section 1.2.1). The models for gravity perception are discussed in later chapters.

Possible mechanisms leading to gravitaxis have been discussed since the end of the nineteenth century. Different hypotheses were established either proposing pure physical or physiological ones (for reviews, see Barlow, 1995; Bean, 1984; Hemmersbach et al., 1999b; Machemer & Bräucker, 1992; Roberts, 1970). A physical mechanism assumes a passive alignment of the cell (e.g., caused by the heavier posterior cell end). In contrast, the existence of a physiological receptor is postulated for the detection of deviations from the gravity vector, thus initiating a sequence of sensory transduction events that finally result in an active course correction [e.g., controlling the ciliary/flagellary beat pattern (cf. Chapter 9)]. There is no fundamental morphological difference between cilia and flagella (cf. Section 4.1.1), but they are semantically differentiated between the two taxonomic groups.

1.2 Definitions

Microorganisms, as well as multicellular plants and animals, respond to environmental stimuli by a multitude of responses. There have been several attempts

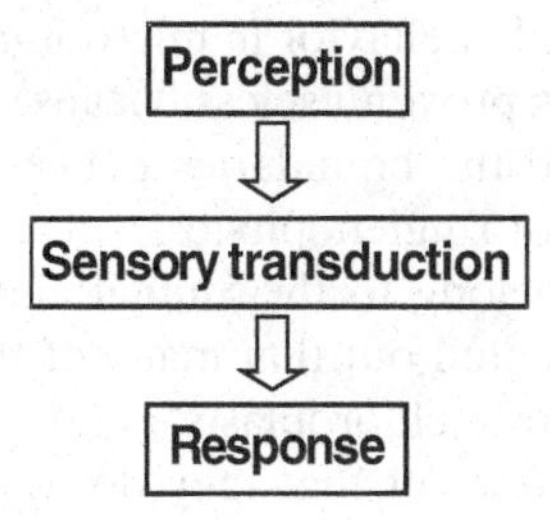

Figure 1.1. Signal transduction chain in living organisms that respond to environmental stimuli (for details, see text).

to develop a unified system of definitions for these responses. Unfortunately, there are two main reasons that have, up to now, prevented generally accepted definitions. One is due to our incomplete knowledge and understanding of the underlying mechanisms. In these cases, we need to wait for a detailed analysis of the receptor and the signal transduction chain resulting in the ultimate motor response to classify the reaction correctly. The second problem stems from the nonuniform usage of terms by different groups and researchers, and noncompliance with definitions already established in the literature.

A true behavioral response of an organism to an environmental stimulus requires the presence of several components. First, the organism needs to possess an appropriate receptor for the physical or chemical nature of the stimulus. This is followed by a signal transduction chain that is often regarded as a "black box," since the individual elements are often not known (Fig. 1.1). Usually, there is a transformation of the energy of the stimulus (e.g., light, gravitational field, or chemical gradient) into a different form of energy (conformation change of a molecule, electric gradient, biochemical reaction) often accompanied by an amplification of the signal. The internal signal is relayed to the actuator (e.g., flagellum or cilium), which may be located in the same cell or in a different one in multicellular organisms, thus generating the visible response. For instance, the gravireceptor for gravitropism in higher plant roots is located in statocytes (statoliths containing cells) organized in a tissue called **statenchym** located in the root columella, while the growth response occurs in the elongation zone of the root several millimeters above the root tip. Amplification can be achieved by gating ion fluxes across a membrane (Nowycky & Thomas, 2002) or a cascade of enzymatic reactions as in vertebrate and invertebrate vision (Müller & Kaupp, 1998; Nagy & Stieve, 1995). This definition of the responses to environmental stimuli excludes passive orientation of organisms by external forces, such as magnetic or electric field lines or acceleration, etc. For instance, magnetic orientation of bacteria is not a taxis in the strict sense, because the cells are passively aligned by the Earth's magnetic field lines due to the presence of small magnetic particles within the cells.

We will attempt to summarize the definitions for responses of organisms to external stimuli, mainly responses to gravity, accepted by most researchers. We will also describe aberrant usage by some groups. We will divide the discussion into one part on motile microorganisms and a second part on higher plants. We will consider multicellular animals only in passing.

For the description of motile behavior in microorganisms, the distinction between plants and animals has proven useless because the behavior can be classified for both kingdoms and the boundaries between them are flexible. One typical example is the taxon of Euglenophytes, which has been claimed by both zoologists and botanists to belong to the animal and plant kingdoms, respectively. The botanists have pointed out that many of the members in this group are photosynthetic – having true chloroplasts – and consequently are plants. In contrast, the zoologists pointed out that they do not have a typical plant cell wall and possess cellular features often found in animals. In this group, there are a number of heterotrophic organisms that lack chloroplasts. Even the same species can be photosynthetic or lose its chloroplasts and live heterotrophically. Based on these considerations, the term "protists," first used by Verworn, may be appropriate to describe these microorganisms (Verworn, 1889b). We will use the term microorganisms in a loose sense for small unicellular and multicellular organisms, including prokaryotes and eukaryotes (i.e., bacteria, cyanobacteria, protists, small algae, and fungi). This usage makes it difficult to draw a line between a larger multicellular microorganism and a small animal or lower plant. But, this distinction may be irrelevant for the current discussion.

1.2.1 Responses of motile microorganisms to environmental stimuli

Motile microorganisms can be powered by different mechanisms, including swimming by flagella or cilia (Melkonian, 1992) or crawling along surfaces by means of membranella or cirri; they can glide, which is described as sliding, twitching, bending, jerking, or ameboid movement (Häder & Hoiczyk, 1992). They can orient the direction of their movement with respect to the source of an external stimulus. This behavior is called **taxis**. The direction of movement can be toward the source of the stimulus (**positive taxis**) or away from it (**negative taxis**). It can also be at an angle with respect to the stimulus direction (**transverse** or **diataxis**), resulting in a bimodal splitting of a population of microorganisms. In any case, the orientation is with respect to a vectorial stimulus, and the cells respond continuously by eliciting course corrections if they are offset from the chosen direction by mechanical forces or random deviations.

The organisms can respond to a multitude of environmental stimuli that is indicated as a prefix to the appropriate term. The directional response to light is called **phototaxis** (cf. Section 7.2). Cells move toward the light source (*positive phototaxis*), away from it (*negative phototaxis*), or at a certain angle (*dia-* or *transverse phototaxis*). The response to a source of a chemical is called **chemotaxis** (cf. Section 7.3). This can be an attractant such as a source of food (e.g., sugar) or a sexual pheromone or a repellent (e.g., phenol). One terminological problem is that the organism cannot perceive the location of the source of the chemical. While a microoganism can detect the location of a light source and move toward it, there is no way of knowing where the source of the chemical is. Therefore, the organisms are limited to a random walk that is biased by a chemical gradient. An organism will move in a randomly chosen direction and continue on this path if it detects an increase over time in the concentration of the (attractant) chemical. It will undergo either organized course corrections or random reorientations if it is going in the wrong direction and senses a decrease in the attractant concentration (cf. Section 7.3). The opposite behavior is observed for repellents. The term **chemotaxis** can be specified by the chemical the organism is responding to, such as in the case of oxytaxis or aerotaxis when the organism orients its swimming path with respect to an oxygen gradient within the water column.

Likewise, the response to a thermal gradient is called **thermotaxis** (cf. Section 7.4), which can be positive (toward the warmer side of the gradient) or negative (toward the cooler side). Some organisms follow the direction of electrical field lines (**galvanotaxis**), even though this behavior may be not a natural phenomenon and induced by a secondary effect (cf. Section 7.6). Bacteria, but also many eukaryotic organisms, have been found to orient with respect to the magnetic field lines of the Earth (**magnetotaxis**). Although it is obvious that magnetotactic orientation is advantageous for higher animals – such as migratory birds – this behavior is somewhat surprising for motile microorganisms: Why would a small cell with a limited swimming range navigate due North or South. The ecological advantage may not be found in the directional movement with respect to the magnetic poles of the Earth, but rather in a mechanism to adjust the vertical position in the water column (cf. Sections 6.4 and 7.5).

The directional orientation with respect to a gravitational field is called **gravitaxis**. In the earlier literature, it was termed **geotaxis**. However, since it is not a specific response of organisms on Earth (greek: *geia*), but can be elicited by any kind of acceleration (e.g., on other celestial bodies or by centrifugation), this term was replaced by the wider term gravitaxis. There is a long and still ongoing discussion among researchers whether the orientation of a given microorganism is due to an active physiological response, including active reorientation (which would be regarded as a true gravitaxis) or by a passive alignment of the cell in the water column, because it might be tail heavy and passively adjusted by the gravitational field (cf. Section 1.1 and Chapter 9). Further details for various organism groups are given in later chapters.

A taxis is a vectorial, long-term response to a given stimulus direction; therefore, the definition of a minimal exposure time is difficult. Usually, an organism orients itself to the stimulus as long as this is present. In contrast, a **phobic response** is a transient reaction to a change in the stimulus strength independent of the direction of the stimulus source (Diehn et al., 1977). This can be best exemplified by the photophobic response of motile microorganisms. When a swimming microorganism experiences a sudden change in light intensity, it responds with a transient behavioral program which, depending on the species, can be a sudden stop, a change in direction, a reversal of movement, or a more complicated pattern of responses. The direction of the response is independent of the stimulus direction; it is irrelevant whether the actinic (stimulating) light beam impinges from above or below or from the side. Thus, the eliciting stimulus can be described as dI/dt, indicating that the change in light intensity has to occur fast. The cells will not respond to slow changes (e.g., during sunrise or sunset). The duration of the response is also a built-in property of the behavioral response of the organism (Doughty, 1991). After the transient reaction, the organism resumes its normal swimming behavior. The photophobic response can be elicited by a sudden increase in the light intensity (step-up photophobic response) or a sudden decrease (step-down photophobic reaction). Both reactions can be observed in the same organism, but at different absolute irradiances (Doughty, 1993; Doughty & Diehn, 1984). Similar phobic responses can be induced by changes in other stimulus qualities and are consequently defined, for example, as chemophobic, mechanophobic, thermophobic, etc., reaction. Graviphobic reactions are not easily observed, because the gravitational field on the Earth is rather constant, but sudden acceleration or deceleration may induce graviphobic responses in motile microorganisms (Machemer, 1998).

The third type of motile responses in microorganisms is called **kinesis**. According to the classical definition (Diehn et al., 1977), this reaction is a steady-state dependence of the movement velocity on the stimulus intensity. This response is strictly independent of the stimulus direction. **Photokinesis** describes the dependence of the swimming speed on the light intensity, compared with that in the dark control. If a cell swims faster at a certain irradiance than in the dark, the behavior is described as positive photokinesis; if it moves slower, this is defined as negative photokinesis. The organism may become motionless in darkness (termed **Dunkelstarre** by the earlier German authors) or immotile in bright light (**Lichtstarre**). Kinesis has also been found to be induced by other stimuli qualities (Dinallo et al., 1982; van Houten et al., 1982; Zhulin & Armitage, 1993).

In some ciliates, a kinetic behavior is found to be induced by gravity, which has been defined as **gravikinesis**. However, unfortunately, this term does not comply with the accepted definition of the term kinesis, since it is not independent of the direction of the stimulus (in this case, the direction of the gravitational field). In a flat, horizontal cuvette, microorganisms are forced to swim horizontally. Without any other vectorial stimulation, the cells will move at the same velocity V_1 (with some statistical deviation), irrespective of their direction (Fig. 1.2a). In a vertical cuvette, the cells will simultaneously swim and sediment, since they usually have a higher intracellular density than the surrounding medium (Fig. 1.2b). The force F that acts on the cells is

$$F = \frac{4}{3}\pi a^3 \Delta\rho g, \tag{1.1}$$

where a = radius of the object (assumed to be spherical), $\Delta\rho$ = the difference in specific density between object and surrounding medium, and g = gravitational acceleration. The resulting sedimentation velocity is (Stokes law)

$$V_s = \frac{2a^2 \Delta\rho}{9e}. \tag{1.2}$$

In upward swimming cells, the sedimentation velocity V_s will vectorially subtract from the locomotion speed V_1, and in downward swimming cells, add to the individual swimming speed. Horizontally swimming cells should not be affected, except for a downward sedimentation that would slightly bend the horizontal

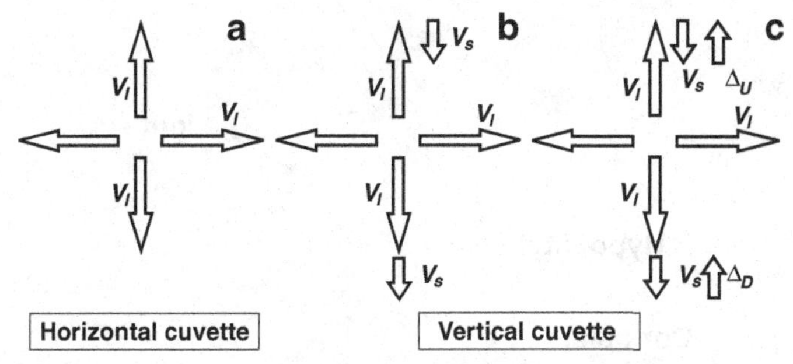

Figure 1.2. Diagram to explain the observed gravikinesis in certain ciliates. In a horizontal cuvette, all cells swim with about the same velocity (**a**). In cells without gravikinesis, the sedimentation velocity adds vectorially to the upward or downward swimming velocity (**b**). In contrast, in cells with gravikinesis, the effect of sedimentation is partially, completely, or overcompensated by an increased upward swimming velocity and a lowered downward swimming velocity (**c**).

path downward (Nagel & Machemer, 2000b). This predicted behavior is found in the flagellate *Euglena* (Häder, 1996b; Vogel et al., 1993).

In certain ciliates – such as *Paramecium, Didinium, Tetrahymena*, and *Loxodes* – computer-aided image analysis and statistical evaluation of a large number of cells indicated a different result (Hemmersbach-Krause et al., 1993b; Machemer et al., 1991; Ooya et al., 1992). In upward swimming cells, sedimentation is compensated by an increased swimming velocity (kinesis); and, in downward swimming cells, a slower swimming velocity is observed (Fig. 1.2c; Machemer et al., 1991). The downward velocity V_D modified by the gravikinetic component Δ_D is

$$V_D = V_1 + V_S - \Delta_D \tag{1.3}$$

and the upward velocity V_U modified by the gravikinetic component Δ_U is

$$V_U = V_1 - V_S + \Delta_U, \tag{1.4}$$

where V_S = sedimentation velocity; Δ_U, Δ_D = gravikinetic components during upward and downward swimming, respectively; and V_1 = swimming velocity independent of the gravitational influence. The gravikinetic component Δ is

$$\Delta = (V_D - V_U)/2 - V_S. \tag{1.5}$$

Compensation of sedimentation can be either complete ($V_S = \Delta$; e.g., *Loxodes*), partial ($V_S > \Delta$; e.g., *Paramecium* and *Didinium*), or even overcompensated ($V_s < \Delta$; e.g., *Tetrahymena*). The same species may even have different compensation rates, depending on the age of the culture (Bräucker et al., 1994). The gravikinetic effect, however, is independent of the swimming velocity, because it is the same in slow and fast swimmers (Machemer & Machemer-Röhnisch, 1996).

It has been speculated that gravikinesis is due to the pressure of the cytoplasm onto the respective lower membrane, which activates mechanosensitive channels. In the rear end, there are hyperpolarizing channels in the membrane and in the front end depolarizing ones. Hyperpolarization of the cell results in an increased

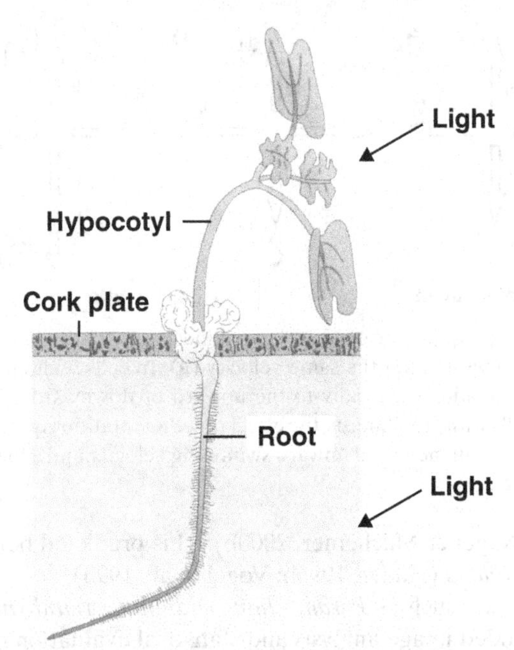

Figure 1.3. Positive and negative phototropism in the shoot and root, respectively, of a mustard seedling (*Sinapis alba*). The leaves are oriented perpendicular to the impinging light beam (diaphototropism) (modified from Nultsch, 2001).

swimming velocity and depolarization in a decreased one (Baba et al., 1991; Machemer et al., 1991; cf. Chapters 4 and 9). Fraenkel and Gunn (1961; Gunn et al., 1934) made a distinction between orthokinesis (which they defined as a change in linear velocity) and klinokinesis (which is an alteration in the frequency of directional changes or angular velocity). These definitions have been used by some zoologists.

1.2.2 Behavioral responses of sessile plants to environmental stimuli

Higher and lower sessile plants also show movement responses to environmental stimuli. Like phototaxis in motile microorganisms, phototropism is a steady-state bending of an organ of a sessile plant with respect to the direction of the light source. Shoots of higher plants bend toward the light source (positive phototropism), while roots either bend away from it (negative phototropism) or are indifferent toward light (Fig. 1.3). Lateral branches and leaves usually orient themselves perpendicular or at a different angle with respect to the impinging light beam (dia- or transversal phototropism). In some plants, leaves, flowers, or shoots follow the course of the sun over the day, which is called sun tracking or heliotropism (Koller, 2001). Also, organs in lower plants bend with respect to the light direction [e.g., the sporangiophores (vertical aerial hyphae that carry the sporangium) in the mold *Phycomyces* (Galland, 2001)].

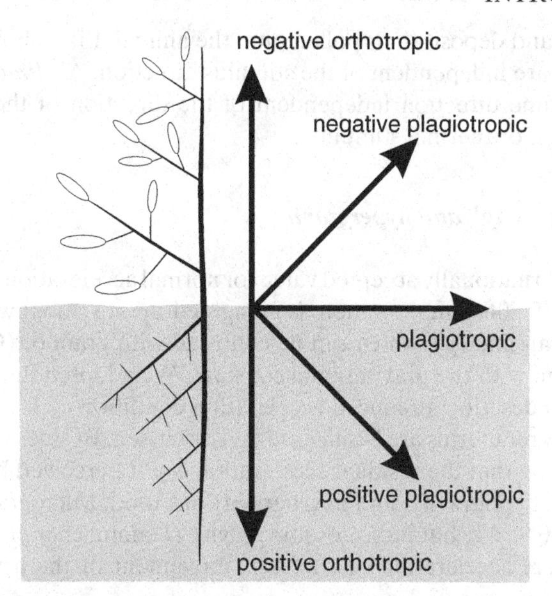

Figure 1.4. Positive and negative gravitropism of primary roots and shoots, respectively, of higher plants. Secondary roots and branches orient themselves at an angle (plagiotropic) to the gravitational force, and tertiary roots and shoots, as well as leaves, are often agravitropic (modified after Lüttge et al., 1994).

Without light stimulation, organs of higher plants respond to gravitactic stimulation. Primary shoots show negative (orthotropic) **gravitropism** and grow upward, primary roots positive gravitropism and consequently grow downward. Lateral branches and roots – as well as leaves – some flowers and fruits orient more or less perpendicular (plagiotropic) to the gravitational field (Fig. 1.4). The so-called setpoint angle (Edelmann et al., 2002) is controlled by two antagonistic forces – negative gravitropism and **epinasty** – which is an endogenous force in the direction opposite to gravitactic bending (Kang, 1979). This can be demonstrated by rotating a plant such as a *Coleus* slowly with its shoot in horizontal orientation (cf. clinostat principle, Chapter 2). By this means, the gravitational stimulus is randomized, and the epinasty causes the leaves to bend downward.

As in motile microorganism, tropisms to other stimuli are described by a prefix such as

- mechano- (response to mechanical forces)
- gravi- (response to gravity)
- thigmo- or hapto- (response to touching an object)
- rheo- (response to streaming water or air)
- seismo- (response to vibrations)
- chemo- (response to a chemical stimulus)
- galvano- (response to an electrical stimulus)

Nondirectional responses to environmental stimuli are called **nasties**. A sudden decrease in solar radiation by a passing cloud can cause some flowers to close (photonasty); a seismic stimulus of a pollinating insect may induce the stamina to bend inward and deposit their pollen onto the animal. Like phobic responses, nastic reactions are independent of the stimulus direction. A *Mimosa* leaf closes always in the same direction independent of the direction of the mechanical, seismic, chemical, or thermal stimulus.

1.2.3 "Microgravity" and hypergravity

Although an internationally accepted value for normal acceleration due to gravity on Earth exists (9.80665 m s^{-2}), there is no agreed-upon symbol when it is used as a unit. One can find (g), which can be confused with gram or (G) that might lead to confusion with the gravitational constant. We adopted the conventional usage "$\times g$." To describe a reduced acceleration condition ($<1 \times g$) – e.g., in a spacecraft – different terms are being used and discussed. Besides weightlessness or $0 \times g$, assuming that the residual acceleration is not perceived by the system, the terms **low *g*, hypogravity**, and **microgravity** are used. Microgravity does not correspond to $10^{-6} \times g$, but indicates low gravity (Hammacher et al., 1987) and takes the residual acceleration due to, e.g., movement of the astronauts, into account.

1.3 Ecological significance

It is an obvious advantage for a higher plant to orient in the Earth's gravitational field so that the shoot grows upward to reach the top of the canopy to harvest solar energy for its photosynthetic apparatus. It is also advantageous for the roots to grow downward into the soil to reach the water table. Likewise, photosynthetic motile microorganisms have an advantage when they swim upward in the water column to reach the surface to have access to sunlight. Also, nonphotosynthetic microorganisms gain from gravitactic orientation (e.g., to find their mating partners at the surface) to find food and oxygen at a certain layer in the water or to find the bottom to settle and become sessile or to find microaerophilic conditions. However, this is a limited view of a much more complex ecological situation.

As we will see in the upcoming chapters, microorganisms respond to a multitude of external stimuli, including light and gravity, temperature and chemical gradients, the magnetic field of the earth, fluid currents, and even electric fields (Berman & Rodhe, 1971; Kamykowski & Zentara, 1977). The final response of the organisms to these multiple stimuli may be a vectorial addition, or one stimulus overrides the response to the others or the answer is the result of a complex network.

As an example, in darkness, photosynthetic flagellates swim upward, guided by the gravitational field of the Earth. This behavior is of ecological advantage, because it will bring the cells to the top of the water column to be near the surface when the sun rises. The upward orientation is supported by positive phototaxis (movement toward a light source), as soon as light at low irradiances is available. In contrast, at high solar radiation, negative phototaxis takes over in many organisms, which leads the cells away from the surface. This is a reasonable response because excessive solar radiation can harm the photosynthetic apparatus and damage cellular structures, including the motor (flagella). Negative phototaxis overrides gravitaxis and responses to other stimuli (cf. Chapter 7). In other organisms, the direction of gravitactic orientation can be reversed. Also, cyanobacteria with a gliding motility have been found to escape excessive radiation by vertical migration (Garcia-Pichel et al., 1994).

In some cases, the ecological significance of graviperception may not be obvious. Why do certain ciliates show negative gravitaxis and thus swim upward while others swim downward in the water column? One possible answer may be found in the modulation of gravitactic orientation by other external factors. *Paramecium* shows a pronounced negative gravitaxis at low oxygen tensions, whereas orientation becomes more random at higher oxygen tensions (Hemmersbach-Krause et al., 1991a). This behavior can be interpreted by assuming that the cells swim upward toward the water surface, where their food is, and they do not care about graviperception when they have reached surface waters, signaled by higher oxygen concentrations. That their behavior is really guided by gravity and not by a chemical gradient (aerotaxis) has been shown by experiments in sealed cuvettes under controlled oxygen conditions. In contrast, the ciliate *Loxodes* shows a pronounced positive gravitaxis at high oxygen tensions and becomes less well oriented at low levels. This behavior will guide the cells downward into the sediment of its habitat, where these microaerophilic ciliates find adequate oxygen concentrations (Fenchel & Finlay, 1986, 1990; Finlay et al., 1986a,b).

Not all microorganisms are capable of active movements. These pelagic organisms can also adjust their position in the water column. Several cyanobacteria and diatoms can modify their buoyancy by producing gas vacuoles or oil droplets, respectively (Kinsman et al., 1991; Overmann & Pfennig, 1992; Reynolds et al., 1987). Buoyancy can also be modified by other external stimuli, including chemicals, temperature, and light (Spencer & King, 1985; Thomas & Walsby, 1986; Walsby, 1987).

The response to the environmental stimuli can be governed by endogenous circadian rhythms. Dinoflagellates are known to undergo vertical migrations of up to 30 m (Taylor et al., 1966; Tyler & Seliger, 1978; Yentsch et al., 1964). The ecological significance of this may be that at night the cells do not need to be near the surface, and it may be rather advantageous to be further down in the water column [e.g., to escape predation (Kuhlmann, 1994; Kuhlmann et al., 1998)]. Gravitactic orientation in the flagellate *Euglena* is also modified by an endogenous rhythm (cf. Chapter 5).

Primary and secondary consumers are also known to respond to external stimuli (Buchanan & Goldberg, 1981; Davison & Stross, 1986) and to undergo vertical migrations (Leech & Williamson, 2001; Rhode et al., 2001; Ringelberg, 1999). This behavior is modified for example, by ultraviolet irradiation (Rhode et al., 2001). However, in some cases, the vertical migration in primary or secondary consumers is not a response to external stimuli. Rather, the organisms follow their food during their movement (Gliwicz, 1986).

2

Methods in Gravitational Biology

To vary the influence of the unique stimulus gravity, different experimental and technical approaches have been followed and developed. Today, we are in the ideal situation to perform gravitational biological experiments on the ground – by means of, for example, clinostats and centrifuges – and in real microgravity using different facilities in dependence of the time of free fall needed. Comparative studies between simulated and real microgravity reveal similar results, though the response in actual microgravity appears to be more pronounced and faster. The methods that are used to answer questions in gravitational biology are presented, explained, and discussed within this chapter.

In their natural environment, swimming microorganisms are confronted with a large number of interacting stimuli, which, after a complex signal processing, result in behavioral responses. To understand the impact of a **single** stimulus, such as gravity, the behavior has to be studied under controlled and defined conditions. Experimenters had to learn that the behavior of microorganisms also depends on parameters such as geometry of the observation chamber, cell density, thermoconvection, time of measurement with respect to circadian rhythm, and that the responses are "full of surprises" (Kessler, 1985b). To study the responses with respect to gravity, an observation chamber should be completely filled and air bubbles should be excluded to avoid chemotactic responses (cf. Section 7.3) and shearing forces. The observation light should not stimulate the cells (cf. Section 7.2), thus gravitaxis experiments are normally performed in red or infrared light. It would be ideal to observe the swimming behavior without any spatial restriction. However, such a three-dimensional (3D) analysis is rather complicated and expensive; therefore, most observations are performed under the partial restriction to two dimensions (cf. Chapter 3). The classical behavioral studies in vertically positioned capillaries restrict the behavior of swimming cells in horizontal directions so that reorientation maneuvers become difficult. Bean (1977) describes that a several hundred micrometer extension in the horizontal direction is necessary to allow reorientation in *Chlamydomonas* and the observation of negative gravitaxis. Taneda (1987) reported negative gravitaxis in *Paramecium* in glass tubes of >1.8 mm diameter. Using larger observation chambers creates the problem of inducing convection currents. Bioconvective phenomena (= behavioral pattern swimming) can be observed in large test tubes (1.3 × 10 cm) filled with a rather dense cell culture. Due to their negative gravitaxis, the cells accumulate at the top of the sealed tube and, after reaching a high cell density, dynamic, finger-like clouds can be observed falling downward (Hemmersbach-Krause et al., 1991b; Kessler, 1985a).

With respect to the previously mentioned points, square or circular observation chambers with a depth exceeding the diameter of a cell several times (Winet, 1973) – but still following cells within the optical focus – have been used for behavioral studies. Focusing on the central layer of water is used to avoid hydrodynamic wall effects (Wu, 1977).

In the past, studies of the graviresponses of swimming microorganisms have been predominantly nonquantitative, analyzing primarily the qualitative or semiquantitative net orientation of a whole population. The development of quantitative methods for the characterization of the graviresponses has led to a deeper understanding of the underlying mechanism (Bean, 1977; Taneda et al., 1987; Taneda, 1987 – cf. Chapter 3).

The influence of $1 \times g$ (9.81 m s^{-2}) is a unique stimulus condition on Earth, because it has been constant and permanent during the evolution of life. Over the years, the question of how gravity influences the development and functions of organisms has fascinated scientists. Thus, different experimental approaches have been followed to manipulate the effects of gravity. In addition to ground-based methods, experimenters obtained the opportunity to perform experiments under real free-fall (microgravity) conditions (Cogoli & Gmünder, 1991). Ground-based studies should be a prerequisite before performing an expensive experiment in space. They are essential for the preparation of space experiments, such as testing and optimization of flight hardware and for verification of results obtained in microgravity.

Due to the presence of masses in our universe, true zero gravity does not exist. Even for a typical shuttle mission at an altitude of, for example, 350 km, the level of gravity is 9.04 m s^{-2} (i.e., only 8% less than the gravitational field at the Earth's surface). To achieve the condition of reduced gravity, the spacecraft or the space station has to be accelerated. As a consequence, the gravitational force and the centrifugal force are equal in amount, but with opposite vectors, resulting in weightlessness or a free-fall condition. This acceleration condition is normally termed *microgravity*, as a residual acceleration in the range of 10^{-4}–$10^{-6} \times g$ cannot be excluded (cf. Section 1.2.3). Compared with the "gravity environment", various physical phenomena behave differently in the theoretical state of zero acceleration. Whereas sedimentation, hydrostatic pressure, and convection are linearly proportional to gravity, being zero in weightlessness, diffusion persists. Regarding the effects of gravity on a cell, one should distinguish direct and indirect effects. Whereas direct effects act on a cell and its components

(= "functional unit" according to Briegleb, 1988), indirect ones may act on the cellular surrounding, thus outside the "gravity-sensitive functional unit" (Briegleb, 1988; Albrecht-Bühler, 1991). Sometimes a discrimination between indirect and direct effects is difficult to make and might lead to a misinterpretation of the results. Is an enhanced proliferation rate of cell cultures in microgravity induced by a direct effect of microgravity on the cellular level or due to the optimal distribution of food particles (lack of sedimentation) and thus by an indirect effect? The absence of convection could also result in an accumulation of metabolic waste products around the cells and tissues, affecting growth and differentiation as a secondary effect of microgravity on cell metabolism. The physical background of microgravity studies on the cellular level were evaluated by Pollard (1965), regarding phenomena like Brownian motion, convection, hydrostatic forces, stress, and temperature. He concluded that a cell with a diameter larger than 10 μm should experience gravity.

2.1 Horizontal microscopes and clinostats

A simple way of studying the impact of gravity on biological systems is to observe their behavior in a horizontally positioned observation unit. Such an experimental setup ("horizontal microscope") can be used to show how gravity influences, for example, plant growth and the behavior of free-moving organisms.

A device that enables the rotation of, for instance, a plant or a chamber filled with a cell culture around an axis perpendicular to the force of gravity is called a *clinostat* (= constant inclination). The classical clinostat uses rotation speeds of 1–2 rpm and was mainly used by plant physiologists to investigate the gravitropism of plants (Pfeffer, 1897; Davenport, 1908). The aim of this device was to identify gravisensitive processes. Due to the relatively long response time of plants in the range of minutes, the gravitropic response is abolished to a certain extent under these conditions. However, later research revealed that the slow-rotating clinostat produces a stress situation due to the omnilateral gravitational stimulation – and thus signal overload – causing disturbances at the ultrastructural and molecular levels [e.g., development of lytic compartments in roots (Hensel & Sievers, 1980), an increase in ethylene production in roots (Hensel & Iversen, 1980), and protein degradation via the ubiquitin system in leaves of *Vicia faba* (Hunte et al., 1993)].

Muller (1959) proposed to rotate clinostats much faster, at such an angular velocity that the system would no longer perceive the rapidly turning gravity vector. He predicted that a human being exposed on a fast-rotating platform would experience "weightlessness." This idea was adopted by Briegleb to investigate the effect of "simulated weightlessness" on small living systems (Briegleb, 1988, 1992; Klaus et al., 1997a, 1998; Schatz et al., 1973). The theory is explained in Fig. 2.1. Under $1 \times g$ conditions (Fig. 2.1a), particles will sediment and take a spatial position that is determined by their weight (G) and buoyancy. Thus, a heterogeneous particle distribution is found under $1 \times g$ conditions. Under free-fall conditions, particles homogeneously distribute as sedimentation is abolished (Fig. 2.1b). A homogeneous particle distribution can also be achieved on ground by rotating a suspension (Fig. 2.1c). Under this condition, particles will still fall

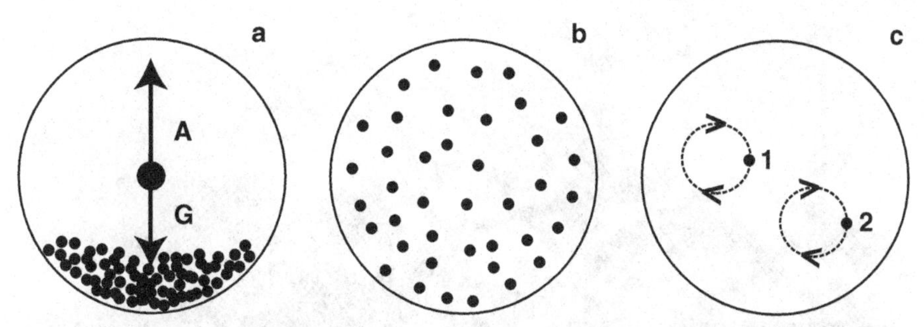

Figure 2.1. Scheme to explain how functional weightlessness can be achieved by means of a fast-rotating 2D clinostat. **(a)** Regarding a suspension of particles with a density higher than that of the surrounding medium, sedimentation occurs under $1 \times g$ conditions. The spatial position of the particles is determined by their weight (G) and buoyancy (A). **(b)** Under microgravity, particles homogeneously distribute due to the lack of sedimentation. **(c)** This condition can be also achieved on the ground by rotating a suspension (shown for two particles). Thus, particles will be forced on circular paths, whose radii are determined by the speed of rotation. Increasing the revolution will decrease the diameter of the circles, finally approaching zero. Simultaneously, a centrifugal force is induced, depending on the speed of rotation and the distance of the particles from the rotation axis, thus driving the particles to the periphery of the system. (Courtesy of A. Schatz, DLR, Cologne, Germany.)

and be forced onto circular paths. The radii of the circles decrease the faster the system is rotated. If the rotation speed is too high, particles will be centrifuged. At an appropriate rotation speed, particle movement – due to sedimentation and centrifugation – is kept within the limits of the Brownian motion (Klaus et al., 1997a).

The movement of particles within a cell and the effect of clinorotation on them can be observed in *Elodea canadensis.* In this water plant, intracellular particles (calcium oxalate crystals) with a size of about 1 μm normally settle at $1 \times g$, but show a random distribution at 100 rpm on a clinostat (Briegleb, 1992). Commonly, visible sedimentation of cell organelles (e.g., nucleus) will not occur due to their fixation within the cytoskeleton. In this case, changes of the influence of gravity might modify tension forces of the cytoskeleton and in turn ion channel functioning and finally cellular behavior. The role of the cytoskeleton in preventing sedimentation has, for example, been shown for the *Chara* rhizoid (Braun et al., 2002) and the statocytes of roots (Hensel, 1985).

Although gravity is still present, its vector is continuously being reoriented during rotation on a clinostat. Whether signal cancellation is achieved finally depends on the relationship between time for one rotation and response time of the system, as well as the effective radius and the threshold for accelerations of the perceiving system (Klaus et al., 1997a). If a tube with a radius of 1 mm is rotated at a speed of 60 rpm, the centrifugal force is $0 \times g$ in the center and $4.2 \cdot 10^{-3} \times g$ at the outer perimeter. The Coriolis acceleration, a force acting on moving objects within a rotating system (e.g., clinostats, centrifuges, and even Earth), reaches no more than $2 \cdot 10^{-23} \times g$ (worst case) under this experimental condition (Briegleb, 1992) and can be neglected. Discussions of a possible induction of circular motion, electric currents and mechanical vibrations, and the influence

Figure 2.2. View of a clinostat microscope. The optical parts are mounted on a horizontal bank. The optical axis is adjusted to the center of rotation of the whole apparatus. During rotation, microscopic observation of the samples is possible. (Courtesy of DLR, Cologne, Germany.)

of the Earth's magnetic field demand a repetition of the experiments in real microgravity (Albrecht-Bühler, 1991; Ayed et al., 1992).

With respect to the scientific demands, different kinds of fast-rotating, two-dimensional (2D) clinostats have been developed. They are all characterized by one horizontal rotation axis, exact positioning of the sample on the rotation axis, and revolution rates in the range of 40–100 rpm. To produce no further acceleration forces than the centrifugal force, they are rotated at constant speed. By equipping such clinostats with a microscope, direct observation and in vivo analysis of the objects during the experimental run are possible (Fig. 2.2; Block et al., 1986b). For long-term cultivation and processing of higher cell numbers, test-tube clinostats have been constructed, where, for example, a tube with cells can be inserted along the rotation axis (Fig. 2.3). Clinostats with submerged cuvettes have been used for developmental studies with aquatic systems.

The validity of the clinostat approach can be shown by comparison of data obtained during the simulation experiment and under real free-fall conditions. Some experimenters already had this possibility (Cogoli, 1992; Kordyum, 1997). Some examples include the following:

- Human T-lymphocytes in suspension show 90% inhibition of activation in microgravity (Cogoli et al., 1984) and 50–75% on the clinostat (Cogoli et al., 1980).

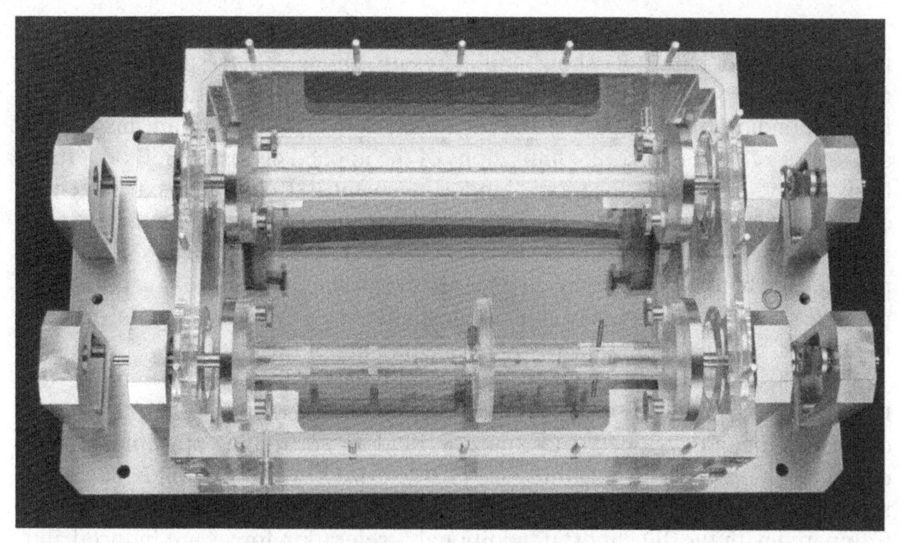

Figure 2.3. View of two cuvette clinostats, for example, for cultivation of cells within test tubes that rotate around their longitudinal axes. (Courtesy of DLR, Cologne, Germany.)

- A two-fold increase in cell numbers and a shortening of the duration of the lag phase of cell division in *Escherichia coli* have been stated in space (−5.8 h) and on the clinostat (−2.1 h) (Klaus et al., 1997a; Klaus, 2002).
- The growth rate of *Paramecium tetraurelia* increased by 100% in microgravity (Planel et al., 1981, 1982; Richoilley et al., 1988) and by 20–30% on the clinostat (Ayed et al., 1992).
- While *Paramecium biaurelia* cells show negative gravitaxis under 1 × *g*, they show a random distribution in microgravity and on the clinostat. The cells swim faster on the clinostat than in microgravity, giving indication for a still persisting mechanostimulation and hyperpolarization during clinorotation (Hemmersbach-Krause et al., 1993a,b).
- *Euglena gracilis* cells show negative gravitaxis in 1 × *g* and a random distribution in microgravity and on the clinostat. Compared with microgravity, an increase in the number of spontaneous directional turns of the cells on the clinostat has been observed (Vogel et al., 1993).
- The velocity of cytoplasmic streaming of *Physarum polycephalum* increased by 120% in microgravity and by 40% on the clinostat (Block et al., 1986a,b).
- The amyloplasts in the statocytes of roots no longer sediment and show a similar distribution pattern in microgravity and on the clinostat (Sievers & Hejnowicz, 1992; Volkmann & Sievers, 1979).
- The displacement of statoliths in the rhizoids and protonemata of *Chara* in microgravity and on the clinostat was comparable, though the transport of the statoliths was faster in microgravity (Braun et al., 2002). For the parameters tested, fast rotation offered a better simulation of the microgravity condition than slow rotation of the clinostat (Cai et al., 1997).

These examples show that both experimental approaches (space experiment and fast-rotating 2D clinostat) show similar results, though the response in actual microgravity appears to be more pronounced and/or faster. Studies with very fast responding free-swimming and highly mechanosensitive organisms – such as gravitactic protists – reveal limitations of the applied clinostat method. Obviously, the systems experience some kind of mechanical stress by the experimental condition. Nevertheless, one can state so far that the clinostat is an efficient tool to detect gravisensitive mechanisms, and studies on this instrument should be a prerequisite before performing an expensive experiment in space. Further comparative studies in functional weightlessness under clearly defined conditions and in real microgravity will finally prove the suitability of this method.

With the assumption to produce a "gravity-vector averaging situation" for larger samples, clinostats with two rotation axes oriented at right angles to each other have been developed (3D clinostat, random positioning machine; Hoson et al., 1995, 1997). These facilities are characterized by the randomly changing rotation speed (e.g., 2 rpm) and direction every 30 or 60 s (Hoson et al., 1997). Compared with the 2D clinostat, additional acceleration forces are induced and shearing forces will not approach zero. Whether irritation and permanent gravitational stimulation or functional weightlessness is achieved for the object depends on its threshold and reaction time.

There is a need for comparison of experimental data obtained on the different facilities (e.g., fast, slow-rotating, 2D, 3D clinostats), with results from space to prove the suitability of the methods. This has been done with respect to lymphocytes, revealing that – at least for this cell sytem – comparable results were obtained on the 2D clinostat and the 3D clinostat, but not by using the free-fall machine (FFM; Schwarzenberg et al., 1998; Cai et al., 1997). It should be kept in mind that, with respect to gravity, random distribution must not be identical with a physiological stimulus-free condition.

2.2 Free-fall machine

The condition of apparent (near) weightlessness can be achieved on or near Earth by exposing an object to the condition of free-fall, which can be orchestrated using different approaches. Based on the hypothesis that there is a minimal time threshold for sensing a change in gravity, a "Free-Fall Machine (FFM)" has been constructed. Biological samples are inserted on a platform that can fall in a vertical shaft. At the lower end of the shaft, the sample is stopped and ejected upward (Mesland et al., 1996). The free-fall condition approximately lasts for 900 ms and is interrupted by an acceleration of about $20 \times g$ for 20–80 ms. Although the cell cycle of *Chlamydomonas* on the FFM is shortened similar to microgravity conditions (space experiment), compared with the $1 \times g$ control (Mesland et al., 1996), the depressed activation of lymphocytes found on the clinostat and in space were not obtained on the FFM (Schwarzenberg et al., 1998, 1999). Knowledge of thresholds and further experimentation under free-fall conditions in comparison with microgravity will show whether gravisensitive systems really experience continuous free-fall and not the periodic changes of acceleration within such a FFM.

2.3 Drop facilities: towers, shafts, and balloons

Short-term experiments performed in drop facilities provide experimental times of only a few seconds (towers and shafts: 2.1–10 s; balloons: 30–60 s), with residual accelerations of 10^{-4}–$10^{-5} \times g$. These facilities are available at several locations all over the world [e.g., at ZARM (Zentrum für Angewandte Raumfahrttechnologie und Mikrogravitation, Bremen, Germany), at the NASA Glenn Drop Tower in the United States, or at JAMIC (Japan Microgravity Center, Kamisunagawa, Hokkaido, Japan). Balloons to drop objects from high altitudes (MIKROBA) are launched, for example, near Kiruna (Sweden) and Beijing (China). A free-fall period of about 60 s is achieved by this method. The 110-m drop tower located at ZARM is an evacuated chamber in which the experimental hardware is dropped inside a pressurized capsule. This facility delivers 4.74 s of free-fall with a quality of $10^{-5} \times g$. In the future, the microgravity time will be doubled by the construction of a catapult system installed at the base of the shaft. Whether a biological system will perceive the maximum acceleration of about $30 \times g$ depends on different parameters, as in the case of the FFM.

Though the experiment time is rather short, drop facilities offer the big advantage that normally no acceleration force $>1 \times g$ is applied to the system before the onset of free-fall conditions, except the sample is catapulted up first. The facilities are mainly used for experiments in material sciences. However, regarding fast biological reactions – such as the excitation of nerve cells – this method might also be of interest. Fast-reacting gravisensitive protists have been exposed to free-fall in a drop tower, showing a relaxation of their graviresponses during this short experimental time (cf. Section 4.1.6).

2.4 Parabolic flights

2.4.1 *Aircraft*

Reduced gravity can also bc achieved during parabolic flights on aircraft (Fig. 2.4). These flights are used for the training of astronauts, testing of space hardware, and also for short-duration experiments. Series of parabolic maneuvers result in about 22-s periods of reduced gravity (about $10^{-2} \times g$). From the normal horizontal flight, the aircraft climbs at 47° (pull-up) to 23,000 ft over an interval of about 20 s. Then, the engine thrust is considerably reduced up to the point where it just overcomes the aerodynamic drag and the pilot ends the lift. The aircraft is then under free-fall conditions for about 20–25 s. Subsequently, a $1.8 \times g$ pull-out phase is executed on the down side of the parabola to bring the aircraft

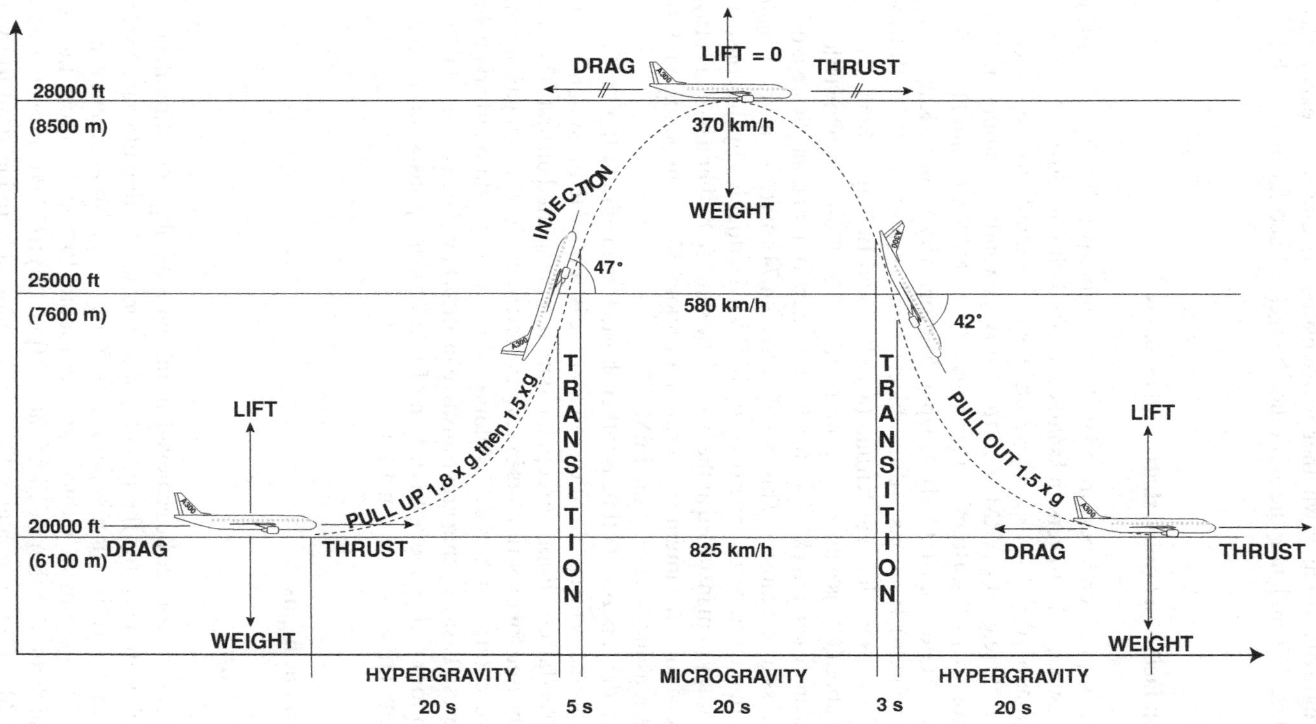

Figure 2.4. Typical flight parabola of an aircraft (A 300), providing alternating periods of hypo- and hypergravity.

back to its steady horizontal position within the following 20 s. Two parabolas are separated by a time interval of 2 min. Within one campaign, normally thirty parabolas each are flown on three consecutive days, thus providing a large amount of data. Cultures of *Euglena gracilis* cells were exposed to the changing acceleration profile during the parabolic flights to understand the mechanism of gravity-sensing by measuring the intracellular calcium concentration. A transient increase in the intracellular calcium concentration was detected during a transition from lower to higher accelerations (1.0–1.8 $\times$ g or microgravity to 1.8 $\times$ g), but not in the opposite direction (Richter et al., 2002), indicating a pronounced calcium influx during reorientation of the cells, which is regarded as the primary event in gravity sensing.

2.4.2 *Sounding rockets*

Sounding rocket activities in Europe are dominated by the German and Swedish programs: TEXUS (Technologische Experimente unter Schwerelosigkeit), Mini-TEXUS, MASER (Materials Science Experiment Rocket), and MAXUS (long duration sounding rocket programme). The launch campaigns take place at the ESRANGE launch site, near Kiruna (Sweden).

Depending on the type of rocket and the apogee of the flight, microgravity conditions of 3–4 min (Mini-TEXUS), 6–7 min (TEXUS, MASER), or 13 min (MAXUS) are provided. Between 1977 and 2003, 40 TEXUS, 6 Mini-TEXUS, 9 MASER, and 5 MAXUS flights have taken place. Their flight profiles are shown in Fig. 2.5. The ascent phase of a TEXUS rocket (Fig. 2.6) lasts for about 6 s, with a mean acceleration of 5 $\times$ g. For stabilization of the accelerated ascent, a spin of the rocket is induced. After burnout of the last stage, only the spin acceleration is effective and stopped within 1 s by releasing ropes tethering a weight, followed by the free-fall period of 360 s, which is terminated by the reentry, with an acceleration of about 26 $\times$ g. Due to the prolonged experimental time in microgravity, a broad spectrum of biological samples has been exposed so far. As the samples are exposed to hypergravity and vibrations during launch, a possible influence of these effects has to be tested. Thus, ground-based studies have to be performed using vibration equipments and centrifuges before the space experiment. Does the biological system react to the launch conditions? If it does, how long does this effect persist? If desired by the experimenter, an in-flight 1 $\times$ g reference centrifuge can be used. The advantage of sounding rocket experiments with living systems are late access (30 min before lift-off) and early retrieval (30 min after landing) of the samples, and the high, nearly undisturbed quality of microgravity. Different modules for the scientific needs of the experiments have been developed for observation, fixation, and stimulation of the samples (for technical descriptions, see Cogoli & Friedrich, 1997; Huijser et al., 1995). A direct manipulation and observation of the experiment during the flight is provided by means of telecommands and video downlink. Mainly fast responses have been investigated in sounding rocket experiments (for reviews, see Cogoli

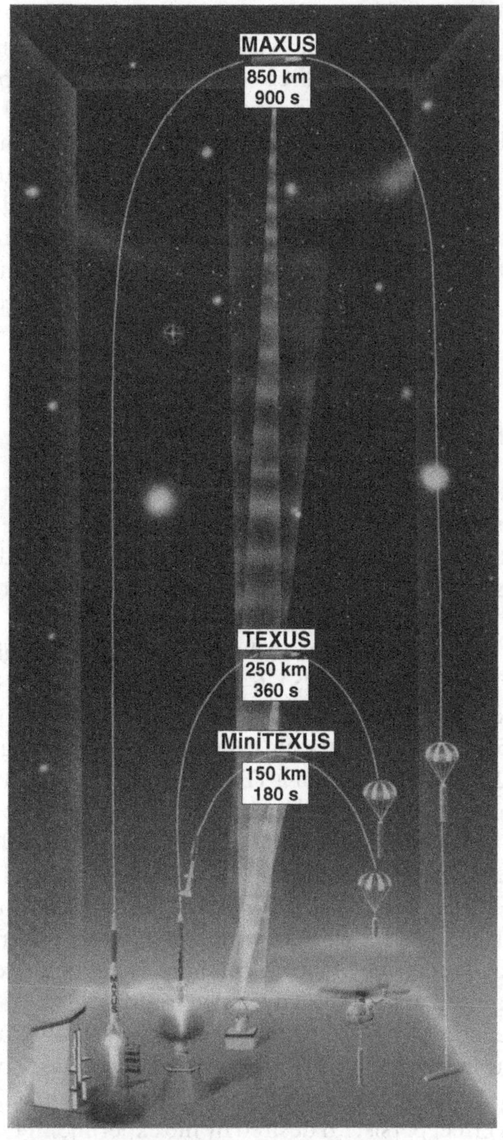

Figure 2.5. Flight profiles of sounding rockets. For further details, see text. (Courtesy of Astrium, Bremen, Germany.)

et al., 1997) enhancing, for example, our understanding of the physiological characteristics of the graviresponses of different protists (*Paramecium, Euglena*, and *Loxodes*), the capacity and task of the cytoskeleton in being an active force against sedimentation (Braun et al., 2002), and the influence of gravity on distinct steps in signal transduction chains, such as early gene expression (De Groot et al., 1990).

Figure 2.6. Launch of a TEXUS (Skylark-7) rocket from Kiruna in Sweden. (Courtesy of W. Engler, Kayser-Threde, Munich, Germany.)

2.5 Centrifuges

Because reactions to gravity are normally rather weak, the response can be increased under elevated accelerations by centrifuges. In addition, centrifuges are useful equipment for experimentation in space. There, they are used to simulate an onboard $1 \times g$ environment, thus providing a comparison with ground controls to ensure that the obtained results depend on gravity and not, for example, on

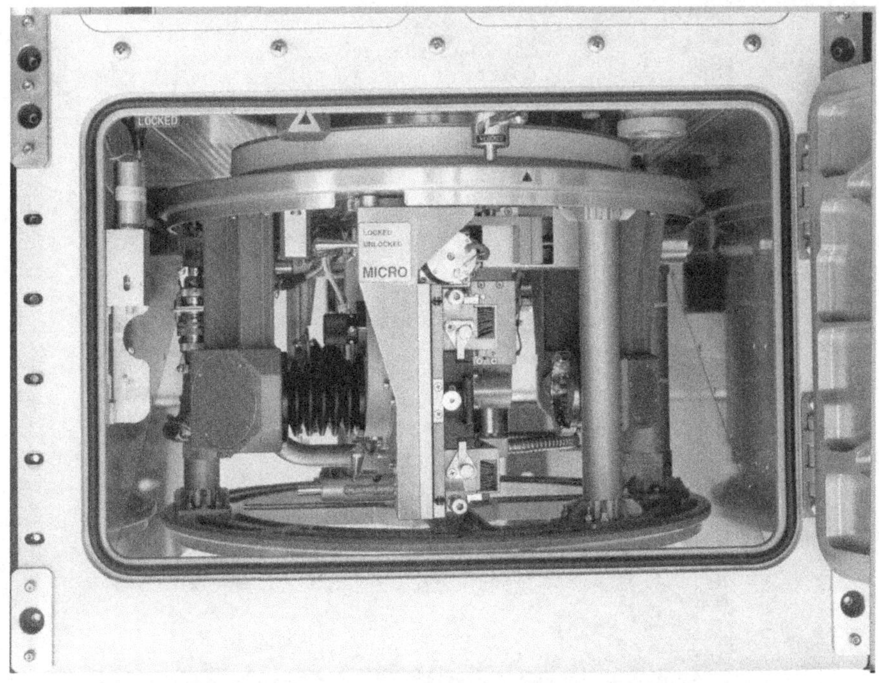

Figure 2.7. Scheme of NIZEMI – a slow-rotating centrifuge microscope. Centrifuge turntable with microscopic (**left**) and macroscopic observation units (**right**). (Courtesy of Astrium, Friedrichshafen, Germany.)

the changed radiation condition in space. Centrifuges in microgravity also enable the experimenter to study samples under defined acceleration profiles, which are a necessity for threshold studies.

The first centrifuge reported for biological experiments was constructed by Knight (1806), using a rotating platform driven by a waterwheel. Placing different kinds of seeds at various positions on the centrifuge radius, he observed that roots and shoots developed along the resultants of gravity and centrifugal force. Over the years, further devices have been constructed for biological research (Brown, 1992; Smith, 1992), reaching high standards of centrifuge microscopes (Hiramoto & Kamitsubo, 1995; Hemmersbach-Krause et al., 1992; Friedrich et al., 1996; Bräucker & Machemer, 2002).

The centrifuge microscope NIZEMI (Niedergeschwindigkeits-Zentrifugen-mikroskop; a low-speed, rotating centrifuge microscope) has been successfully used for ground-based studies and was also flown on the shuttle mission IML-2 (Fig. 2.7). This facility allows the observation of small biological and material science samples during stepwise and defined accelerations from $1 \times g$ to $10 \times g$ on ground and from $10^{-3} \times g$ to $2 \times g$ in space, thus being ideally suited for studying adaptation processes and determination of gravity-dependent thresholds (Hemmersbach et al., 1996a,b). The system consists of a microscope and a macroobservation unit, both mounted tangentially on a rotating platform. Seven life sciences experiments were performed on NIZEMI during the Spacelab mission IML-2 in 1994. The objects were protozoa, algae, slime molds, and plants – for the study of thresholds, dose-response curves, kinetics, behavior, and developmental processes (results are summarized by Cogoli, 1996). For this mission, the samples were prepared by the scientists on the ground, then stored and incubated within the BIORACK (a biological rack in the shuttle), and exposed at dedicated times to variable and defined acceleration steps in the observation unit of NIZEMI. Close contact between the well-trained astronauts and the scientists guaranteed the success of the experiments. The images were recorded by a black and white CCD camera. Data were stored on video and/or downlinked and evaluated. A centrifuge microscope is also available on sounding rockets, providing the opportunity for threshold studies in fast-reacting systems, such as gravitactic protists.

In addition, experimenters were rather creative in providing hypergravity conditions for various kinds of samples [e.g., the CECILIA (= centrifuge for ciliates) centrifuge, for the study and direct observation of graviresponses of swimming protists on ground – housing up to six experimental setups and video recorder units (Bräucker & Machemer, 2002); the STATEX (STATolithen-EXperiment) centrifuge (Neubert et al., 1991), originally developed for microgravity experiments on statoliths of frogs; the BOTEX centrifuge, for experiments with plants; or centrifuge designs developed to incubate several 1- to 2-ml samples (cell culture) in parallel].

2.6 Shuttles, satellites, and space stations

Long-term biological studies in microgravity were enabled by the development of automatic satellites, platforms (e.g., Eureca), and human-tended space laboratories [e.g., space shuttles, Spacelab, Spacehab, the International Space Station (ISS), and MIR; for review of missions and the national research programs in the life sciences of the European Space Agency (ESA) member states between 1980–1996, see Fitton & Moore, 1996; for current missions, see the homepages of ESA, Deutsches Zentrum für Luft- und Raumfahrt (DLR), and NASA]. However, before going into space and creating space laboratories, it had to be tested whether living systems can survive under space conditions. Thus, during the period between 1949–1952, animals – mainly dogs – were exposed to short-term hypergravity ($5.5 \times g$ for 4 min). In 1957, the dog Laika was launched on Sputnik 2, and, for the first time, physiological functions – such as heart activity and food uptake – were measured under microgravity conditions. In 1961, Yuri Gagarin was the first man in space and fulfilled a dream of mankind. He circled the Earth and was exposed to microgravity for 89 min. The increasing stay of humans in space revealed that we can live there better than expected. Adaptation to microgravity conditions is managed rather fast. The duration of the stay – limited by the fact that, for example, bones and muscles reduce and lates readapt to Earth conditions – is the problem. By means of altered nutrition and countermeasurements, this problem is addressed. Since the early 1970s, recoverable and unmanned capsules of the Bion and Foton type were derived from the Vostok spacecraft, which carried Yuri Gagarin into space. By means of automatically performed experiments in dedicated facilities (e.g., the Biobox), questions about how the space environment (microgravity and radiation) acts on living systems have been investigated.

Space Station MIR fulfilled the dream of scientists and offered for the first time the fascinating possibility of a long-term stay of humans in microgravity. The station was permanently expanded over 10 years, from 1986 on. For 15 years, MIR circulated at a height of 300–400 km above ground, with a speed of 28,000 km/h. The station was visited by 109 spacecrafts and more than 100 astronauts. The experience of a long-term stay of humans in space and experimentation in microgravity are utilized for the ISS.

The ISS is located about 400 km above ground. The assembly began in November 1998, and the completion is planned for 2006. Various elements are added using a number of supply vehicles, primarily the Space Shuttle. Pressurized modules and external platforms provide the living and working accommodations for up to seven astronauts. Specific racks offer experimental conditions and equipments (e.g., BIOLAB, MCS; cf. Chapter 11) for dedicated and systematic studies in microgravity.

2.7 Direct manipulation of gravisensors

Manipulation of the density of the medium offers an experimental approach to identify the site of graviperception. The density of a gravitactic protist exceeds the density of the surrounding medium by about 2–5%, depending on species and feeding status (Taneda, 1987; Lebert & Häder, 1996). An artificial increase of the cytoplasmic mass in ciliates can be achieved by means of iron feeding (Köhler, 1921, 1922; Watzke et al., 1998). If gravity is perceived by an intracellular gravireceptor, this mechanism works independently of the density difference between cell and medium. As a consequence, experiments under isodensity conditions will not affect the graviresponses, as shown for the gravitaxis of *Loxodes*. If, however, the gravisensory mechanism is located at or in the cell membrane – or in the case of a pure physically guided mechanism – adjustment of the densities of cell and medium will switch off the graviresponses, as shown for *Paramecium* and *Euglena*. Common methods to achieve neutral buoyancy are to use Percoll or Ficoll (Lebert & Häder, 1996; Hemmersbach et al., 1998; Wayne et al., 1990).

To identify the role of receptor candidates in the gravity-signal transduction chain in single cells, different micromanipulation techniques have been applied:

- Destruction of the sensor by means of a laser beam [e.g., the statocyste organelle in *Loxodes* (Hemmersbach et al., 1998; cf. Chapter 4)]
- Manipulation of the position of the statoliths by means of optical tweezers and centrifugal forces (Braun et al., 2002)
- Displacement of gravisensors (amyloplasts) by magneto-mechanic forces in high-gradient magnetic fields (Kuznetsov & Hasenstein, 1995).

Great efforts have been undertaken to block specifically mechanosensitive ion channels or steps in the postulated signal transduction pathway by means of chemical inhibitors, quenchers, and blockers (Lebert & Häder, 1996; Nagel & Machemer, 2000a; Millet & Pickard, 1988; Wayne et al., 1990), by environmental tools such as ultraviolet radiation (Häder & Liu, 1990a,b) or temperature (Block et al., 1999), and to study the consequences for graviresponses. Unfortunately, the effects of these manipulations might be rather complex for cellular physiology, thus making interpretation of the obtained results more difficult. A promising step forward will be the induction and isolation of mutants with respect to gravisensing (Lebert & Häder, 1997a).

3

Image Analysis

Motility and orientation of motile microorganisms can be quantified basically by two approaches: population methods and individual tracking. Individual tracking analyzes the movement parameters of single organisms and subsequently averages over the behavior of a statistically significant number of individuals to evaluate the behavior of a population. Population methods operate on the assumption that the movement of the individual organisms will lead to a translocation of the whole population. Both methods have their advantages and drawbacks. Individual tracking is tedious and can be prone to error or bias, and may only be applicable for restricted path segments. Reaching a significant conclusion for a whole population may strain the patience of an experimenter. On the other hand, population methods may measure something different than the experimenter assumes. For example, a population may seem to be moving in a certain direction and end up, for instance, at the top of a water column, which may be interpreted as the result of negative gravitactic orientation. However, the behavioral result may be due to the organisms moving in random directions, but become immobile near the water surface. Other phenomena leading to a displacement of a population, erroneously interpreted as the result of gravitaxis, include sedimentation, phobic responses, kinetic effects, or the response to other stimuli – including light, magnetic field lines, or chemical gradients (such as oxygen, carbon dioxide, or nutrients). In recent years, fluorescence imaging turned out to be one of the most powerful tools to address questions regarding functions of certain proteins, second messengers, and ions in living cells. In this respect, we will discuss the possibilities to monitor *in vivo* changes in calcium concentration, as well as the membrane potential. The discussed strategies can be applied to related questions. Finally, experiments performed on the ground and in space are presented.

3.1 Introduction

In addition to being tedious, manual tracking of motile microorganisms may be biased by the expectation of the experimenter. In a class experiment, students were asked to analyze the directional movement of flagellates. One group of students was handed a videotape with cell tracks and asked to draw short movement vectors on an acetate sheet overlaying a video monitor. They were told that the cells show negative gravitaxis. At the end of the day, the students had produced a large amount of data clearly indicating a preferential upward orientation of the cells. The experiment was repeated with another group of students, who were asked to evaluate the same tape and were given the same explanation, with the only difference being that the organisms were said to show positive gravitaxis. Again, the results showed a high statistical significance for the expected result. Even in double-blind experiments, it is difficult for an experimenter to arrive at an objective analysis and to not involuntarily exclude organisms moving in an unexpected direction. The state-of-the-art way around this dilemma is to use fully automatic image analysis, which avoids human biases. In addition, automatic real-time analysis can determine large numbers of tracks and – due to simultaneous tracking of hundreds or thousands of organisms – in a rather short time (Alani & Pan, 2001; Häder, 1994b).

The recent improvements in video technology and the even more dramatic developments in the computer hardware allow real-time analysis of large numbers of tracks that provides high statistical significance (Clark & Nelson, 1991; Häder & Vogel, 1991; Häder, 1991c, 1992). In addition, the development of mathematical approaches and computer algorithms facilitate a more and more detailed extraction of relevant movement and orientation parameters (Taylor et al., 1997; Bengtsson et al., 1994).

Whereas the human eye and brain are still superior to even the fastest computers, the latter have a number of advantages that make machine vision an attractive tool. In contrast to human researchers, automatic cell tracking instruments do not get tired or bored, and are characterized by high objectivity. However, the accuracy highly depends on the experimental setup.

A recently developed tracking software (WinTrack 2000, Real Time Computer, Möhrendorf, Germany) has been timed to track more than 30,000 cell tracks per minute (Häder & Lebert, 2000). This speed facilitated the analysis of as many as possible cell tracks from recent experiments on parabolic airplane and rocket flights with limited duration on the order of seconds to a few minutes (Häder, 1994a,b, 1996a,b; Häder et al., 1996). The movement parameters extracted from the cell tracks include the sizes, shapes, and positions of the objects (Serra, 1980; Kokubo & Hardy, 1982), their linear and angular velocities, and their deviations from a predefined direction.

3.2 Hardware

Early systems used photographic cameras and long-term exposure to document the tracks of organisms. Bacteria, ciliates, or flagellates were observed by dark-field microscopy, and the camera recorded the tracks over several seconds or minutes with the shutter open. Variants of this method used stroboscopic monitoring light, which resulted in dashed tracks on the photo (Fukui & Asai, 1985).

Objects of interest can be recorded by various devices. The traditional tool is a video camera, with a lens of suitable magnification. Since a number of organisms may directly or indirectly respond to the measurement light beam, infrared sensitive cameras and infrared light sources can be used (Häder, 1986). Today, mostly charge-coupled devices (CCDs) are used for recording, which have a number of advantages. They are less bulky than a video camera and can be even smaller than a match box, and they are rugged and withstand vibrations of up to $100 \times g$ and impact accelerations in excess of $40 \times g$ as encountered during unmanned space flights. They can be built with a high spatial resolution and are relatively inexpensive. The spectral sensitivity of CCDs extends far into the infrared (Vogel et al., 1993).

To monitor organisms of microscopic dimensions, the recording device is mounted on a light microscope (Häder 1990). Both bright-field and dark-field microscopy can be used to record the organisms of interest. The latter one is used to increase the optical contrast and to increase the apparent size of very small objects, such as bacteria by the halo around the actual organism. In this technique, cellular details are lost; but, in most cases, only the position and movement of cells are of interest (Häder, 1987d; Häder & Häder, 1988a, 1989a). Phase-contrast, interference contrast, and polarization microscopy are less useful for image analysis, because by using these techniques the objects of interest may change from darker (compared with the background) to brighter or the other way round, which makes them difficult to detect by recognition algorithms (cf. Section 3.3.1). Larger microorganisms, such as ciliates, or multicellular organisms can be tracked with a macro lens.

The spatial resolution should be at least as high as that of the subsequent digitizer. The temporal resolution is limited by the operating frequency. In Europe and other parts of the world with 50-Hz power frequency, full frames are recorded at a video frequency of 25 Hz (30 Hz in the U.S. standard). Each full frame consists of two half-frames. In CCD cameras, the image is composed of an array of pixels in x and y directions. The gray value of each pixel is read out from the pixel matrix and also transmitted line by line.

If the video image is to be analyzed on line, each video frame must be digitized and analyzed in no more than 40 ms (33 ms for the U.S. standard), which is the interval between two subsequent full frames. Thus, the analysis is limited to events that are slower than this frequency. At that speed, it is not possible to follow the beating pattern of flagella or cilia, which in flagellates operate at a frequency on the order of about 50 Hz. For special purposes, high-speed CCDs can be employed, which have a much higher temporal resolution than the standard video signal that can be in excess of 1,000 Hz. If the analysis cannot follow this high-recording frequency, the images can be stored either electronically or on tape prior to analysis, which is then performed offline from the recorded images.

Another limitation of interest is the dynamic range of the recording device (i.e., the difference between the brightest and darkest parts in the image). Most CCD cameras have an automatic gain control (AGC), which electronically changes the amplification when the image is too bright or too dark. This automatic adjustment fails (e.g., when a few dark objects are recorded on a bright background). In this case, the AGC circuit lowers the amplification so that some structure is seen in the bright background, which may not be desired, while details in the dark objects of interest are lost. The use of the AGC is also counterproductive if the gray level of the objects or the background is to be quantified. This is for example, of importance in fluorescence imaging, where the gray value encodes the fluorescence intensity and must be known precisely to quantify e.g. the calcium concentration in a cell (Lebert & Häder, 1999b).

The next step in automatic image analysis is digitization of the image. For this purpose, the image is divided into an array of equally sized elements called pixels. Typical spatial resolutions range from 512 × 512 to 2,048 × 2,048 pixels. Even though the CCD image is divided into distinct pixels, it is not a digital image, because the gray value of each pixel is an analog value. During digitization, the brightness of each pixel is assigned a digital value (gray level). This is performed by a video analog/digital (A/D) flash converter (digitizer). Depending on the hardware, the digitizer has a dynamic range of, for example, 64 or 256 gray values (Castillo et al., 1982). The darkest black is arbitrarily set to a gray value of zero and the brightest to the maximal value (e.g., 255). Color images can be digitized by splitting the information of each pixel into three color channels, which results in three times larger image files (Knecht, 1993).

Today, digitizers have shrunk to the size of a card that fits into the slot of a standard PC (Fig. 3.1). Complete image analysis systems are available from, e.g., Real Time Computer (Möhrendorf, Germany), Universal Imaging Corporation (West Chester, PA, U.S.A.), Clemex (Quebec, Canada), AI Tektron (Düsseldorf, Germany), and AVS/UNIRAS (Birkerod, Denmark).

Image intensifier cameras allow the detection of very low light levels, such as fluorescence signals. They contain a photocathode close to a microchannel plate electron multiplier, and the image is generated on a phosphorescence output screen. The amplification and thus the sensitivity can be adjusted over a wide range. The disadvantage of these cameras is the high thermal noise from the photocathode plus the electron multiplication noise from the microchannel plate that, in most cases, limits the A/D conversion to 10 bit.

Digital images can be enhanced by a number of techniques that rely on their numerical representation in memory. The gray value of each pixel can easily be manipulated mathematically. The brightness can be increased or decreased. However, this can only be done within the range of the possible gray values. Numbers exceeding the highest or lowest possible gray values are arbitrarily set to the highest number or zero, respectively. Multiplication or division of each pixel with a constant factor enhances or reduces the contrast. Thresholding is used to suppress the background below or above a predetermined value that removes noise in these regions.

Many mathematical filters have been developed to improve image quality (Malik, 1980; Malik & Huang, 1988). High pixel noise is reduced by smoothing. In this procedure, each pixel is replaced by the average gray value of a certain

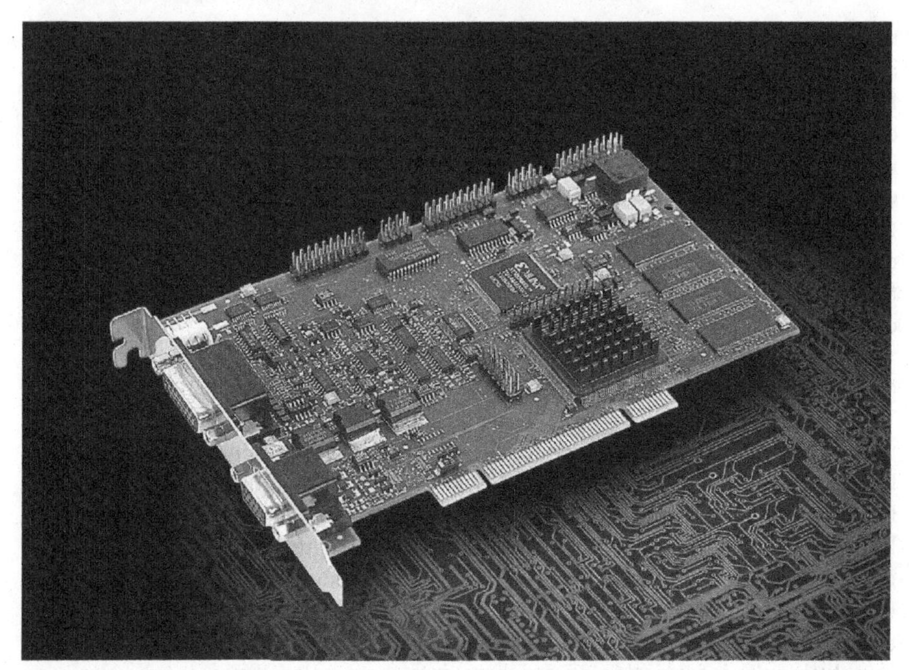

Figure 3.1. The MVtitan-G1 framegrabber card. (Matrix Vixion GmbH, Oppenweiler, Germany, with permission.)

matrix around the pixel. In the 3×3 matrix, the mean value is determined from the original value of this pixel and those of its eight neighbors. This calculation is repeated for each pixel in the image. The technique smoothes sharp boundaries and reduces pixel noise (Marangoni et al., 2000; Julez & Harmon, 1984). The operations can be performed on any size of the calculation array (but usually it has odd numbers for x and y for calculation simplicity), which has, of course, an impact on operational speed. A variant of this technique is a hat filter, which weights the pixels before smoothing.

Laplace and Sobel algorithms are advanced mathematical techniques based on calculations of 3×3 (or larger) matrices of pixels gliding over the image (Voss, 1990). Each pixel in the derived image is calculated from its original value and that of its neighbors. In Laplace filtering, the gray value of the central pixel is multiplied by a specific number (kernel), and the values of all its surrounding eight neighbors are subtracted from this value. For the most common Laplace transformation, the kernel is eight. The result of Laplace filtering (with a kernel of eight) is a general reduction in brightness of all areas with similar gray levels independent of their original value (bright or dark), whereas sharp boundaries show up as bright lines. The Sobel algorithm determines the difference between the gray values opposite pixel pairs (left and right, upper and lower neighbors, and those of the diagonals) and replaces each pixel by the highest calculated difference in the gray values of opposite pixels. This algorithm highlights not only horizontal and vertical, but also diagonal edges.

The numerical values in the pixel matrix can also be manipulated by replacing the individual numbers by those read from a look-up table (LUT; Adler, 1996) defined in software. One application is to convert a positive image into its negative: each 0 is substituted by a 255, each 1 by 254, etc. In most cases, the LUT is defined in software, but the manipulation of the gray values is done by hardware to allow real-time manipulation. Another example is binarization: each gray value below a certain threshold is set to zero (black), and the rest is set to 255 (white). This algorithm increases the contrast to its maximum and can be used to segment the image [i.e., extract the objects (bright) from the background (dark)]. The human eye and brain cannot easily distinguish between similar gray values. Therefore, a certain range of gray values of interest can be spread out, whereas background gray values are set to 0 or 255.

The human eye can distinguish colors more easily than gray values. Therefore, often a pseudo-color technique is used. In this approach, different gray values are mapped to the different color channels of the monitor. The color ranges can overlap to produce yellow or orange. Using an 8-bit definition for each color channel allows a combination of more than 16 million colors. One method with visually pleasing results is a so-called rainbow LUT, which maps increasing gray values to increasing wavelengths in the visible spectrum from blue via green and yellow to red. True color digitization requires separate treatment of the three colors. Of course, these can also be manipulated by LUTs, which are used to modify the color presentation of photographic images using specific image software.

3.3 Software

The example of a modern tracking software (WinTrack 2000, Real Time Computer, Möhrendorf, Germany) is used to discuss the current possibilities of these complex tools (Fig. 3.2). This software package was developed in Visual Basic taking advantage of the Windows environment, including 3D appearance, communication boxes, buttons, DirectX elements, etc.

Most of the screen is occupied by the digitized image, which can display the life incoming video or a snapshot frozen from the video stream at any time. In the snapshot mode, the image can be binarized to test which areas of the image are currently recognized as objects and which as background. Another toggle allows selection of the bright-field mode (the software detects dark objects on bright background) or dark-field mode (bright objects on dark background).

Before operation, the image is adjusted in focus and brightness using the optical controls on the microscope. The video settings allow a fine tuning by sliding the brightness contrast controls or by entering numerical values. The result of each manipulation of the settings can be followed visually on screen. In addition, an online gray value histogram can be calculated from the current gray value distribution in the image. Because this histogram is updated online, the user can see the change in the histogram as the brightness and control sliders are adjusted.

A zoom function facilitates to visualize details of the objects. The image can be magnified by integer factors of up to 16 times or reduced in size by the same

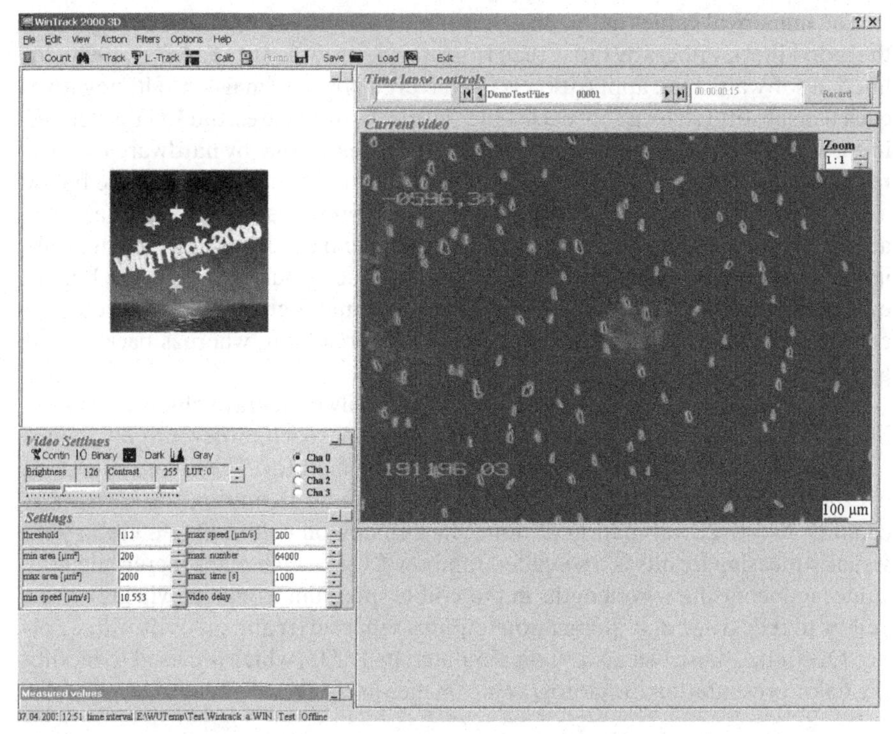

Figure 3.2. Initial screen of WinTrack 2000 (Real Time Computer, Möhrendorf, Germany).

factor. This procedure neither changes the pixel resolution nor does it affect the determination of movement parameters, which is performed on the full image regardless of whether it is displayed fully or only in part. Images from the incoming video stream can be stored on the hard disk of the computer and later retrieved for subsequent analysis.

Usually, image analysis systems operate on the basis of individual pixels (Saxton, 1994). To quantify sizes and velocities in real physical units, the system needs to be calibrated. The WinTrack 2000 software system has a built-in calibration tool. It remembers the calibration parameters for subsequent sessions until a new calibration is performed. In addition, the depth of the measurement cuvette can be indicated, so that the program can determine, for example, true cell densities per ml. A scale bar is projected into the digitized video image.

3.3.1 Identification of objects

Segmentation is a simple, fast, and – especially for real-time image analysis – successful method that is based on a pixel-oriented threshold technique. The underlying assumption is that there are two regions in the image that can be attributed to the objects and background, respectively. Ideally, the gray values of the two regions each follow a normal distribution and show two local maxima in the gray value histogram separated by a well-defined trough (Fig. 3.3).

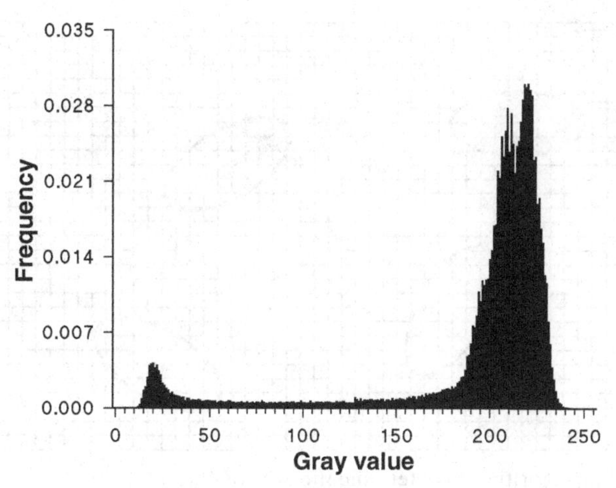

Figure 3.3. Histogram of the gray value distribution in an image of dark organisms on a bright background.

A threshold is defined in this minimum between the two maxima to distinguish between objects (low gray values in bright-field mode) and the background (high gray values). All pixels with gray values below the threshold are regarded as belonging to objects and all pixels with gray values above the threshold as belonging to the background. In the dark-field mode, the algorithm is inverted. The thresholding operation results in a binary image in which all object pixels can be assigned the value 1 and all background pixels the value 0. This technique can be applied to all images with an even background and clear object boundaries. If the background is not homogeneous and cannot be improved, a dynamic background allocation is utilized. A human observer does not have this problem, because that individual recognizes an object by its contrast to the background in its close vicinity. This human behavior can by built into software algorithms. If necessary, the homogeneity of a region can be improved by smoothing operations. If the uneven background remains constant for a given video sequence, an initial image can be recorded and subtracted from the subsequent images. By this technique especially, moving objects are emphasized.

When a gray value threshold has been defined on the basis of the gray value histogram, the image is scanned line by line from the top left corner until a first pixel is hit, the gray level of which is above the defined threshold value (Vaija et al., 1995). Once an object has been identified, there are several methods to identify its area and contour in the image (Häder, 1987a, 1988; Whalen, 1997; Wu & Barba, 1995). The first detected object pixel is an edge pixel. Its eight neighbor pixels are tested for being object pixels. Each found object pixel belongs to the same object, unless two objects touch each other – a complication which we will discuss at a later stage. The *x* and *y* coordinates of all those pixels identified as object pixels are recorded, and for each new object pixel the above operation is repeated (Fig. 3.4). This procedure warrants that only those pixels are identified that are connected forming one unique object. Especially with larger objects, this brute-force method is time consuming.

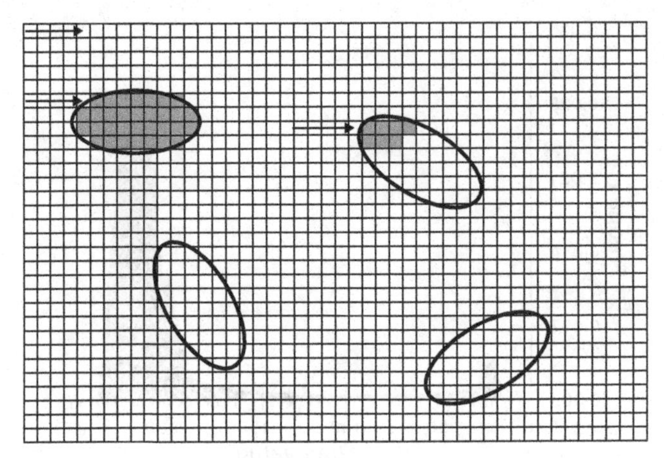

Figure 3.4. Filling algorithm to determine the area of objects.

As an alternative, an edge detection algorithm can be utilized (Grant & Reid, 1981; Pavlidis, 1982; Jähne, 1989). A simple technique also starts by scanning the image line by line until an organism pixel has been found. The x and y coordinates of this edge pixel are stored. Then, the next edge pixel is searched in a clockwise direction by scanning the neighbors of the initial edge pixel. This procedure is repeated until the first edge pixel is hit again and thus the outline has been determined. During calculation, the minimal and maximal x and y coordinates in the outline are memorized from which the geometric center of gravity can be calculated (Jobbagy & Furnee, 1994). A more elegant variation of this approach is the chain code algorithm (Freeman, 1980). A chain code is a sequence of short contour elements coded in numerical form, which uniquely describe the outline of an object in a digital image. The advantage of this robust technique is that it is very fast and reduces the amount of information drastically. One disadvantage is that, in contrast to the object fill method, "holes" in objects are not detected.

Freeman (1980) has provided the mathematical tools to determine important parameters of the identified objects from the chain code, including the length of the outline, the area of the enclosed object, and the coordinates of the center of gravity. The area A of the two-dimensional (2D) projection of the organism can be determined from the integration over the chain elements with respect to the x axis.

An alternative strategy uses a pattern recognition algorithm. Once the software has detected an object, the form and size are stored in memory. During cell tracking, the image is screened for this pattern. This can be done even when the pattern is turned around its center or changes in size; however, in this case, the computational overhead and the time needed are larger.

3.3.2 Cell counting and area determination

In many medical and biological laboratories, cell counting is routinely performed to enumerate microscopic or macroscopic objects – such as cells or colonies – and determine their areas (Häder & Griebenow, 1987). The WinTrack 2000 software contains a module for this purpose. Motile cells do not need to be fixed or immobilized unless they move very fast, because digitization is done in real time. The sample may not only contain the objects to be counted, but also may be contaminated by smaller or larger objects, which the user wants to exclude from the analysis. For this purpose, the software provides a "Settings" window in which the user can define lower and upper limits for the area of identified objects. Objects that do not fit in this size range are ignored. The software determines a size distribution of the identified objects in ten equal size classes. Repetitive counting is possible with a user-defined number of counts. The measured parameters are the number of objects, the mean area, and the mean form factor shown in the "measured values" window. The form factor is a parameter to describe the shape of an object and to distinguish, for example, round from long objects. It is calculated from the ratio of the squared perimeter L and the area A adjusted to the area of a circle.

If two or more objects are in direct contact with each other, errors in counting occur (Häder & Griebenow, 1988). When the average size of the objects is known, bigger ones can be ignored or the area can be divided by the average area to correct the count. This procedure may still yield inaccurate results, especially in dense suspensions. To solve this problem, the erosion technique has been developed: shells, one pixel wide, are removed along the outline of each object and set to a specific gray value. This process is repeated under visual control of the operator until all connected areas are separated at the isthmus. Subsequently, the shells are blown up (dilated) again to their original value, leaving a gap at least one-pixel wide between adjacent objects. In some software applications, objects can be separated manually before analysis using the cursor controls, a mouse or digitizing tablet, or a touch-sensitive screen.

Cell counting can be used to analyze the result of gravitactically moving organisms in a population. Increasing cell numbers at the top of a swimming cuvette would indicate negative gravitactic orientation of the organisms. The method was also used in an ecological approach to determine the densities in populations of freshwater and marine phytoplankton (Häder & Griebenow, 1988; Eggersdorfer & Häder, 1991a,b). Cell suspensions were filled in a Plexiglas column (1 m long for freshwater organisms and 3 m for marine phytoplankton), and samples were taken from eighteen evenly spaced lateral outlets using a peristaltic pump that could handle eighteen samples in parallel. Samples were taken at regular time intervals to determine the density distribution of the populations over the day (see Fig. 5.19). The position of the cells was found to be controlled by light and gravity, and also to follow an endogenous rhythm (see Section 5.5).

3.3.3 Organism tracking

In the following discussion, we will concentrate on black and white images. However, the same algorithms can be used for color images, with some modifications. In 2D image analysis, the detected objects are only a 2D representation of three-dimensional (3D) organisms. Movement perpendicular to the plane of projection cannot be detected in 2D image analysis. Analysis of very fast moving objects can be difficult, because translocation of the objects can be so far that the identification of the object fails, especially at high densities of moving objects. Correspondence between an object in subsequent frames can only be established if the movement vectors are smaller than the mean distance between the objects. If the object distances are statistically distributed, misinterpretations cannot be excluded.

Another difficulty occurs when objects hit or partially or totally occlude each other. A related problem occurs when formerly touching objects move apart; in this case, no correlation can be found to previous frames. These problems can be aggravated by shadow images or nonuniform irradiation in the image.

The position (centroid), area, and form of each object are determined in each frame of a sequence and stored for the following analysis (Häder, 1990, 1994a,b). Complex search algorithms are used to find the correspondence between objects in subsequent frames so that movement vectors can be determined (Häder, 1991a). In a method developed by Noble and Levine (1986), the positions of objects in subsequent frames are stored in an array. Then, the movement vectors are calculated by assuming that the corresponding object in the subsequent frame is the nearest neighbor.

Using subsequent frames recorded at frequent and regular time intervals makes it possible to follow movement vectors of organisms in the time domain (Allen, 1985; Amos, 1987; Coates et al., 1985; Rikmenspoel & Isles, 1985; Burton et al., 1986; Gordon et al., 1984). Using the same technique, the growth or change in form of an organism or organelle can be followed over time (Omasa & Onoe, 1984; Omasa & Aiga, 1987; Jaffe et al., 1985; Popescu et al., 1989; Häder & Lebert, 1998). The objects can range from individual cells such as flagellates or ciliates (Dusenbery, 1985; Häder, 1997a; Häder & Hemmersbach, 1997; Hemmersbach et al., 1996a; Häder et al., 1988) to multicellular organisms (Sanderson & Dirksen, 1985).

The following discussion concentrates on the movements of whole organisms and disregards the analysis of ciliary and flagellar beat patterns (Baba & Mogami, 1985; Omoto & Brokaw, 1985; Cantatore et al., 1989; Häder & Lebert, 1985). The track module in the WinTrack 2000 software determines the movement parameters of motile objects on the basis of short track segments (minimal 160 ms). A few (e.g., five) subsequent frames from the incoming video sequence are captured and stored in memory. All objects in the first image are identified using the algorithms described above for object detection. Next, the positions of the cells are determined in the second image. This and the following images are not scanned line by line, rather the organisms are searched starting at the corresponding gravicenters found in the first image. If an object has been identified at or near the position of the corresponding object in the first image, correspondence is assumed – provided there are no large discrepancies in the areas. If there is no or more than one object in the search area, the organism is regarded as lost and excluded from the rest of the calculation. The procedure is repeated for the third, fourth, and fifth frames. Movement vectors are calculated for each successfully tracked

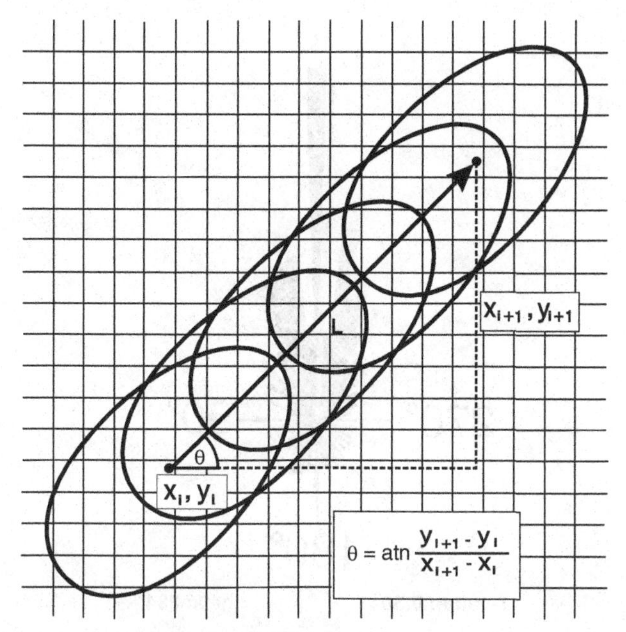

Figure 3.5. Track analysis from a short sequence of video frames. The length of the vector connecting the centroid in the first and final images can be used to determine the linear velocity of the object. The angular deviation from a predefined direction can be used to determine the precision of orientation of a population.

object using the x and y coordinates of the corresponding gravicenters in the first and fifth frames (Fig. 3.5). These vectors are used to calculate the direction and velocity for the moving objects.

The area (again with the correct physical dimension, e.g., square micrometer) and form factor of each object are calculated, and all parameters recorded. After each analysis cycle, the results are updated on screen indicating the determined parameters (i.e., the number of evaluated tracks, the mean area and mean form factor of the detected objects, the mean velocity of movement, the elapsed time, and the percentage of motile cells are displayed). All values are based exclusively on the objects which meet the selection criteria (minimal and maximal area and speed), with the exception of the percentage of motile cells.

A histogram is plotted and updated after each cycle to show the direction of movement of the tracked organisms binned in sixty-four sectors (Fig. 3.6). The upward direction in the histogram (0°) corresponds with the upward direction in the video window. A statistical value for the precision of orientation, the r-value, is calculated according to the following equation (Mardia, 1972; Batschelet, 1981):

$$r = \frac{\sqrt{\left(\sum \sin\alpha\right)^2 + \left(\sqrt{\sum \cos\alpha}\right)^2}}{n}, \tag{3.1}$$

where n is the number of tracks and α the angular deviation of each track segment defined above. The r-value runs from zero for a completely random orientation to

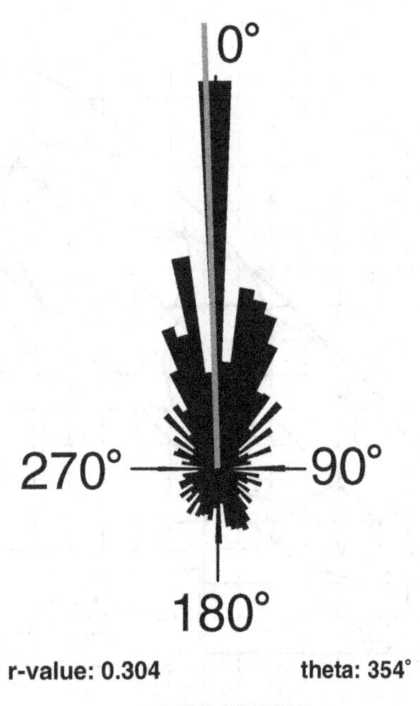

Figure 3.6. Orientation histogram showing the direction of movement of a population of microorganisms. The inserted bar to the left of 0° indicates the mean direction of the population (theta = 354°). The *r*-value is a measure of the precision of orientation. The Rayleigh test is used to determine whether the distribution is random or not.

one for an ideal orientation of all objects in the same direction. A Rayleigh test is applied to determine whether the distribution is random or not. The critical value for the error probability *P* is read from a table. For measurements with more than 100 tracks, the value $2nr^2$ is calculated from the *r*-value and the number of tracks *n* and compared with the table entry.

The alignment is a measure of how well the tracks are aligned with the *x* or *y* axis and calculated from $(\Sigma|\sin\alpha| - \Sigma|\cos\alpha|)/n$, where α is the angular deviation of each track and *n* the total number of tracks. The numerical value ranges between +1 (all tracks parallel to the *y* axis) and –1 (all tracks parallel to the *x* axis); a movement on the 45° line results in a value of 0. Another orientation parameter is the *k*-value, which is calculated by (Σupward swimming cells – Σdownward swimming cells) ∗ 100/*n*. This value runs between 100 (all cells swim in the upward semicircle) and −100 (all cells swim downward); zero indicates random orientation or a split in the population, with one-half of the organisms moving upward and the other downward.

The mean direction of movement (θ) is calculated, printed numerically, and indicated by a bar in the histogram. The direction histogram can be replaced by a velocity histogram that shows the velocity distribution of the motile objects in dependence of their movement direction (with the correct physical units,

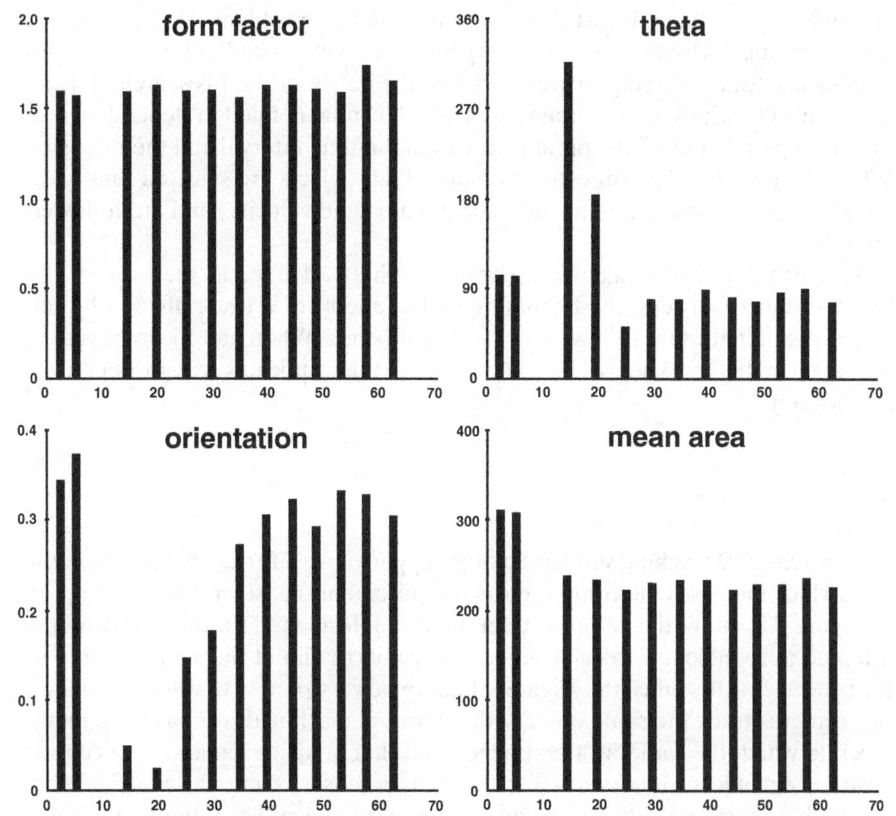

Figure 3.7. Diagrams showing as examples the kinetics of the mean form factor, mean direction of movement (theta), precision of orientation, and the mean area in a population of flagellates during gravitactic orientation.

according to the calibration procedure). The mean velocities are calculated for the upward, downward, and sideward moving objects (moving in 120° sectors each). This is of interest when objects move with different velocities in dependence of their direction (e.g., gravikinesis; Machemer & Bräucker, 1992). The Windows technology allows performance of the calculations in parallel to the video frame acquisition (multithreading).

An advanced option of the tracking module is to display the kinetics of the orientation parameters over the time of the experiment. In this mode, the kinetics of the precision of orientation, mean direction of movement, mean speed in the upward, downward, and sideward direction, overall mean speed, as well as of the mean form factor and of the mean area can be selected to be displayed (Fig. 3.7).

Whereas the track module follows organisms over limited time intervals, it may be of interest to follow the tracks as long as they are in the field of view. This type of analysis is required, for example, for the determination of the free path length of the organisms (Glazzard et al., 1983). During long tracking, chances are high that an object is lost due to correspondence problems, occlusions, and objects hitting each other, as well as organisms moving in and out of focus. If no organism is found at the previous position, the neighborhood is searched using a dynamically adjusted area, the size of which can be defined in dependence of the mean path length of the organisms within the time interval and their density. When the long tracking procedure is started, all objects are selected that meet the selection criteria (minimal and maximal area and velocity) and are followed until they are lost.

Linearity is another calculated parameter that is defined as the ratio of the linear distance between the beginning and endpoint of a track divided by the actual path length that follows all curves and bends. When an object moves on a straight path, the linearity is one; a small value indicates a high degree of meandering.

3.3.4 3D tracking

In most cases 2D tracking suffices. In other applications, 3D tracking is indispensable, which increases the computational requirements considerably.

Figure 3.8 shows the setup to track motile objects in 3D using two cameras oriented perpendicular to each other. The cameras aim at the same volume in the central cuvette. Since the alignment has to be very precise to warrant that the two cameras track the same objects, the device is machined from a single metal block to which the cameras are attached and that holds the cuvette in a central position. An object with a precisely machined tip is inserted into the central cavity, and both cameras are adjusted to produce a sharp image. Then, the cameras are aligned horizontally and vertically by a set of four precision screws each, so that the two images of the object are congruent. The output of the two black and white cameras is routed to two of the three color channels of the digitizer card, so that both aspects of the same object can be analyzed in the same video image.

The software is designed to first identify an object in one camera view and determine its x and y coordinates. Then, it tries to identify a corresponding object in the other color plane at the same y coordinate and determine the z coordinate. The following analysis determines whether the two objects in the two camera views are valid corresponding images of the same object. The subsequent calculation of the direction of movement is done in a 3D space using similar mathematics as for 2D analysis expanded for three dimensions.

3D tracking has been used, for example, to analyze the tracks of the ciliate *Stentor coeruleus* before and after a flash of light (Kühnel-Kratz & Häder, 1993, 1994). The cells were automatically tracked, and when they entered a central volume of their swimming chamber, a flash was elicited. It was found that the cells, in addition to showing photophobic responses, performed a gravitactic orientation moving downward in the water column. This would be advantageous for the cells because they are easily damaged by high light irradiation (Häder & Häder, 1991b).

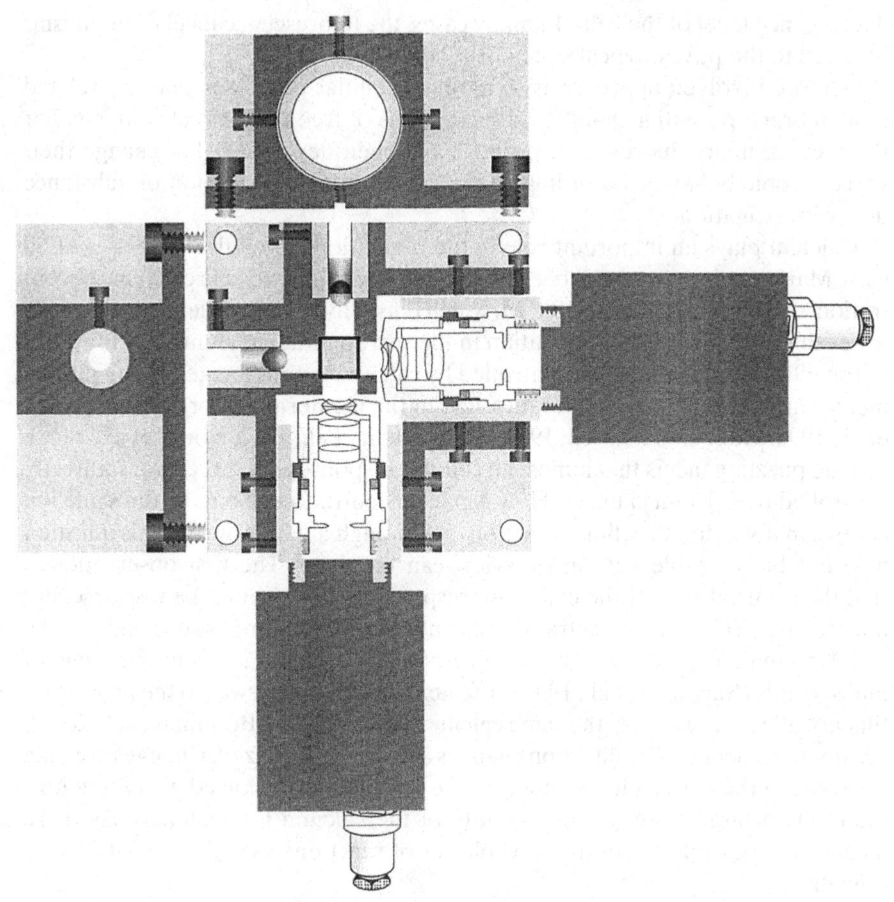

Figure 3.8. Schematic diagram for a 3D tracking device. Two cameras positioned perpendicularly to each other focus on the same spot in a swimming cuvette. The two video images are routed to two of the three color channels of the digitizer card, and the software compares these two partial images.

3.4 Fluorescence imaging

The previous sections summarize the methods for automatic tracking of organisms and a quantitative analysis of behavioral responses to single or multiple stimuli. In the following, we will concentrate on a more specialized topic in motion analysis: fluorescence imaging. In a very basic approach, it is easily possible to track photosynthetic organisms by the autofluorescence of their photosynthetic apparatus. For this specific task, the setup described above has to be slightly modified. Instead of a bright-field or dark-field illumination, an excitation wavelength of approximately 670 nm is used. An emission filter of 680 nm will transmit only the emitted chlorophyll fluorescence radiation. From this point on, tracking is performed as described. An additional advantage of this method is to monitor the fluorescence level of the cells. In many cases, the fluorescence level of organisms is linked to the physiological state of the cell.

A more involved approach is to monitor cellular processes that are related to membrane potential changes or fluctuations of free cellular calcium, etc. For these cases, many fluorescence probes have been developed that change their spectroscopic behavior according to the concentration of the ion or substance under investigation.

Calcium plays an important role in the regulation of cellular processes of all cells. Many mechanisms have been evolved to keep the cytosolic calcium concentration at a low level of about 10^{-7} M, which is usually several orders of magnitude lower than the calcium concentration in the surrounding medium. This low level of the intracellular calcium is controlled by calcium pumps located in the plasma membrane, in the ciliary membranes, and in the membrane of organelles (Webb et al., 1996; Gangola & Rosen, 1987; MacLennan et al., 1997; Norris et al., 1996).

One puzzling fact is that almost all cellular responses are directly or indirectly controlled by calcium changes. How can a concentration change in the same ion control many if not all cellular reactions? Although a full answer to this question may not be available yet, an overview can be given. The first observation is that the total calcium of the cell is not responsible for triggering a response, but only the free calcium concentration (i.e., not bound or stored somewhere in the cell; Bootman et al., 2001). The lifetime of a free calcium ion is in the range of milliseconds (Sugimori et al., 1994). The next part of the answer to the problem is that not all reactions show the same calcium dependences (Bootman et al., 2001). Finally, the calcium concentration changes are strictly localized. Changes are only observed in the close neighborhood of the ion channels involved. Consequently, the calcium targets are in close vicinity of the calcium ion channels. By these means, it is possible to control a whole set of reactions with the same trigger – calcium.

Changes in the calcium concentration, mediated by activated calcium channels in the plasma membrane or in the membrane of internal calcium stores, lead to defined responses of the cell (Golovian & Blaustein, 1997). Several mechanisms can trigger the opening or closing of different calcium channels. One example are voltage-gated calcium channels, which can be triggered by changes in the membrane potential of only a few millivolts and that are located in the plasma membrane, in the T-tubule membrane of muscle cells, or in the cilia membrane. Other calcium channels are activated by the binding of specific ligands – such as hormones, ions, or cyclic nucleotides – at binding sites at or near these channels. In these cases, calcium acts as intracellular second messenger. In some cases, calcium channels are triggered by mechanical stimulation (Petrov & Usherwood, 1994; Sachs & Morris, 1998) [e.g., mechanosensitive channels in membranes of ciliates and flagellates and transduction channels in the hair bundle of vertebrates (Machemer & Bräucker, 1992; Machemer & Teunis, 1996; Yoshimura, 1996; Howard et al., 1988)]. Stretch-activated channels can also be found in plant cells or retinal pigment epithelial cells (Shimmen, 1997; Stalmans & Himpens, 1997). The pharmacology of mechano-gated channels is well known and reviewed elsewhere (Hamill & McBride, 1996).

The principle of all fluorescence techniques is the excitation of molecules with suitable light qualities and the detection of the resulting considerably weaker emission (Haugland, 2002; Tevini & Häder, 1985). Fluorescence probes are specially designed molecules with a specific binding site (e.g., ion-specific chelators or antibodies) coupled to a fluorescent dye (mostly a polyaromatic hydrocarbon or heterocyclic molecule (Haugland, 2002) detected with fluorescence techniques. Binding the target substance leads to conformational changes of the dye molecule resulting in changes of its absorption or fluorescence properties.

Another tool in cell biology is the development of photoactivatable (caged) components. Substances like ions, chelators, hormones, second messengers, or nucleotides are bound to special groups (mostly derivates of *o*-nitrobenzylic compounds). As long as they are bound to this "cage," they are biologically inactive (Haugland, 2002). The photolysis of the caged substances can be elicited by the application of low-energy ultraviolet (UV) flashes (Gilroy et al., 1990). The internal calcium concentration can be manipulated with photolabile calcium chelators, such as nitrophenyl-ethyleneglycol-bis(β-aminoethyl ether)-*N*,*N*,*N*′,*N*′-tetraacetic acid (NP-EGTA) or dimethoxy-nitrophene, which show a drastic decrease in their affinity to calcium after UV irradiation (about 12,500-fold in the case of NP-EGTA; Haugland, 2002). In addition, there are primarily inactive calcium chelators that show an increase in their affinity on illumination (Haugland, 2002; Kaplan et al., 1988; Graham & Kaplan, 1993). It is even possible to combine fluorescence measurements with electrophysiological techniques (Mason et al., 1995; Helmchen et al., 1997). By this means, one can simultaneously measure membrane potential and calcium concentration, which is very advantageous, for instance, in the investigation of voltage-dependent calcium kinetics.

Confocal microscopy allows 3D reconstructions of fluorescence-labeled cell structures (Braun & Wasteneys, 1998) or the detection of local changes in ion activity (Stricker, 1997; Nitschke et al., 1997; Lipp et al., 1996).

Fluorescent calcium indicators with different chemical or physical properties are commercially available (Molecular Probes, Eugene, OR, U.S.A.). The calcium binding indicators are normally derived from the calcium chelators BAPTA [1,2-bis(*o*-aminophenoxy)ethane-*N*,*N*,*N*′,*N*′-tetraacetic acid] or EGTA, which form dynamic complexes with calcium ions (Gurney & Bates, 1995). The resulting conformational change of the binding sites on calcium binding affects the spectroscopic properties of the fluorophore. With one-wavelength (wavelength stable) indicators like Calcium Crimson (Fig. 3.9), an increase in the amplitude of the fluorescence can be observed. In contrast, calcium-binding, dual wavelength or wavelength shifting indicators such as Fura-2 show a shift of the fluorescence excitation and a corresponding change in the fluorescence emission intensity when the excitation wavelength is kept constant. In wavelength-stable dyes, the fluorescence intensity is related to the calcium concentration. In contrast, the response of wavelength-shifting dyes is determined by the ratio of the emission signal at two different excitation wavelengths. One excitation wavelength corresponds to the absorption maximum of the calcium-bound form of the dye and the second to the calcium-free form. In Fura-2 measurements, the emission at 510 nm is monitored after excitation at 340 nm (calcium-bound) and 380 nm (calcium-free).

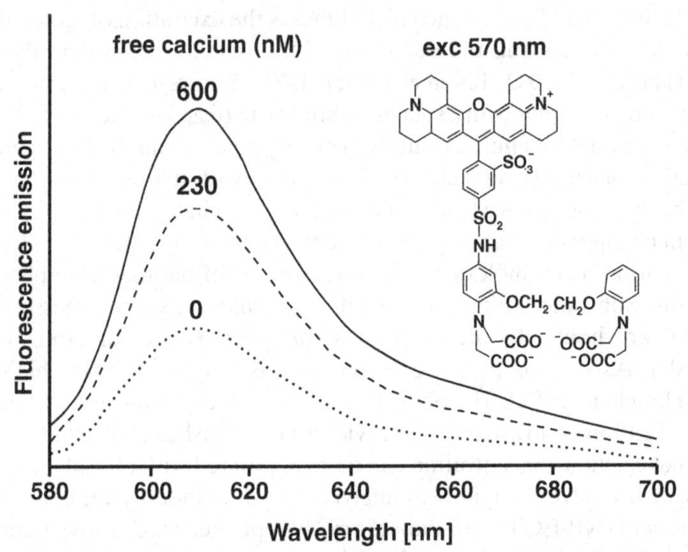

Figure 3.9. Fluorescence emission spectra and chemical structure of the wavelength-stable fluorescent calcium indicator Calcium Crimson. Maximal absorption at 590 nm, and maximal emission at 615 nm. The amplitude of its fluorescence intensity increases on calcium binding. (Modified from Haugland, 1997, 2002.)

The ratiometric indicator Indo-1 is a dual emission dye that is excited at 338 nm. It changes the emission maximum from 480 nm (calcium-free) to 405 nm.

A list of common fluorescence indicators is given in Table 3.1. Several factors have to be considered for the right choice. First, the K_d value. Then, the toxicity to the cell is an issue. In addition, a possible interference between excitation wavelength and photo-controlled behavioral reaction, like phototactic or photophobic responses (cf. Chapter 7 for details) might be a problem. In flagellates, most calcium indicators cannot be used due to the absorption properties of the dyes that would lead to an interference between gravitaxis and phototaxis, as well as photophobic events that might have unpredictable influences on the calcium concentration.

Due to the charge of the carboxylic acid groups at its binding site, the dye molecule is lipophobic and can hardly pass the cell membrane (Fig. 3.9). In contrast, the acetoxy methyl (AM) ester form of the dye, in which the carboxylic groups are uncharged by coupling to an acetoxy methyl, can easily enter into the cell. To avoid accumulation of the small dyes inside compartments (like chloroplasts or mitochondria), some dyes are available as large dextran-bound molecules (see below for loading details).

To measure the near-membrane calcium concentration, which is most likely different from the cytosolic calcium concentration, some calcium indicators are conjugated with a lipophilic alkyl chain (Etter et al., 1994, 1996). Membrane-bound C_{18}-Fura-2 responds several times faster to transient changes in the calcium concentration upon depolarization than cytosolic Fura-2 (Etter et al., 1994).

Table 3.1. *Summary of commonly used fluorescence dyes*

Indicator	Ex/Em (nm)	F_{Ca}/F_{free}	K_d
Single wavelength dyes			
Fluo3	490/530	~100–200	390 nM
Rhodamine Red	550/580	>100	570 nM
Calcium Green-1	506/531	~14	190 nM
Calcium Green-2	503/536	~100	550 nM
Calcium Green-5N	506/532	~38	14 μM
Calcium Orange	549/576	~3	185 nM
Calcium Orange-5N	549/582	~5	20 μM
Calcium Crimson	590/615	~2.5	185 nM

Indicator	Ex (nm)	Em (nm)	K_d
Wavelength shifting dyes			
Quin-2	330 + 350	490	60 nM
Fura-2	330 + 350	510	145 nM
Mag-Fura-2	340 + 380	510	25 μM
Fura Red	420 + 480	660	140 nM
Indo-1	360	405 + 480	230 nM

Note: Ex = excitation wavelength; Em = emission wavelength; F_{Ca} = fluorescence level of calcium-bound dye; F_{free} = fluorescence level of calcium-free dye; K_d = dissociation constant.

Another method of calcium measurements is the use of bioluminescence based on calcium indicators like aequorin (Borle, 1994; Haugland, 2002). The aequorin complex consists of an apoprotein (22,000 molecular weight) and the luminophore coelenterazine. The presence of oxygen and calcium results in oxidation of the luminophore and release of a blue light photon. Because in this case there is no need for excitation light, the systems are very suitable for the indication of calcium dynamics in cells with pronounced autofluorescence, such as photosynthetic cells (Cubitt et al., 1995; Schaap et al., 1996; Johnson et al., 1995; Knight et al., 1991).

Another sensitive approach for calcium measurement is the use of calcium-sensitive microelectrodes. There was good coincidence of the signal obtained with an electrode, compared with simultaneous Fura-2 measurements in *Sinapis alba* (Felle & Hepler, 1997).

Excitation and detection of the fluorescence emission of single cells, as well as for multicellular organisms, can be performed with a microscope equipped with a suitable filter set. The excitation light can be produced by a xenon lamp combined with an excitation filter with suitable transmission properties. The excitation light is deflected by a dichroic mirror (reflection in the excitation wavelength range and transmission in the emission wavelength range) through the objective onto the sample. The emitted fluorescence light will pass through the dichroic mirror. The emission light is selected with an emission filter transmitting in the fluorescence wavelength range (Fig. 3.10). Because the intensity of the fluorescence signal is

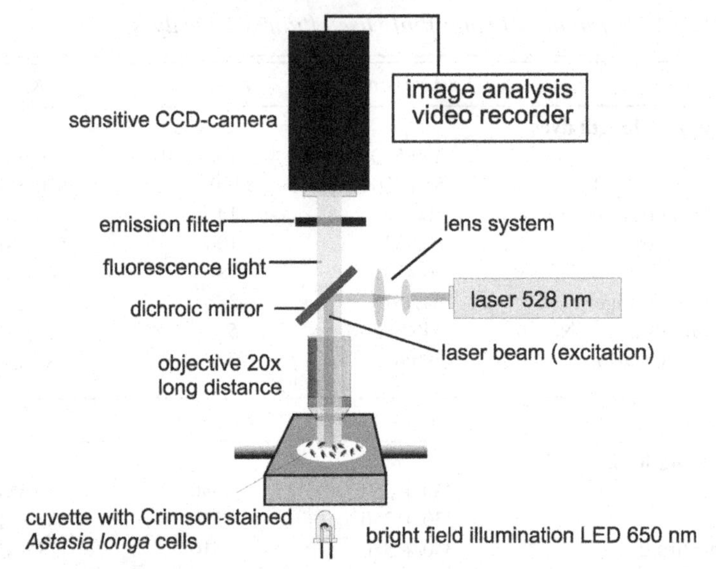

Figure 3.10. Schematic drawing of the experimental setup for calcium measurements in the motile flagellate *Astasia longa*.

normally very low (10^{-5}–10^{-6} lx; Mason et al., 1999), images need to be recorded by a sensitive camera (intensified CCD camera, integrating camera or single-photon camera) that is mounted on top of the microscope. Alternatively, confocal microscopy can be used to record the fluorescent images. Moving the focus also in the z direction allows 3D image reconstruction.

The chemical and structural properties of the dye determines the loading procedure. The basic question is whether the dye is charged under physiological conditions or not. Charged dyes are essentially impermeable for the cell membrane. In these cases, it is possible to apply the indicator substances, such as an AM (acetoxy methyl) ester. The intracellular concentration of the trapped dye is several times higher than the concentration applied to the medium. However, when applying an indicator substance to a cell culture, the dye concentrations will vary considerably from cell to cell.

Alternatively, non-acetoxy methylized dyes can be loaded noninvasively into pH-tolerant cells when the medium is adjusted to a low pH (acid loading). In this case, carboxylic acids are protonated, resulting in an uncharged form of the dye. After membrane permeation, the cellular neutral pH will result in a deprotonation of the substance (Legue et al., 1997).

High molecular dextran conjugates of calcium indicators were developed to overcome compartimentation. These dyes are conjugated to dextran molecules (10,000–2,000,000 molecular weight) and are generally loaded by the help of invasive techniques, such as electroporation, scrape loading, microinjection (Stricker, 1996, 1997), or membrane perforation with UV laser microbeams (Stricker, 1996; Greulich & Pilarczyk, 1998). Electroporation and scrape loading result in transient pores in the plasma membrane, which allows molecules to enter the cell from the surrounding medium. In the case of electroporation, this short time permeabilization of the membrane is achieved with the application of short electric pulses (Weaver, 1993), in the case of scrape loading by application of mechanical forces (Borle, 1994; Behrenfeld et al., 1993; El Fouly et al., 1987; McNeil et al., 1984; Schlatterer et al., 1992).

In ciliates, some experiments with calcium indictors were performed to address questions regarding trichocysts discharge (for an overview, see Plattner & Klauke, 2001). As incubation of ciliates in Fura leads to an uptake of the substancelike food via the cytopharynx and thus predominant loading of the food vacuoles of the animals, the cells had to be loaded with Fura by microinjection. To the best of our knowledge, the only case where calcium indicator measurements in protists were done in the context of gravity-related motility is in the case of *Euglena gracilis* and *Astasia longa.*

In the following, we will describe the experimental approach to monitor calcium concentration changes during reorientation of free swimming cells in space and ground experiments (Richter et al., 2001b). The experiments were performed with *Astasia longa*, a close relative of *Euglena gracilis*. The chlorophyll-free *Astasia* cells show only a negligible autofluorescence – in contrast to the green *Euglena* cells – and are not influenced by the excitation light of the green laser as were the colorless *Euglena* strains (e.g., 1F or 9B; Lebert & Häder, 1997a), which showed a strong photophobic response due to the laser illumination.

The cells were concentrated and incubated for several hours in the presence of Calcium Crimson dextran. The loading was performed by electroporation.

Figure 3.11. Experimental setup of the MAXUS 3 space experiment. Crimson-loaded *Astasia longa* cells were excited with laser light at 538 nm. The fluorescence signal from the cells is detected with an image intensifier CCD camera. The cuvette with the cells **(top right)** can be turned to bring cells back into the camera view field. (Photo courtesy of Astrium, Bremen, Germany.)

However, loading of cells is more the result of the combination of scrape loading and electroporation. Subsequently, the cells were washed several times in fresh medium following the electroporation procedure. After a final test for sufficient fluorescence levels, the cells were transferred into custom-made, stainless-steel cuvettes and mounted into an experimental module of a sounding rocket (Fig. 3.11). The space experiment consisted of two identical setups. Two epifluorescence microscopes, equipped with 20× long distance objectives, were mounted on a centrifuge. The excitation was performed with lasers at 538 nm. The intensity of the lasers was reduced with neutral glass filters, and the red part of the laser radiation – resulting from the pump diode – was removed with suitable filters. A beam splitter deflected the laser beam onto the sample. The fluorescence of the samples was additionally filtered through an emission filter combination. The fluorescence signal of the two microscopes was recorded with an intensified CCD camera, each transmitted to the ground control center and recorded with video recorders. Weak bright-field illumination was possible. Telemetry commands were available to control the experiment from the ground. The centrifuge could be adjusted to defined accelerations. During the launch, the cuvette was constantly turned around its short axis to avoid cell sedimentation or swimming out of the camera field of view.

4

Ciliates

Ciliates can be regarded as swimming sensory cells. Their ion channels in the cell membrane have been extensively studied, and a direct correlation between particular ion currents, the membrane potential, the control of ciliary activity, and the swimming behavior of the cells was established. Conclusively, changes in ion fluxes can be identified by corresponding changes in swimming velocity and swimming direction. Thus, ciliates represent suitable model systems to study with noninvasive methods the effects of changes of environmental stimuli on the cellular level. By studying distinct graviresponses (gravitaxis and gravikinesis) under different gravitational stimulations, new results were found indicating that different mechanisms for graviperception have been developed. Uniquely, in the ciliate family Loxodidae, specialized gravireceptor organelles exist, whereas in other species, common cell structures seem to be responsible for gravisensing. Based on the fact that in many ciliates mechanosensitive ion channels are arranged in a bipolar manner and thus ideally suited for perception of the linear stimulus gravity, the old "statocyst hypothesis" was renewed. In the current hypothesis, gravity (e.g., in *Paramecium*) is perceived by sensing the mass of the cell body via distinct stimulation of mechanosensitive ion channels. Signal amplification by cytoskeletal elements, as well as involvement of the ubiquitous second messenger cAMP, seems likely.

4.1 *Paramecium*

4.1.1 Morphological aspects

The biology of the holotrich ciliate *Paramecium* has fascinated scientists over decades, and a large amount of knowledge of its behavior and structure is

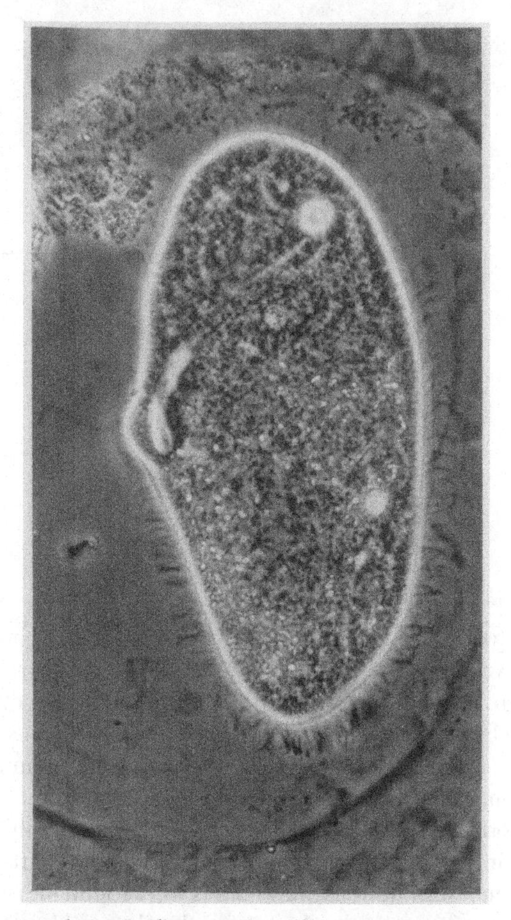

Figure 4.1. *Paramecium biaurelia* (length 170 μm) seen in phase contrast.

available. From the literature, some basic works should be mentioned, which are the famous ones by Jennings (1906) *Behavior of Lower Organisms* and more recent books such as *Paramecium* (Wagtendonk, 1974; Görtz, 1988). *Paramecium* is a typical eukaryotic, unicellular, aerobic organism (Fig. 4.1) bearing some specializations, such as contractile vacuoles for osmoregulation, trichocysts (= extrusive organelles), a dual nuclear system (macronucleus and micronucleus), the alveolar system (comparable with the sarcoplasmatic reticulum in animal muscles), and specialized regions for food uptake (cytopharynx) and defecation (cytoproct). (For reviews on cytology, see Wichtermann, 1986.) As the name "slipper animal" indicates, *Paramecium* has a slipper-like shape that is maintained by the so-called pellicle. The pellicle consists of three layers: the plasma membrane, the alveolar system (membrane-bound sacs) and the epiplasm, an amorphous or filamentous layer (for a review, see Allen, 1988). The pellicle is molded into ridges, forming patterns of hexagons and parallelograms. From these depressions, cilia emerge – single or in pairs – which are used for motility. Intracellular, complex, cytoskeletal structures are organized in cilium–basal body complexes, bundles, ribbons, and meshworks – responsible for maintenance of cell shape and intracellular movements. Cilia with the typical "9 + 2 pattern" [nine peripheral double microtubules (= doublets) and two separated ones (= singlets) in the center] grow from basal bodies. Although the A-microtubule of the doublets consists of thirteen protofilaments, the attached B-tubule has ten of its own filaments and shares three with the A-tubule. The A-tubule bears the dynein arms, which are the force-generating systems. Cilia are arranged in rows on the cell's cortex. They are covered by the plasma membrane (= ciliary membrane; for further details, see Hausmann & Hülsmann, 1996).

4.1.2 **Paramecium** *– a swimming sensory cell*

Paramecium has been established as a model for studying the physiology of excitable cells (Hinrichsen & Schultz, 1988) and signal transduction processes (Pech, 1995). It shows well-characterized behavioral responses to various environmental stimuli, such as temperature, nutrients, or chemicals (cf. Chapter 7). The transduction of a stimulus into a proper cellular signal is coupled to bioelectric events (graded receptor potentials), resulting in a change of the membrane potential (de- or hyperpolarization). In the case of chemosensory transduction, it was shown that signal molecules might either directly pass the cell membrane or bind to receptor molecules, thus inducing intracellular second messenger production (cf. Section 7.3). Identification and understanding of the cellular signal transduction mechanisms offer a great challenge for neurobiological, biochemical, and electrophysiological studies (Hinrichsen & Schultz, 1988; Machemer & de Peyer, 1977; Pech, 1995).

An undisturbed *Paramecium* swims forward on a left helical path with a constant velocity; occasionally, it stops and sometimes even swims backward. When the front end of a cell touches an object, transient backward swimming is observed, which was termed **avoiding reaction** (Jennings, 1906). The complex mechanism of modification of the swimming pattern has been identified by means of electrophysiological and biochemical studies, and the identification of a large collection of behavioral mutants.

4.1.3 Ion channels

The membrane of *Paramecium* has distinct membrane channels (eleven types identified so far), which differ in their mode of activation (e.g., stimulus mode, voltage/ion dependence) and ion permeability. Several calcium-dependent channels have been reported. In addition, two types of potassium-selective channels and also a calcium calmodulin-activated sodium channel are described (for a review, see Machemer, 1988a).

An unequal distribution of ion concentrations across the cell membrane generates an electrochemical potential that can be calculated according to the Nernst equation. Combination of the two major equilibrium potentials, E_{Ca} and E_K, and their corresponding ion conductances determine the membrane potential

V_m of a ciliate, which is maintained by calcium and potassium pumps. The cell membrane of freshwater paramecia has a low permeability for calcium, and the resting potential of about −40 mV mainly depends on potassium and, to a smaller degree, on calcium. The negative membrane potential leads to an uneven distribution of other ions at both sides of the cell membrane. However, because either concentration differences of these ion species (i.e., chloride and magnesium) or their membrane conductances are low, they only contribute marginally to the membrane potential.

In addition to channels for regulation of the resting potential, the cell membrane bears receptor channels, which are activated by a specific stimulus. Sensing involves small deflections of the membrane potential due to an increase or decrease of ion conductances. Receptor potentials are generated, graded by the number of activated channels, and result in de- or hyperpolarization – depending on the ions involved. Depolarization results from a sudden increase in calcium conductance of the cell membrane and thus the opening of voltage-sensitive calcium channels in the ciliary membrane allowing a passive calcium influx, driven by the electrochemical gradient. Thus, calcium can flow through these channels, increasing the calcium concentration in the intraciliary space. The calcium influx is limited by the increasing intraciliary calcium concentration (Brehm & Eckert, 1978; Machemer, 1988b). Free calcium in *Paramecium* has been reported to be in the range between 30 and 600 nM at resting potential, rising between 10- and 100-fold upon ciliary reversal. These values differ, depending on the method applied and the temporal and spatial (intracellular and intraciliary calcium concentrations) resolutions (for a review, see Plattner & Klauke, 2001). The calcium concentrations published so far for *Paramecium* are intracellular ones, which should differ from the intraciliary ones. Hyperpolarization results from changes in potassium efflux and reduction in the calcium concentration in the ciliary space.

Considerable knowledge exists on the mechanoreception in ciliates based on the activation of mechanoreceptor channels in the cell membrane (Machemer & Deitmer, 1985). Mechanosensitive ion channels are located in the somatic plasma membrane, whereas voltage-sensitive ion channels can also be found in the ciliary membrane. The mechanosensitive ion channels are distributed in a characteristic manner. At the anterior pole of a *Paramecium* cell, mainly calcium mechanoreceptor channels are found, at the posterior part hyperpolarizing mechanosensitive potassium channels. In between, both types of channels occur (de Peyer & Machemer, 1978; Ogura & Machemer, 1980; Deitmer, 1982). Distinct mechanical stimulation of different parts of the cell lead to characteristic behavioral responses, depending on the ion channels involved.

4.1.4 Regulation of the ciliary beat pattern

To perform a coordinated swimming of a *Paramecium* cell, 3,000–6,000 cilia have to beat in a coordinated way (Fig. 4.2). Bending of an individual cilium occurs by sliding of adjacent peripheral doublets along each other. The force generating elements of cilia and flagella are the dynein arms, which have an ATPase activity.

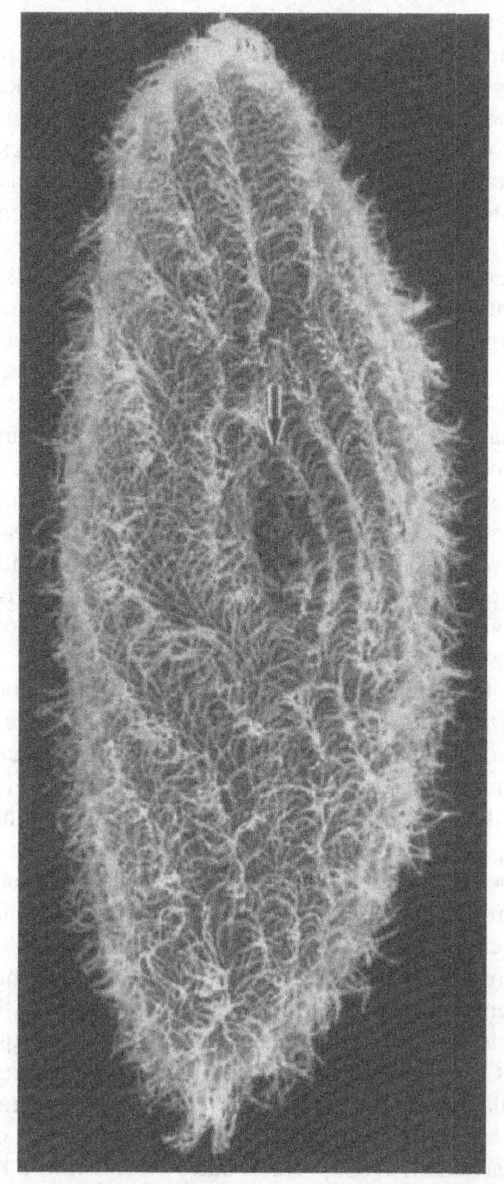

Figure 4.2. View of *Paramecium* in scanning electron microscopy. (Courtesy of W. Foisner, Salzburg, Austria.)

The complex cycle of attachment, detachment, shortening, shifting, and reattachment of the dynein arms is controlled by ATP and its hydrolysis (Satir et al., 1981). As different dynein arms act on different regions of peripheral doublets in a specific temporal pattern, the movement of a cilium results in two distinct phases: the effective power stroke (= propelling phase) and the recovery phase. During the effective stroke, the cilium is stiff and straight and bends near to its basal region (dynein arms of the upper doublets are active) and consequently moves fast in a plane perpendicular to the surface of the cell. During the recovery stroke, bending of the cilium from the base to tip leads it back to the starting position by a counterclockwise rotation in a plane nearly parallel to the cell's surface (dynein arms of the lower doublets are active). It is still unclear whether the central tubules rotate passively and have a stabilizing function and/or whether they have a regulatory function with respect to the ciliary movement.

The ciliary beat in *Paramecium* is regulated in two ways: by a change in the direction of the power stroke and by changing the beat frequency. Both mechanisms determine the swimming behavior of a cell: the swimming direction (direction of the effective stroke) and the swimming velocity (beat frequency) – both are controlled by the intraciliary calcium concentration and the membrane potential. Depolarization or hyperpolarization of the cell membrane can be induced by mechanostimulation, chemical stimuli, or altered extracellular ion concentrations (Naitoh & Eckert, 1968; Eckert, 1972). Depending on the degree of depolarization caused by the increasing intraciliary calcium concentration, the beat frequency increases (depolarization beyond 5 mV), decreases (depolarization above 5 mV), or even induces a reversal of the power stroke and thus backward swimming (depolarization reaches action potential). A cell swims backward until the resting potential is reestablished and calcium is pumped out of the intraciliary space (for a review, see Machemer, 1988b).

The frequency of the ciliary beat, and thus the swimming velocity, is increased by hyperpolarization, which is caused by changes in potassium efflux and reduction in the calcium concentration in the ciliary space. As a result, the frequency of the ciliary beat increases, and the beat direction changes further to the posterior – thus reducing the diameter of the helix and finally increasing the swimming velocity (Machemer, 1974; Naitoh & Eckert, 1969).

The firm link between membrane potential and ciliary function is well documented (= electromotoric coupling; Machemer, 1986). But, how is the signal transduced to the cilium and how is a graded membrane signal translated into a fine-tuned motor response? Naitoh postulated that calcium ions act directly on the intraciliary elements (Naitoh & Eckert, 1968). "Assuming that a Ca^{2+} concentration transmits the membrane signal to the axoneme, the question arises as to how a uniform Ca^{2+} concentration at a particular time can act to induce nine pairs of microtubular doublets to perform different sliding programs in sequence, and to perform theses sequences differently at different levels of the axonemal cross section" (Machemer, 1988a). Travis and Nelson showed that calcium has no direct effect on the ciliary ATPase and the phosphorylation of dynein (Travis & Nelson, 1988). Its influence might be transmitted by calcium-binding proteins, such as calmodulin (Blum et al., 1980) and calcium-dependent enzymes, such as protein kinases (Preston et al., 1990; Noguchi et al., 2001). High concentrations of calmodulin were identified in the cilia (Maihle et al., 1981; Evans & Nelson, 1989) and the soma (Walter & Schultz, 1981). Blocking of the calmodulin action by antibodies inhibited a reversal of the ciliary beat, thus demonstrating its regulatory function (Hinrichsen et al., 1992).

Cyclic nucleotides are controversially discussed with respect to their role in regulating the ciliary beating (Hennessey et al., 1985; Bonini et al., 1986; Schultz et al., 1986; Majima et al., 1986; Nakaoka & Machemer, 1990; Pech, 1995). The cAMP and cGMP second messenger systems have been shown to operate in *Paramecium*, and adenylate cyclases and guanylate cyclases are located in the ciliary membrane (Schultz & Klumpp, 1980, 1988). Changes in membrane potential might be correlated with changes in the concentration of second messengers, though the time course of events is still under discussion (Hinrichsen & Schultz, 1988; Preston & Saimi, 1990; Pech, 1995). While an increasing cAMP concentration is correlated with hyperpolarization of the cell membrane (Bonini et al., 1986; Schultz & Klumpp, 1988; Schultz & Schönborn, 1994), cGMP is linked to depolarizing stimuli (Schultz et al., 1986; Majima et al., 1986). To study regulatory events in signal transduction, *Paramecium* cells were permeabilized with detergents and the cilia of these permeable "models" or isolated cortical sheets reactivated by Mg-ATP. These cell "models" allowed investigation of the role of messengers and ions on the ciliary beat. It was shown that cyclic nucleotides are involved in the fine regulation of the ciliary beat with respect to beat frequency, which determines the swimming velocity. In addition, orientation of the power stroke, which determines the swimming paths, is controlled by cyclic nucleotides (Bonini et al., 1986; Noguchi et al., 1991, 2000, 2001; Okamoto & Nakaoka, 1994a,b; Pech, 1995). Noguchi and coworkers (1991, 2000) demonstrated that the outer dynein arms contribute to the control of the ciliary beating direction in response to calcium and cyclic nucleotides. Furthermore, the response to second messengers differs, depending on localization of the cilia on the body. Cilia on the left-hand side of the body changed the orientation of their power stroke at lower concentrations of cAMP or cGMP, but not cilia in the right-hand field. The intracellular differentiation of this second messenger-regulated mechanism for regulation of the ciliary beat is unknown (Noguchi et al., 1991; Okamoto & Nakaoka, 1994a, 1994b). Schultz et al. (1997) postulated that increasing cGMP might affect the mechanical and electrical properties of the cell membrane (e.g., via regulation of nucleotide-gated ion channels). In this context, they might stabilize the cell "under changing environmental conditions."

In addition to the regulation of the beat pattern of one cilium, several thousands of them have to be coordinated to achieve a controlled movement. The coordination obviously occurs by hydrodynamic effects. In *Paramecium,* the cilia parallel to the beat direction are in the same beating phase, and their bending state is shifted in phase with respect to the neighboring cilia (metachronal waves). Waves of beating cilia can be observed, moving from the posterior left to the anterior right, resulting in a left helical movement of the whole cell (for further description, see Hausmann & Hülsmann, 1996).

4.1.5 **Paramecium** ***mutants***

Different behavioral mutants have been isolated in *Paramecium*, characterized by defects in their electrical properties. For example, the mutant *pawn* of *Paramecium* lacks functional voltage-sensitive calcium channels, thus having little or no voltage-gated calcium current. As a consequence, the mutant can neither depolarize to the level of an action potential, nor perform ciliary reversal nor swim backward (Kung & Eckert, 1972). In contrast, the mutant *dancer* allows an increased calcium flux into the cell, due to a reduction in inactivation of voltage-gated calcium channels: thus, a prolonged response to stimulation can be observed. Due to a decreased calcium-dependent potassium current, the mutant *pantophobiac* shows long periods of backward swimming upon depolarizing stimuli. Other mutations seem to be useful to identify molecular mechanisms in signal transduction chains. Examples are the barium-shy mutant (*baA*), characterized by affected ion conductances due to altered lipid compositions of the membrane, or mutants with deficient chemotactic responses (for reviews, see Kung et al., 1975; Hinrichsen & Schultz, 1988; Ramanathan et al., 1988).

With respect to our question, how gravisensation is managed in single cells, no mutants with respect to mechanosensitivity and gravisensitivity are known so far. However, nature offers some exciting alternatives for comparative studies: though bipolar mechanosensitivity is obviously displayed in many ciliates (Machemer & Deitmer, 1985), exclusively anterior mechanoreceptor responses are found in *Stentor* (Machemer & Deitmer, 1987; Wood, 1982), whereas *Didinium* only has depolarizing mechanoreceptor responses (Hara & Asai, 1980). Comparative, behavioral studies between, for example, *Paramecium* and *Didinium*, gave fruitful insights with respect to gravisensation (cf. Section 4.3).

4.1.6 Graviresponses of **Paramecium**

Under normal gravitational conditions ($1 \times g$), *Paramecium* shows a negative gravitaxis whose degree of orientation varies in dependence of different factors, such as feeding status, culture age, temperature, and oxygen concentration of the medium. In 1889, Verworn described this "negative geotaxis," as he observed that the majority of cells swim upward in a flask (Verworn, 1989b). In addition, *Paramecium* uses regulation of its swimming speed (deceleration during downward swimming and acceleration during upward swimming) to compensate at least part of its sedimentation velocity (82–89 μm/s; Hemmersbach-Krause et al., 1993b; Machemer et al., 1991; Bräucker et al., 1994). This phenomenon was called gravikinesis (for definitions, cf. Chapter 1).

Experiments in hypergravity. To study the underlying mechanism leading to gravitaxis and gravikinesis, the behavioral responses of *Paramecium* were investigated under increased gravitational stimulation (cf. Section 2.5). Köhler (1922) already stated that, during low-speed centrifugation, the cells swim toward the center of the centrifuge (**centrotaxis**), indicating that they maintain their negative gravitaxis. Centrifuge microscopes revealed further details. Higher accelerations of *Paramecium* cells within a density-graded medium placed in a horizontally positioned cuvette showed a random distribution up to $100 \times g$, whereas a gravitactic (centripetal or centrifugal) orientation of the cells was only noted at very high accelerations of $300–400 \times g$ (Kuroda & Kamiya, 1989; Kuroda et al., 1986). However, using another experimental setup to observe cells at "physiological" accelerations (up to $10 \times g$) showed that, under these conditions, *Paramecium* increased the precision of gravitaxis and enhanced gravikinesis – that means their capacity to counterbalance sedimentation. Furthermore, centrifuge experiments under slightly increased acceleration allow a precise determination of the sedimentation rates of immobilized cells (Machemer et al., 1997). This kind of experiments was performed in a vertically positioned cuvette within a slow-rotating centrifuge microscope [NIZEMI (Niedergeschwindigkeits-Zentrifugenmikroskop; cf. Section 2.5)]. No adaptation of the graviresponses to hypergravity was observed, and even repetitive stimulation between $1 \times g$ and hypergravity did not change the corresponding value of gravikinesis and the precision of gravitaxis (Hemmersbach-Krause et al., 1992, 1996; Bräucker et al., 1994).

The fact that gravitaxis and gravikinesis are maintained even at $5 \times g$ shows that the underlying mechanism also works at higher accelerations. The response to hypergravity is rather fast and can be measured at the behavioral level within seconds, which is useful for a free-swimming organism. Interestingly, the time courses of gravitaxis and gravikinesis toward hypergravity differed; the same was observed in studies at low temperatures. Cultivation of *Paramecium* cells at 10°C decreased their swimming velocity by about 75%, whereas gravikinesis was maintained, but gravitaxis was lost after about 1 week. This phenomenon raises the question of whether the two kinds of graviresponses are regulated by two different mechanisms (Hemmersbach et al., 2001).

Experiments in microgravity. To learn how *Paramecium* reacts to microgravity conditions, its swimming behavior was investigated on sounding rockets. With late access (45 min before lift-off), the observation chambers were installed within the rocket on the stage of a horizontal microscope (cf. Section 2.4). At $1 \times g$ conditions on ground, *Paramecium* showed a precise negative gravitaxis (Fig. 4.3a,b). At the beginning of the microgravity phase, 60 s after lift-off, the first video signal was transmitted from the rocket. At that time, the cells showed a precise orientation, however, with an even higher precision than at $1 \times g$. The direction of orientation was determined by the centrifugal spin acceleration that was the last acceleration acting on the cells (cf. Section 2.4.2). Thus, it can be concluded that the cells behave during hypergravity of the launch as on the centrifuge on ground by making their negative gravitactic response more precise. The gravitactic response declines slowly, and, after 80 s in microgravity, a random distribution of the paramecia was registered (Fig. 4.3c), giving evidence that the term "gravi" is justified for the observed orientation at $1 \times g$. After retrieval of the payload (45 min after landing of the rocket), cells showed negative gravitaxis again (Hemmersbach-Krause et al., 1993a). Also, with respect to gravikinesis, a transient response to microgravity was registered during the 7-min lasting microgravity experiment. The differences between upward, downward, and horizontal swimming velocities disappeared in microgravity. Compared with $1 \times g$, an increased mean swimming velocity of the paramecia was measured for up to 3 min, then it approached the value of the former horizontal swimming velocity (Hemmersbach-Krause et al., 1993b). Analysis of the linearity of the swimming paths showed no significant differences between $1 \times g$ (Fig. 4.3, top right-hand panel) and microgravity conditions (Fig. 4.3, bottom right-hand panel), indicating

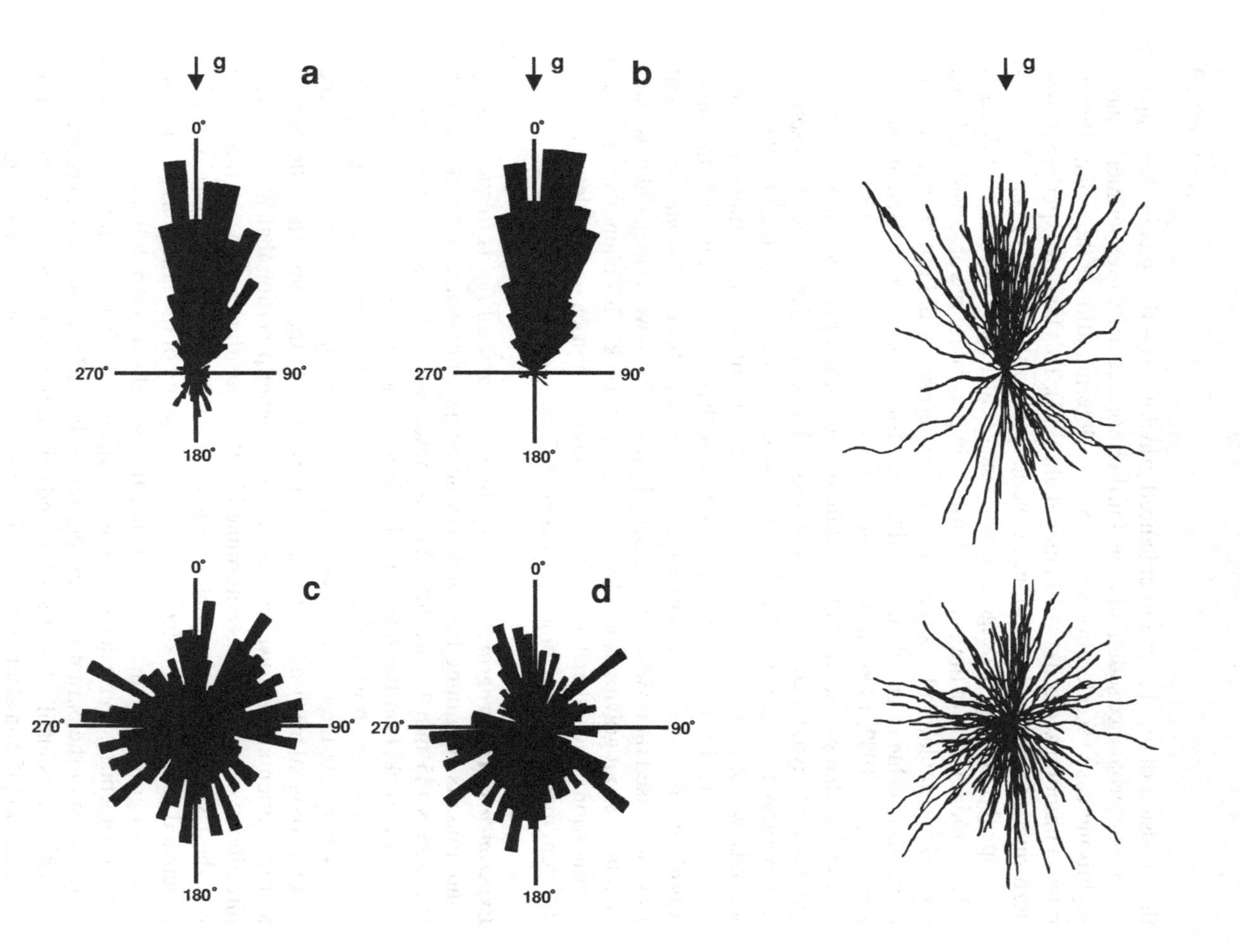
g
a
0°
270°
90°
180°
g
b
0°
270°
90°
180°
g
c
0°
270°
90°
180°
d
0°
270°
90°
180°

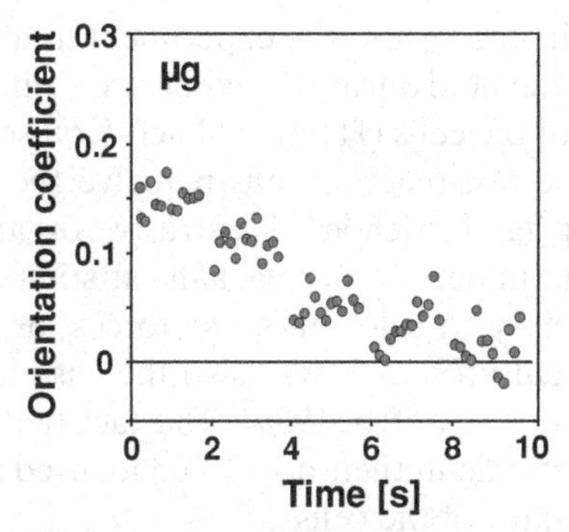

Figure 4.4. Time course of the relaxation of the gravitactic orientation of *Paramecium caudatum* after a step transition from 1 × *g* to microgravity in the drop shaft. The degree of orientation is represented by the orientation coefficient, ranging from 0 (random) to 1 (high degree of orientation). (Modified after Bräucker et al., 1998, with permission.)

no induction of strong depolarizing effects at the membrane level; otherwise, directional turns should occur (Hemmersbach-Krause et al., 1993a). Furthermore, *Paramecium* had been exposed to drop tower and drop shaft experiments, with a free-fall time of 4.7 s or 10 s – allowing the experimenter to study the initial transient responses after the onset of microgravity, but the subsequent steps could not be studied due to the short experimental time (Machemer et al., 1992). After the step transition from 1 × *g* to microgravity, the gravitactic orientation gradually disappears (Fig. 4.4). Transition to microgravity also induces a coincidence of the gravity-dependent swimming rates. Rocket and drop tower experiments support the assumption that the gravity-unrelated swimming velocity of a *Paramecium* cell corresponds to the horizontal swimming velocity at 1 × *g* (Machemer et al., 1991), which fits with the hypothesis that gravikinesis is based on the distinct arrangement of mechano(gravi-)receptors (cf. Fig. 4.6; Machemer & Bräucker, 1992; Bräucker et al., 2001). Nevertheless, it should be mentioned that equal swimming velocities must not represent identical physiological status.

Comparison of the behavior in microgravity with experiments on a fast-running clinostat revealed similar changes with respect to orientation, but with some delay – a phenomenon also stated for other organisms (cf. Section 2.1). After starting the clinostat, the cells continued to swim, with a residual orientation in the former upward swimming direction. After 120 s, a random distribution was registered that persisted until the clinostat was stopped (Fig. 4.3d). After stopping the clinostat following a 2-h run, *Paramecium* regained its negative gravitaxis

---

Figure 4.3. Histograms of gravitactic orientation of *Paramecium biaurelia* under 1 × *g* conditions (**a**) and (**b**) and after 80 s in microgravity during a parabolic flight on a TEXUS rocket (**c**) and 120 s after starting a fast-rotating clinostat (**d**). The linearity of the individual cell tracks (**right-hand panels**) remained high. (**Top right figure**) 1 × *g* control and (**bottom right figure**) microgravity. (After Hemmersbach-Krause et al., 1993a.)

within the first minute. In contrast to the experiment in microgravity, the swimming velocity remained elevated during the whole experimental run, indicating a mechanical stimulation of the cells (Hemmersbach-Krause et al., 1993a,b, 1994). It seems likely that these fast-reacting cells perceive the rapidly turning gravity vector. On the other hand, trichocysts, extrusive organelles of *Paramecium* that are discharged, e.g., under strong mechanical stimulation, an increase of the membrane permeability to calcium, shear forces, or cell damage was still present in clinorotated cultures, as it was also the case in space-exposed cells (Hemmersbach-Krause et al., 1991b, 1994). The fact that trichocysts were still attached to the cell membrane in their normal condensed morphology indicates the good physiological status of the cells.

Different cellular functions – such as endocytosis (food uptake, vacuole formation) and cyclosis (= transport, digestion, and excretion of nutrient), as well as the general ultrastructure of *Paramecium* – appeared unchanged after exposure on a fast clinostat (Hemmersbach & Briegleb, 1987). The proliferation rate of *P. tetraurelia* was investigated in space and on the fast-rotating clinostat (Planel et al., 1982, 1987), revealing a significant increase compared to the $1 \times g$ controls. Whether this is the result of a direct interaction of gravity with the cell or an indirect effect, such as optimized distribution of food and thus faster growing, remained under discussion. Planel and coworkers (1982) also stated that *P. tetraurelia* cells appeared more spherical in microgravity and that the content of cell protein – as well as the magnesium and calcium concentrations – decreased, compared with the $1 \times g$ controls.

Threshold experiments. The existence of a threshold is a clear indication for a physiological response; a pure physical mechanism would act in a linear fashion, thus allowing an extrapolation from $1 \times g$ and hypergravity to $0\ g$. The question whether a threshold of graviperception exists has to be answered under real hypogravity conditions. As a consequence, *Paramecium* was exposed to defined acceleration steps between microgravity and $1.5 \times g$ using centrifuge microscopes in space. During the 15-day IML-2 mission in 1994, samples were investigated on different mission days. Samples that had been cultivated for 3 days in space on a $1 \times g$ reference centrifuge were transferred to the slow-rotating centrifuge microscope NIZEMI (Fig. 2.7, cf. Section 2.5) and accelerated to $1 \times g$, then stepwise down to $10^{-3} \times g$ (= stop of the centrifuge). Negative gravitaxis of the paramecia was measured with decreasing precision up to $0.3 \times g$ and achieved a random distribution at and below $0.16 \times g$. Alternatively, cells that had been cultivated for 10 days in microgravity were exposed to an increasing acceleration profile. Again, random distribution was measured up to $0.16 \times g$, whereas negative gravitaxis was induced at $\geq 0.3 \times g$ (Hemmersbach et al., 1996b,c). No adaptation phenomena to microgravity concerning the responsiveness were observed. The significant difference between upward and downward swimming velocity disappeared at $<0.6 \times g$. Another experiment on a centrifuge microscope on MAXUS 2 (experimental time 12.5 min; cf. Section 2.4.2) allowed observation of the behavior of *Paramecium* and *Loxodes* (see Section 4.2) in parallel (Hemmersbach et al., 1998). Online computer-controlled image analysis of the video-downlinked microscopic images allowed determination of the responses of the cells to the following acceleration steps on ground. Whereas in microgravity, random distribution and equal swimming velocities in all directions were registered, negative gravitaxis and a discrimination between upward and downward swimming velocities were seen at $\geq 0.35 \times g$ – indicating that the thresholds for gravitaxis and gravikinesis of *Paramecium* are in the same range and confirming data from the IML-2 mission. Interestingly, *Loxodes*, which was studied in parallel on the centrifuge, showed gravitaxis at accelerations as low as $0.15 \times g$. The higher gravisensitivity of *Loxodes* seems reasonable, as this ciliate has distinct gravisensory organelles (see Section 4.2). Threshold values for graviresponses obtained so far indicate that protists are able to detect even 10% of the normal gravitational field strength. From an energetic point of view, a focusing and amplifying system has to be postulated, such as cortical contractile filaments in connection to the gating of gravireceptor channels (cf. Chapter 9, Fig. 9.8; Machemer-Röhnisch et al., 1996).

Identification of the site of graviperception. *Paramecium* was subjected to media of different densities to learn more about the graviperception mechanism. Graviperception should remain unaffected under these conditions if an intracellular gravireceptor is active. In contrast, if graviperception takes place at the level of the plasma membrane or in the case of a passive buoyancy mechanism, the graviresponses should disappear in neutral-buoyancy experiments. With increasing density of the medium, the negative gravitaxis of *Paramecium* became less obvious, and orientation completely disappeared at 1.04–1.05 g/cm^3, which corresponds to the mean cell density (cf. Chapter 9; Fig. 9.6). Higher densities than that of the cells reversed the direction of movement of *Paramecium* (Taneda & Miyata, 1995; Ooya et al., 1992; Hemmersbach et al., 1998; Häder & Hemmersbach, 1997). These results indicate that intracellular sedimenting organelles are not involved in graviperception of *Paramecium* (comparable with the situation in *Euglena*; cf. Section 5.2.3).

Though this result alone does not allow a discrimination between a passive or an active mechanism, additional data from other experiments (such as variable orientation of immobilized paramecia during sedimentation; Kuznicki, 1968; Häder, 1997b) support the hypothesis of a combination of both mechanisms in *Paramecium* (cf. Chapter 9).

In contrast, *Loxodes* maintained its positive gravitaxis independent of the density of the surrounding medium at $1 \times g$, even in combination with hypergravity, where weak gravity-dependent differences become obvious (cf. also Section 4.2.1; Chapter 9; see Fig. 9.6).

To identify the kind of ion channels involved in graviperception, gadolinium and cadmium were used, which are known to affect mechanosensitive ion channels (Millet & Pickard, 1988; Yang & Sachs, 1989; Hamill & McBride, 1996 and references therein). However, behavioral and electrophysiological studies in *Paramecium* showed that, in this species, gadolinium is not specific for mechanosensitive channels and affects also voltage-gated potassium channels, as well as calcium channels. This might be the reason that there was no specific

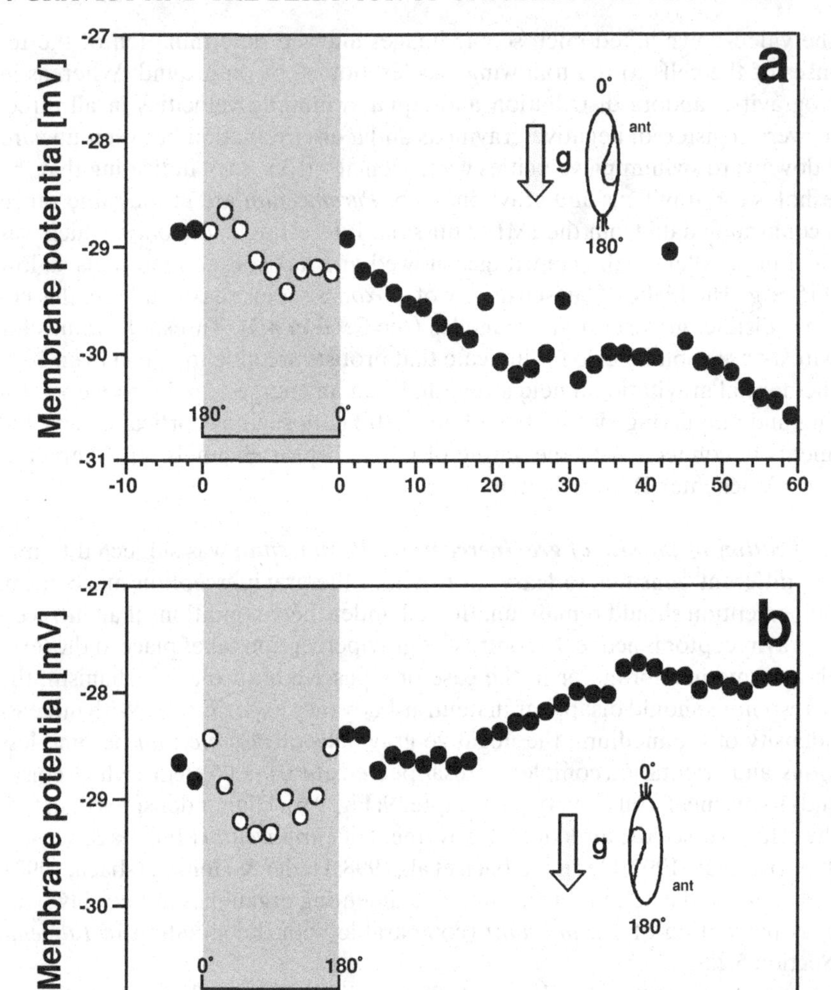

Figure 4.5. The membrane potential of a single *Paramecium* cell was recorded prior, during, and after a reorientation with respect to the gravity vector. A cell with its anterior (ant) pole downward showed a potential of about −29 mV. During and after a 180° turn (**shaded area**), a hyperpolarization can be seen (**a**). A cell fixed in an upright position shows depolarization when turned with its anterior hemisphere downward after the turn (**b**). (Data from Gebauer et al., 1999, with permission.)

effect on the graviresponses of *Paramecium*, in contrast to the findings in *Euglena* (cf. Section 5.2.3; Hemmersbach et al., 2001). Unfortunately, the search of suitable blockers for gravisensitive ion channels in ciliates was so far without success.

Great efforts have been invested to measure directly a gravireceptor potential by means of electrophysiological recording. Depending on the cell's orientation with respect to the gravity vector, a hyperpolarization (Fig. 4.5a) or a depolarization (Fig. 4.5b) of the membrane potential was registered after a 180° turn, indicating the existence of a gravireceptor potential in *Paramecium* (Gebauer et al., 1999) and *Stylonychia* (cf. Section 4.3; Krause, 2003).

The "statocyst hypothesis". Experimental data support the hypothesis that the graviresponses in *Paramecium* result from a sensory mechanism supported by physical parameters of the cell, such as an asymmetric mass distribution toward the posterior part of the cell (cf. also Chapter 9; for reviews, see Bean, 1984; Machemer & Bräucker, 1992; Hemmersbach et al., 1999b; Hemmersbach & Häder, 1999; Hemmersbach & Bräucker, 2002).

The knowledge derived from electrophysiological studies was the basis for a renewed interest in the statocyst hypothesis (Lyon, 1905; Köhler, 1922, 1930). The bipolar arrangement of mechanosensitive ion channels at the cell poles is ideally suited for a distinct gravistimulation in dependence of the orientation of a cell (Fig. 4.6). But, what is the effective mass that interacts with gravity? As the density of a *Paramecium* cell (1.04 g/cm ^{3}) exceeds the density of the medium by about 4% (Kuroda & Kamiya, 1989; Taneda, 1987; Hemmersbach et al., 1998), this density difference results in a pressure of 0.08 Pa (Ooya et al., 1992) to 0.1 Pa (Machemer et al., 1992) acting on the lower cell membrane. Consequently, mechanosensitive hyperpolarizing potassium channels should be opened in an upward swimming *Paramecium,* mechanosensitive depolarizing calcium channels in a downward swimming cell (Machemer & Bräucker, 1992; Bräucker et al., 2001; Hemmersbach & Bräucker, 2002). According to the electromotoric coupling – which means that the membrane potential determines the ciliary beat pattern (Machemer, 1988b; Naitoh, 1984) – the swimming velocity is either increased (potassium-induced hyperpolarization) or decreased (calcium-induced depolarization). In turn, due to the changed membrane potential, the direction and frequency of the ciliary beat are regulated. This model is well supported by the data of gravikinesis, as *Paramecium* cells speed up during upward swimming and decelerate during downward swimming. Further evidence came from the ciliate *Didinium,* which is characterized by the lack of hyperpolarizing mechanoreceptor responses (Hara & Asai, 1980). As a consequence, this protist cannot accelerate during upward swimming, but due to the existence of depolarizing mechanoreceptor responses, decelerates during downward swimming (Bräucker et al., 1994; Machemer et al., 1992).

Whether gravitaxis of *Paramecium* is regulated by the same mechanism as gravikinesis is still under discussion. According to the hypothesis of Baba et al. (1991), different receptor populations are activated in the course of the helical swimming path of *Paramecium*: in the downward swimming phase of the helix, the cell is depolarized by stimulation of the anterior gravireceptors, resulting in a decreased swimming velocity. During the upward phase, stimulation of the posterior, hyperpolarizing gravireceptors results in an increased swimming velocity of the cell. Integration of these changes in swimming rate over several helical turns leads to upward curved swimming paths in graviorienting *Paramecium* cells (Fukui & Asai, 1985; Taneda, 1987; Taneda & Miyata, 1995). According to this model, the resulting curvature is smaller the more the swimming track is parallel to the gravity vector, thus resulting in negative gravitaxis.

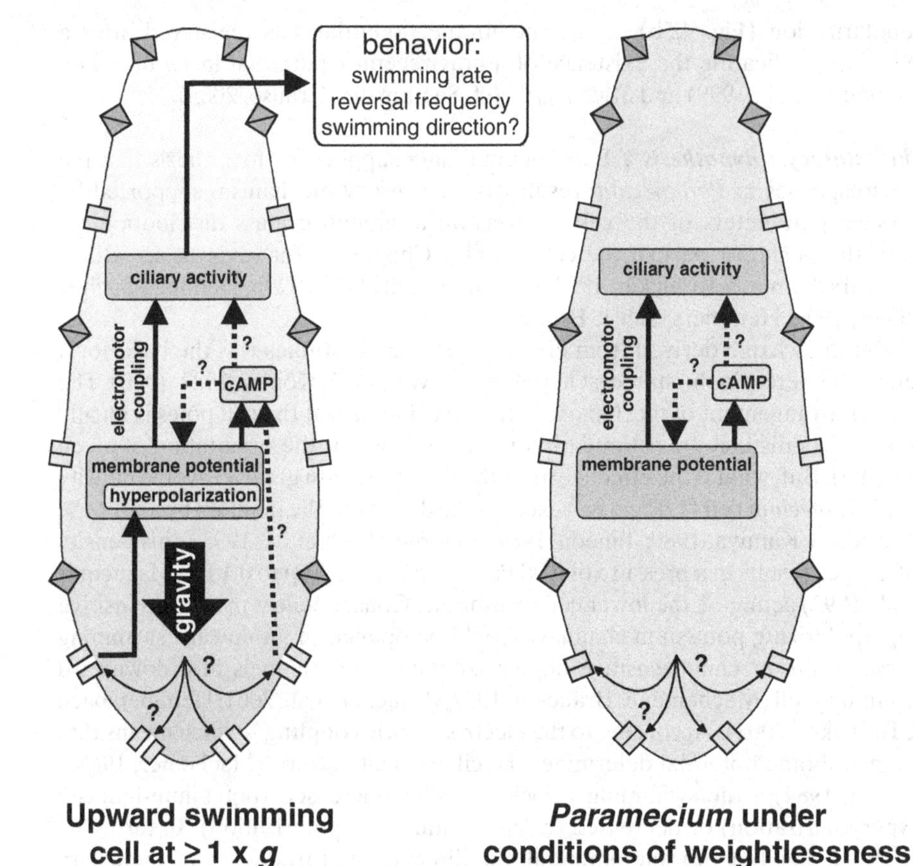

Figure 4.6. Scheme of *Paramecium* demonstrating the polar distribution of mechanosensitive ion channels – a prerequisite for the renewed statocyst hypothesis. The **left cell** demonstrates the situation during upward swimming (negative gravitaxis). Pressure of the cytoplasm on the lower membrane leads to the opening of mechanosensitive potassium channels, hyperpolarization, and thus increased forward swimming. The mechanosensitive calcium channels in the anterior part of the cell are closed in this situation. The involvement of cAMP and cytoskeletal elements is under discussion. In weightlessness (**right cell**), the pressure on the cell membrane no longer exists, and cAMP might decrease (for further details, see text).

As the cyclic nucleotides cAMP and cGMP are obviously responsible for the reorientation and fine regulation of the ciliary beat direction (cf. Section 4.1.4), an involvement in the spatial reorientation of *Paramecium* can be postulated. In the case that gravitational stimulation is coupled with second messenger production – cAMP during hyperpolarization (upward swimming) and cGMP with depolarization (downward swimming) – the following can be postulated: if a negative gravitactic (upward swimming) cell is exposed to hypergravity conditions, where the graviresponses even increase, the pressure on the lower membrane should increase, increasing the number of open potassium channels and in turn the concentration of cAMP. On the other hand, exposure of negative gravitactic

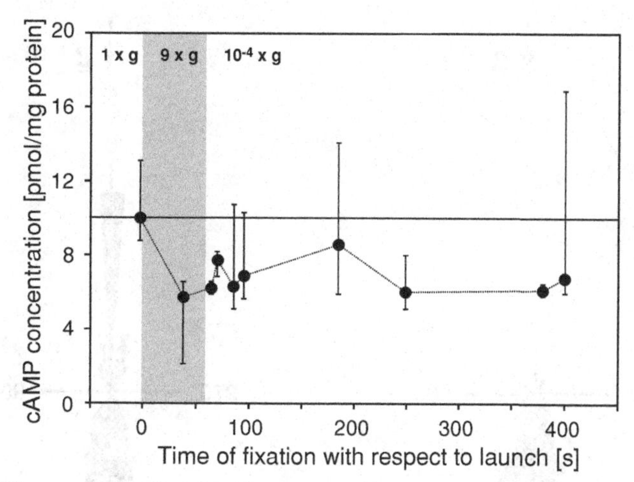

Figure 4.7. Time- and acceleration-dependent changes in cAMP levels of *Paramecium biaurelia* cells during a TEXUS flight (mean, minimal, and maximal values).

cells to microgravity should neutralize the pressure of the cytoplasm on the membrane and thus decrease the cAMP concentration (Fig. 4.6). Our current data support this hypothesis: 4 h after insertion and equilibration within a centrifuge, the cells were accelerated for 45 s at $9 \times g$, then fixated immediately during centrifugation or at defined time intervals after stopping the centrifuge. The mean values of cAMP concentrations showed a 300% increase in hypergravity, which persists for even 2 min after centrifugation, followed by a transient decrease to the range of the $1 \times g$ values. Corresponding cGMP values showed high fluctuations, however, without recognizable relation to the applied accelerations. During the parabolic flight of TEXUS 39, cells were fixated at dedicated times. Data showed an immediate decrease in cAMP level during the hypergravity phase of the rocket, persisting during the total period of microgravity (Fig. 4.7). Again, cGMP concentrations showed gravity-independent variations (Hemmersbach et al., 2002). Unexpectedly (compared with centrifuge data), the decrease in cAMP already occurred during the launch of the rocket. This may be due to less defined acceleration conditions in the rocket (linear and spin accelerations) and the different positioning of the syringes in the different experimental setups.

Data suggest an involvement of cAMP in the gravity signal transduction chain of negative gravitactic cells, though the time course of events remains to be clarified. The question arises whether adaptation to a new environment is coupled with shifts in the basal second messenger levels. An answer should be found by multigeneration experiments on the Space Station.

4.2 *Loxodes*

Karyorelictean ciliates, such as *Loxodes*, are characterized by the inability of their nearly diploid macronucleus to divide during asexual reproduction. Macronuclei develop by differentiation from micronuclei (Raikov, 1985). Although the majority of these ciliates are restricted to marine habitats, the genus *Loxodes* is found

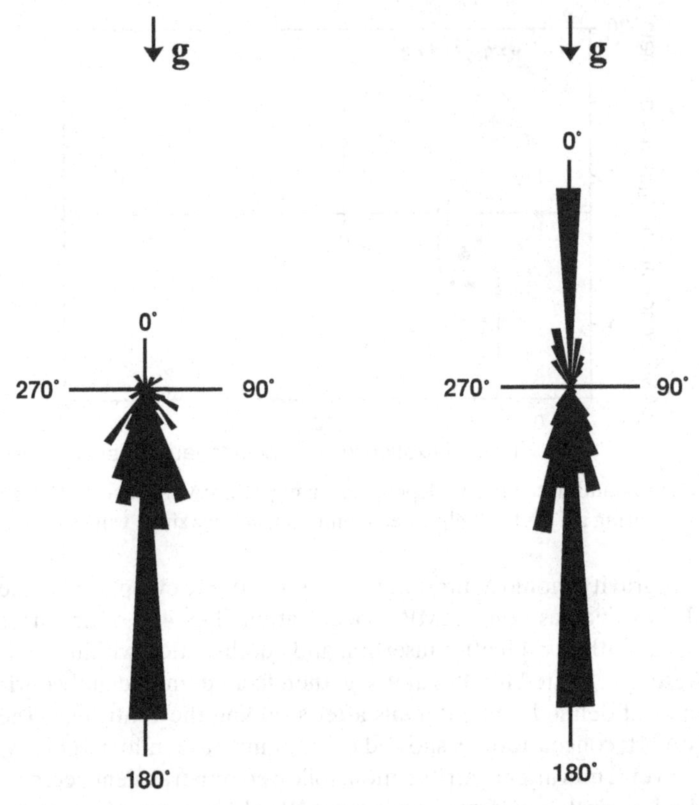

Figure 4.8. Histograms of gravitactic orientation of *Loxodes striatus* under 1 × *g* conditions showing either a positive gravitaxis (**left**) or a bimodal distribution (**right**).

in freshwater habitats. *Loxodes striatus* is a flattened ciliate with eleven rows of cilia on its right side and with only two rows of cilia on its left side (Engelmann, 1862; Penard, 1917). The ciliated side is directed to the substratum when the cell is gliding. Free-swimming cells follow a counterclockwise helical path. Occurrence and aggregation of the microaerophilic *Loxodes* strongly depends on oxygen. It is an ideal model system to study the interaction of light, oxygen, and gravity on a cell and the corresponding signal transduction pathways. Sensitivity to oxygen explains why *Loxodes* cannot be cultivated in Petri dishes, but in vertically positioned test tubes, where they can react to the oxygen gradient. The oxygen tension modifies the sign of gravitaxis and thus the swimming direction. *Loxodes* leaves the sediment of lakes and moves in the water column at low oxygen concentrations. Whereas it shows positive gravitaxis at ≥40% air saturation, either negative or bimodal distribution can be observed at <5–10% atm oxygen (Fenchel & Finlay, 1984; Fig. 4.8). It is proposed that, in *Loxodes*, the mechanism that modulates the mode of gravitaxis is based on a sensor for oxygen. Light increases the toxic effect of oxygen and induces the search of the cells for lower oxygen tension. The facts that *Loxodes* cells respire their oxygen, respiration is sensitive to KCN, the cells show faster and random swimming at 10^{-4}–10^{-6} M KCN – obviously

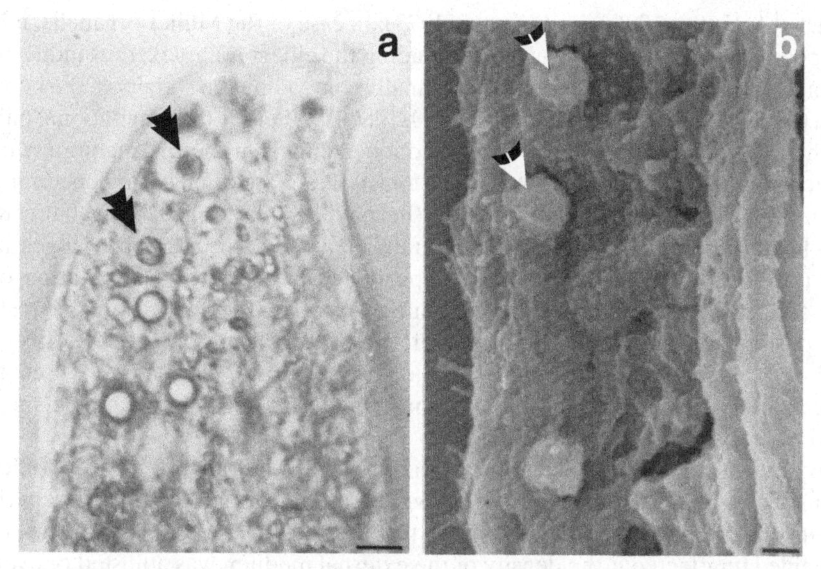

Figure 4.9. Müller organelles at the dorsal rim of *Loxodes striatus*, seen in bright-field (**a**) and scanning electron microscopy (**b**). Scale bars in (a) 10 μm, in (b) 1 μm.

by blocking of oxygen receptor – and finally the accumulation of cells at micromolar oxygen concentrations indicate that the cytochrome oxidase is the oxygen receptor (Finlay & Fenchel, 1986). In addition to oxygen, low temperatures can induce a bimodal distribution in a previously positive gravitactic *Loxodes striatus* culture via a yet unknown mechanism (Block et al., 1999).

4.2.1 Müller organelles of Loxodes *– cellular gravisensors*

Loxodes received attention because it bears statocyst-like organelles. These organelles – discovered in 1786 by O. F. Müller – are a characteristic of karyorelictean ciliates and are called Müller organelles or Müller vesicles (Penard, 1917). The number of Müller organelles varies in dependence of the species and their size between 5 and 25. They are easily visible by means of light microscopy at the dorsal rim of the cell (Fig. 4.9a). Penard gave an astonishingly detailed description despite the poor microscopic techniques at that time. He already saw a vacuole filled with fluid and the statolith connected to a "stick," allowing its positioning in the center of the vacuole. He proposed that this structure functions as mechanoreceptor (statocyst; Penard, 1917).

Electron microscopy and X-ray microanalysis confirmed the findings of Penard, revealing vacuoles of 7–10 μm diameter containing a body of $BaSO_4$ (3–3.5 μm in diameter) fixed to a modified ciliary stick (Rieder et al., 1982; Rieder, 1977; Hubert et al., 1975; Fenchel & Finlay, 1986a; Fig. 4.9b). Interestingly, the lower density of $SrSO_4$ in the Müller bodies of the marine ciliate *Remanella* (Karyorelicta) is compensated by a higher diameter of the mineral body (4 $\pm$ 0.3 μm;

Fenchel & Finlay, 1986a; Bedini et al., 1973). In case of the Müller organelle, it is proposed that gravity is perceived by bending the ciliary complex, thus inducing changes in the membrane potential and finally controlling the activity of the body cilia (Rieder, 1977; Fenchel & Finlay, 1984, 1986a). Whether the gravitational pull of the heavy body is mediated directly to ion channels in the cell membrane or if second messengers are involved is unknown. There are discussions concerning the movement angles of the Müller body (Fenchel & Finlay, 1986a; Neugebauer & Machemer, 1997). Although Fenchel and Finlay (1986a) describe changes in the position of the Müller body in dependence of the spatial orientation of the cell, recent studies revealed maximum excursions of 10° from the resting position (Neugebauer & Machemer, 1997), which in comparison with 6° and −3° deflection of hair bundles in the vestibular ampullae of eel (Rüsch & Thurm, 1989), should produce a saturating stimulus (Neugebauer & Machemer, 1997).

By means of different experimental approaches, the function of the Müller organelles as gravisensor could be shown. First, indications came from experiments under isodensity conditions. The fact that positive gravitaxis of *Loxodes* remained unaffected if the density of the external medium was adjusted or even higher than the density of the cell (1.03 g/cm^3) indicated the existence of an intracellular gravisensor in *Loxodes striatus* (Hemmersbach et al., 1998; cf. Fig. 9.6). Induced by the additional result that gravikinesis was slightly reduced under isodensity conditions – which could be measured in bimodally distributed *Loxodes* cultures – it was concluded that, in *Loxodes*, an intracellular gravisensing mechanism is supported by a membrane-located one (Neugebauer et al., 1998).

To test whether Müller organelles are the gravireceptors, these structures were destroyed by means of laser beams. Individual cells were immobilized and at least one of their Müller organelles was damaged with short series of laser pulses. Subsequently, the cells were transferred into an observation chamber and observed in a horizontal microscope. Analysis of the swimming parameters indicates a loss of gravitactic orientation (Fig. 4.10b), compared with the untreated control cells (Fig. 4.10a), whereas the swimming velocity and the vitality of the cells were unaffected. Destruction of the Müller organelles resulted in random swimming tracks, comparable with the ones observed in microgravity and thus stimulus-free (with respect to gravity) environment (Hemmersbach et al., 1996a). It can be concluded that the Müller organelles serve as cellular gravireceptors, enabling the microaerophilic *Loxodes* to orient in an environment (e.g., sediment), where the density is higher than the one of the cells and where mechanical stimulation at the membrane level will be high. As only isolated cells can be manipulated by the laser beam, the impact of this action on gravikinesis remains open, because this can only be calculated from larger cell populations.

4.2.2 Graviresponses of **Loxodes**

The gravity-dependent behavior of *Loxodes striatus* has been documented in detail. At $1 \times g$, they show a clear gravity-dependent spatial orientation – negative

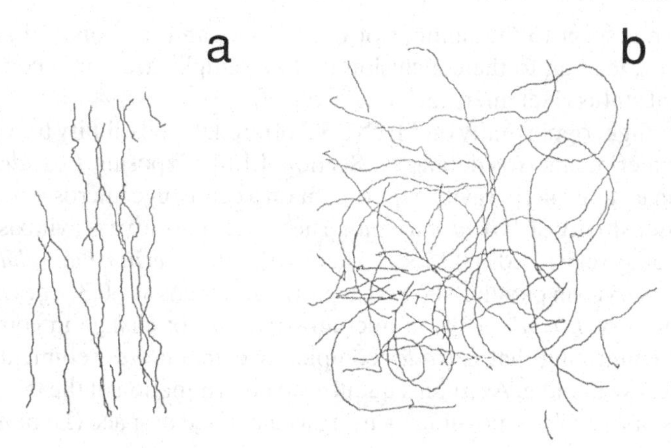

Figure 4.10. Individual cell tracks of *Loxodes striatus* of intact cells (**a**) and after destruction of at least one of the Müller organelles (**b**). This manipulation resulted in random swimming tracks, comparable with the ones observed in microgravity.

or positive gravitaxis, or the occurrence of both within one culture (bimodal distribution; Fig. 4.8). Under hypergravity (tested up to $5 \times g$), swimming *Loxodes* cells maintained their gravitactic behavior, even with an increased precision (Hemmersbach et al., 1996a). Another behavior was stated in the case of gliding *Loxodes* cells under hypergravity conditions, as they reoriented in a counterclockwise direction away from the direction of gravity. According to the authors, this asymmetry of the cell, as well as the asymmetry of the distribution of the cilia, induces a torque. They assumed that, under hypergravity conditions, the cell is no longer able to compensate the increasing torque. As a consequence, the cells are no longer moving exactly in the axis of the resulting acceleration vector (Machemer-Röhnisch et al., 1993). Calculation of the gravikinetic value shows that *Loxodes striatus* completely compensates its sedimentation velocity of 49 μm/s at $1 \times g$ conditions (Bräucker et al., 1992).

Exposing *Loxodes* to the conditions of microgravity resulted in random swimming. During the IML-2 mission, *Loxodes* was cultivated for 15 days in space and exposed to different acceleration profiles by means of a centrifuge. Unexpectedly, they did not react to the variable accelerations, though they showed clear graviresponses in the ground controls from the same culture (performed in parallel) and also in the flight samples after the mission. It remained speculative whether this effect occurred due to temperature problems within the cuvette or due to adaptation to microgravity (Hemmersbach et al., 1996a).

Electron microscopy of the cells from this space experiment revealed no ultrastructural changes of the Müller organelles, though a tendency of a decreased mineralization (barium) was stated. Interestingly, a reduction in barium was also observed in the statoliths of *Chara* rhizoids after cultivation in microgravity (Braun et al., 1999). Due to the fact that the number and distribution of micronuclei, macronuclei, and DNA content of the nuclei vary according to environmental conditions, these parameters were determined after cultivation in microgravity. Staining of cells revealed no difference between flight samples and ground controls with respect to the number of micronuclei and macronuclei and their DNA content, leading to the conclusion that all samples were in a comparable physiological status (Hemmersbach et al., 1999a).

A further flight opportunity on MAXUS 2 offered the possibility to repeat the threshold experiments (for details, cf. Section 4.1.6). Exposing *Loxodes* within this 12.5-min lasting microgravity experiment on a centrifuge microscope showed a high gravisensitivity of *Loxodes striatus*. The onset of positive gravitaxis of *Loxodes* was already registered at $0.15 \times g$, whereas the *Paramecium biaurelia* culture, which was observed in parallel, showed negative gravitaxis at $\geq 0.3 \times g$. The swimming velocities of *Loxodes* cells in microgravity did not change in comparison with $1 \times g$, confirming that *Loxodes* completely compensates sedimentation at $1 \times g$ by faster swimming. As a consequence, no determination of the threshold of gravikinesis of *Loxodes* is possible by using a centrifuge in space (Hemmersbach et al., 1996a).

4.2.3 *Graviperception in* Loxodes *– conclusion*

So far, *Loxodes* is the first known protist that possesses a cellular gravisensor, additional to membrane-located gravisensors, which are found in several other protists. As in other ciliates, the prerequisite of mechanosensitivity is given in *Loxodes*. First, electrophysiological studies indicate that the mechanosensitivity in *Loxodes* is similar to the one in *Paramecium*, at least regarding the receptor channels at the anterior and posterior cell poles (Nagel, 1993). The statocystoid of *Loxodes* may be a specialization due to the living conditions. As this species often creeps on surfaces of detritus or in narrow gaps of the benthos, mechanoreceptors in the cell membrane should be often stimulated by external contacts. This would impede gravisensation and might have favored the development of an intracellular gravisensory mechanism.

4.3 Other ciliates

Different ciliates have been studied with respect to their gravisensitive responses. They differ either in size, ranging from giant cells like *Bursaria truncatella* with a volume of $30 \cdot 10^6\ \mu m^3$ to small ones, such as *Tetrahymena pyriformis*, with a volume of $2 \cdot 10^4\ \mu m^3$. In the case of *Bursaria*, with a sedimentation rate of 923 μm/s, the capacity of gravikinesis is very obvious. Without such a counterbalancing mechanism, this cell would rapidly sink to the ground of a pond. The cell manages this problem by regulation of its swimming velocity to compensate about 60% of its sedimentation rate (Krause, 2003). Exposing this ciliate to the conditions of hypergravity induces a precision of orientation and an increase in gravikinesis, the latter demonstrated by the fact that the upward swimming velocity of *Bursaria* remains virtually constant. Step transition to microgravity in the drop tower of Bremen induced that the two directional-dependent swimming velocities approach the level of the horizontal velocity at $1 \times g$. Intracellular recordings of

Bursaria revealed – as in other ciliated species – an anterior-posterior distribution of calcium and potassium mechanoreceptor channels (Krause, 1999).

The hypotriche ciliate *Stylonychia* crawls along surfaces by means of membranelles and cirri. *Stylonychia* shows negative gravitaxis. Measurement of the movement velocities and calculation of the corresponding gravikinetic values reveal that this ciliate completely compensates its sedimentation velocity. Step transition to microgravity in a drop tower showed that, in *Stylonychia* – as in other ciliates – the graviresponses slowly decreased, indicating the participation of supporting structures (cytoskeleton) in graviperception (Krause, 2003). Comparison with other ciliated species revealed electrophysiological similarities and specifications of *Stylonychia*. The bipolar distribution of mechanosensitive ion channels is existent as in other ciliates. Whereas *Paramecium* shows a transient depolarizing receptor potential, *Stylonychia* has two kinds of voltage-dependent calcium currents, being responsible for a two-peaked action potential – a graded action potential due to activation of calcium channel II (also found in *Paramecium*) and additionally, an all-or-none response (activation of calcium channel I) inducing spontaneous ciliary reversals and thus directional changes of the swimming direction. The two types of channels differ in their thresholds for activation and the capacity to block them by means of drugs (Machemer & de Peyer, 1977; Machemer & Deitmer, 1987). In dependence of the spatial orientation of the cell, a change in membrane potential can be induced due to the pressure of the cytoplasm on the lower cell membrane. Gravireceptor potentials were measured with maximal amplitudes of 4 mV (Krause, 2003).

The ciliate *Didinium nasutum*, a predator of *Paramecium*, is of special interest due to its electrophysiological peculiarities and its impact on the hypothesis of graviperception. *Didinium* can generate depolarizing, but not hyperpolarizing, receptor responses (Hara & Asai, 1980). As a consequence, only depolarizing stimuli can modify the ciliary beat. Analysis of the gravikinetic values reveal that *Didinium* does not actively respond to gravity during upward swimming (lack of hyperpolarizing mechanoreceptor responses), but decreases its swimming velocity during downward swimming. This (monopolar) mechanism is less effective in compensating sedimentation. In horizontally swimming *Paramecium* cells, it has been postulated that, at $1 \times g$ conditions, the bipolar mechanosensitivity cancels gravikinesis – thus the horizontal velocity equals the swimming rate in microgravity and remains more or less unaffected in hypergravity. In contrast, the horizontal swimming velocity of *Didinium* strongly depends on the applied acceleration and increases in microgravity, compared with $1 \times g$ (Bräucker et al., 1994; Machemer et al., 1992) – demonstrating the monopolar mechanosensitivity of this species. Thus, *Didinium* can be regarded as a "kind of mutant with respect to gravisensation." Furthermore, this species shows that **bipolar** mechanosensitivity as a basis of graviperception cannot be generalized.

Due to its small size, studies of the graviresponses of *Tetrahymena* lead to the interesting question concerning the limits of gravisensation with respect to the energetic point of view (cf. Chapter 8). *Tetrahymena* shows negative gravitaxis (Winet & Jahn, 1974) whose degree of orientation increases in hypergravity ($1.8 \times g$), as shown during repeated parabolas of an airplane (Noever et al., 1994).

This ciliate also shows gravikinesis, by which it even overcompensates by about 30% of its sedimentation rate (s = 22 μm/s; Kowalewski et al., 1998). Electrophysiological studies of this species revealed a bipolar distribution of ion channels comparable with the one in *Paramecium* (Takahashi et al., 1980). Using the specific density of the cytoplasm, the cell shape and volume, and the gating distance of a mechanoreceptor channel (here assumed to be 10 nm), a channel gating energy 33 times above the thermal noise level is calculated for *Tetrahymena*. Thus, gravisensation seems likely, even in this small species (Kowalewski et al., 1998).

All ciliated species examined so far show gravitaxis and gravikinesis, enabling these cells to reach and stay in habitats with favorable living conditions. The possession and the polar arrangement of mechanosensitive ion channels seem to be characteristics and prerequisites for gravisensation. The "renewed" statocyst hypothesis – based on mechanosensitive properties of the ciliates – seems reasonable as a model for gravisensation. It might be speculated that the membrane-located gravisensors are early inventions in evolution, whereas the Müller bodies of *Loxodes* are later acquisitions due to their adaptation to special living conditions.

5

Flagellates

Many photosynthetic or heterotrophic flagellates from various taxonomic origins investigated so far have been found to be capable of gravitactic orientation and to orient themselves in the water column by positive, negative, or transversal gravitaxis. Two species can be regarded as model systems, *Chlamydomonas* and *Euglena*, since in these organisms gravitaxis has been studied in more detail. Earlier hypotheses assumed that gravitactic orientation is mediated by a buoy effect, where the cell is tail-heavy, and the flagellum – emerging from the anterior end – pulls the organism upward. A number of observations are in contradiction to this model. Rather, at least in *Euglena*, an active, physiological graviperception mechanism seems to be responsible for the observed orientation. For this organism, the whole cell body is assumed to function as a statolith and exert pressure on the lower membrane. This force is thought to activate mechanosensitive calcium ion channels. During each rotation around its long axis, more calcium enters the cell until a concentration threshold is reached – upon which the flagellum swings out and induces a course correction. Other elements of the sensory transduction chain include changes in the membrane potential, cAMP as secondary messenger, and possibly additional elements. In *Euglena* and other flagellates, gravitaxis is controlled by an endogenous rhythm that also affects the cell form, cAMP concentration, and other physiological parameters. External stress factors – such as excessive solar radiation, high salt concentrations, or micromolar concentrations of heavy metal ions – can modify the response to gravity and invert the direction of movement. The other model system is *Chlamydomonas.* For this organism, it is still discussed whether gravitaxis is mediated by an active physiological receptor or a passive physical phenomenon.

5.1 Introduction

Most flagellates are unicellular heterotrophic or photosynthetic microorganisms. However, they can form aggregates or colonies or even multicellular organisms, such as the green alga *Volvox*. Flagellates belong to many different taxonomic groups, and are listed in the plant and animal kingdoms. Many other groups produce flagellated forms during their life cycle, such as primitive fungi (e.g., Myxomycetes, Chytridiomycetes, and Oomycetes). In mosses and ferns, the flagellated sperm cells remind us of flagellates.

Among the algae, flagellates are found in the Euglenophyceae, Cryptophyceae, Dinoflagellates, Haptophyceae, Chrysophyceae, and Xanthophyceae (Sitte et al., 1998). Among the Bacillariophyceae, no flagellated forms are known. The Chlorophyceae contain unicellular flagellates, flagellated colonies, and multicellular flagellated organisms, as well as flagellated stages in the life cycles of the green macroalgae (van den Hoek et al., 1993). The Phaeophyceae do not contain unicellular flagellates, but all taxonomic groups produce flagellated gametes and/or zoospores. In contrast, the Rhodophyceae do not form flagellated forms at all.

Chlamydomonas is a typical unicellular, green flagellate (Fig. 5.1a,b). Its almost spherical or ovoid body is enclosed by a cellulose cell wall outlined by a cup-shaped chloroplast that contains a pyrenoid involved in starch synthesis. The orange-colored stigma is also located inside the chloroplast near the cell equator and is used in phototactic orientation in light. In the center, there is a single nucleus. The cells are powered by two apical flagella that emerge from basal bodies. The motor responsible for the movement is 520 kDa protein called dynein, which possesses an ATPase function (DiBella & King, 2001). The cells can multiply vegetatively by forming two to sixteen zoospores within the mother cell, which acts as a sporangium. The zoospores are released by rupturing the mother cell wall. Sexual reproduction starts with the fusion of two small gametes, which can be morphologically similar (isogamy) or different in size (anisogamy). In the primitive *Polytoma uvella*, the gametes do not differ from the vegetative cells.

Among the Volvocales, a number of aggregates with increasing specialization of the individual cells are found. In *Oltmannsiella*, four cells form a ribbon, in *Gonium* four to sixteen cells are connected to a flat disk, where all flagella point in the same direction. In *Pandorina*, sixteen cells form a sphere embedded in a common slime mass; and, in *Eudorina* and *Pleodorina*, thirty-two and 128 cells, respectively, form a hollow ball. The most complex organization is found in *Volvox*, in which thousands of individual cells – each following the blueprint of a typical *Chlamydomonas* with two flagella and a stigma – form a macroscopically visible hollow sphere. The cells are interconnected by broad plasmatic bridges, and their flagellar movements are synchronized. This multicellular organism shows a complex specialization with vegetative cells at the front pole and generative cells in the rear half. During vegetative multiplication, individual cells divide into smaller daughter spheres inside the mother organism that eventually bursts open and dies – releasing a number of new individuals that subsequently grow to their typical size.

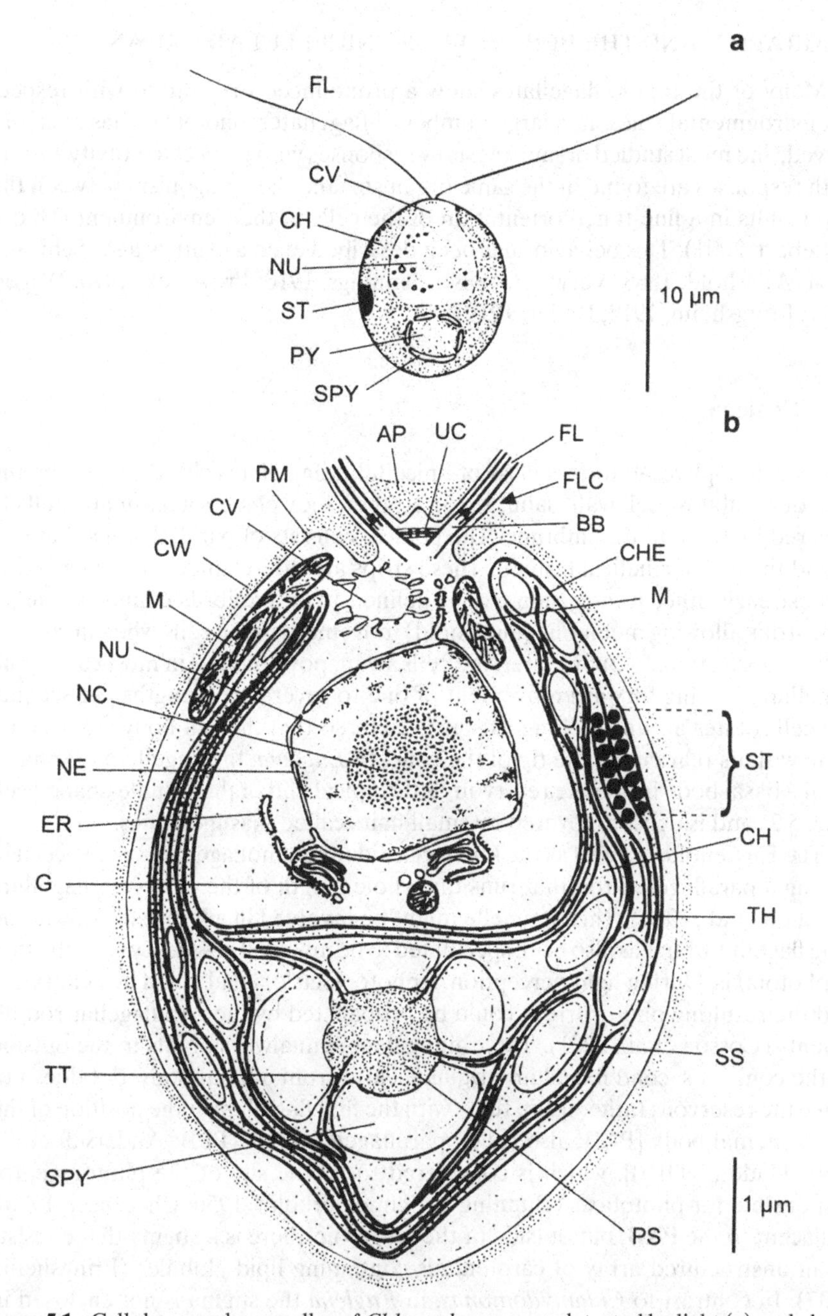

Figure 5.1. Cell shape and organelles of *Chlamydomonas* as detected by light (**a**) and electron microscopy (**b**). AP, apical papilla; BB, basal body; CH, chloroplast; CHE, chloroplast envelope; CV, contractile vacuole; CW, cell wall; ER, endoplasmatic reticulum; FL, flagella; FLC, flagellum channel; G, Golgi apparatus; M, mitochondria; NC, nucleolus; NE, nuclear envelope; NU, nucleus; PM, plasma membrane; PS, pyrenoid stroma; PY, pyrenoid; SPY, starch plate on pyrenoid; SS, starch in stroma; ST, stigma; TH, thylakoid; TT, tubular elongations of the thylakoids into the stroma of the pyrenoid; UC, upper striated connective between basal bodies. (After van den Hoek et al., 1993, with permission.)

Many of the motile flagellates show a pronounced orientation with respect to environmental clues. In a large number of flagellates, phototaxis has been observed, and most studied organisms show responses with respect to gravity. Often, both responses are found in the same organism, and the antagonism between the two results in a fine-tuned orientation of the cells in their environment (Häder & Lebert, 2001). This behavior has been described even a century ago (Schwarz, 1884; Aderhold, 1888; Verworn, 1889b; Jennings, 1910; Prowazek, 1910; Wager, 1911; Pringsheim, 1912; Buder, 1919).

5.2 Euglena

The Euglenophyceae form a class of unicellular flagellates characterized by the lack of a cellulose cell wall (Jahn, 1946). Rather, they possess a plasmatic pellicle covered by the outer membrane. The pellicle consists of parallel strips that surround the cell in a helical fashion. These strips are interconnected, but can glide against each other to a certain extent defined by microfibrils connecting adjacent strips allowing metabolic (euglenoid) movement of the cells when in contact with a substratum. In open water, the cells swim, powered by (in most cases) one flagellum. During forward movement of one to several cell lengths per second, the cell rotates around its long axis at 1–2 Hz, so that the cell body moves on a cone with its rear end at the tip of the cone. In *Euglena*, the flagellum originates from a basal body inside the reservoir at the apical end of the spindle-shaped cell (Fig. 5.2) and is densely covered by small hairs called **mastigonemes**.

The Euglenoids, as well as the kinetoplastids and dinoflagellates, are special in having a paraflagellar rod that runs the whole length of the emerging flagellum (Cachon et al., 1988). This organelle might be involved in adding stiffness to the long flagellum, but has also been speculated to be involved in sensory transduction in phototaxis. During light perception, a photoelectric signal could be generated, and the resulting photocurrent could be propagated by the paraflagellar rod filaments (Cosson et al., 2001). The trailing flagellum always points to the outside of the cone. A second flagellum originates also from a basal body, but does not leave the reservoir; rather its tip fuses with the first flagellum at the position of the paraxonemal body [PAB; also called paraflagellar body (PFB; Andersen et al., 1991; Häder, 1991b)], which is considered to be the site of the photoreceptor responsible for phototaxis (Bünning & Schneiderhöhn, 1956; Checcucci, 1976). Adjacent to the PAB, but outside of the reservoir, there is a stigma that consists of an unstructured array of carotenoids-containing lipid globules (Pringsheim, 1937). In contrast to *Chlamydomonas*, in *Euglena* the stigma is not enclosed in the chloroplast, and it is not the site of photoreception, but assists in light direction perception during helical movement of the cell by periodically shading the PAB in lateral light. Most species contain multiple chloroplasts and perform photosynthesis. However, there are heterotrophic species in this group, such as *Astasia*. The chloroplasts in the green forms can be suppressed or removed by culturing the cells in darkness for several generations or exposing them to chemical inhibitors, such as streptomycin (Pringsheim, 1948).

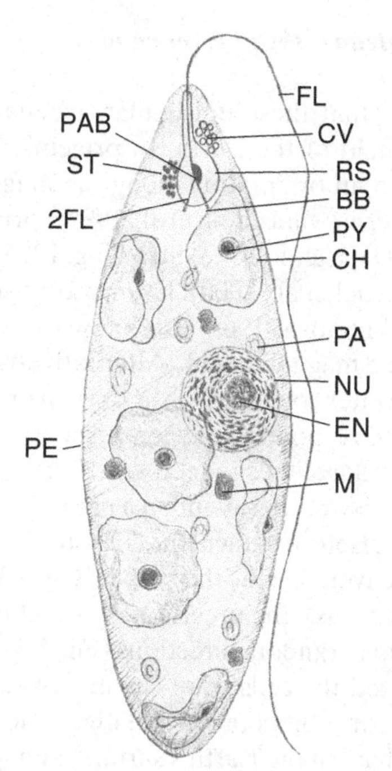

Figure 5.2. Cell shape and organelles of *Euglena gracilis* as seen by light and electron microscopy. BB, basal body; CH, chloroplast; CV, contractile vacuole; EN, endosome; FL, flagellum; M, mitochondrion; NU, nucleus; PA, paramylon; PAB, paraxonemal body; PE, pellicle stripes; PY, pyrenoid; RS, reservoir; ST, stigma; 2FL, 2nd flagellum.

The Euglenophyceae are model organisms to study behavioral responses to environmental stimuli. Most members of this family show a pronounced phototactic orientation (Bancroft, 1913; Mast, 1914; Loeb & Wasteneys, 1916; Oltmanns, 1917; Häder et al., 1981; Colombetti et al., 1982); in some species, a directional movement in chemical gradients, such as oxygen and carbon dioxide, has been described (Bolte, 1920; Porterfield, 1997; Checcucci et al., 1974). Gravitaxis is one of the most prominent responses of high ecological importance and has been found in a number of species (Schwarz, 1884; Jennings, 1910; Pringsheim, 1912; Creutz & Diehn, 1976; Häder, 1987d). Most species show negative gravitaxis, a movement against the gravity vector of the Earth. This is of ecological importance, because it brings them toward the surface of the water column that is vital for light-dependent photosynthetic organisms, especially in the absence of phototactic orientation at night or in murky water. But, also, heterotrophic flagellates profit from this response, because the orientational reaction brings them to a zone with optimal conditions for growth and reproduction. However, some organisms have been described to show positive gravitaxis at least during certain phases of their life cycle. *Euglena gracilis* was found to orient itself positive gravitactically during the first few days after inoculation into a new medium, indicating that cells show this behavior for a short while after cell division (Stallwitz & Häder, 1994).

5.2.1 Gravitaxis in Euglena – the phenomenon

First, it had to be proven that these unicellular flagellates are capable of detecting the gravitational field of the Earth. In principle, upward or downward swimming could be due to an orientation along the magnetic field lines, which has been described in bacteria (Vainshtein et al., 1998; Spring et al., 1998), invertebrates (Lohmann et al., 1995; Camlitepe & Stradling, 1995), as well as vertebrates (Wiltschko et al., 2000; Diebel et al., 2000). Even cyanobacteria (Rai et al., 1998) and ciliates (Kogan & Tikhonova, 1965) are known to orient their movement with respect to the Earth's magnetic field. Alternatively, the cells could follow chemical gradients in the water column, such as oxygen or carbon dioxide. In fact, *Euglena* has been shown to respond to oxygen (oxytaxis; Porterfield, 1997). The first clear proof that these flagellates indeed sense the gravitational field of the Earth was derived from a space experiment on a parabolic rocket flight (TEXUS; Häder et al., 1990b, 1997a). Before launch, the cells showed a pronounced upward movement in the vertical cuvette inside the rocket (Fig. 5.3a). After a few seconds of microgravity (weightlessness), the precision of orientation deteriorated; after about 1 min, the cells swam in random directions (Fig. 5.3b), indicating that under terrestrial conditions indeed the cells were capable of sensing the gravitational field of the Earth. In the near orbit of the rocket flight, the magnetic field lines are not significantly weaker than on the Earth's surface, and gaseous gradients could be excluded because the cells swam in a completely sealed cuvette. From the

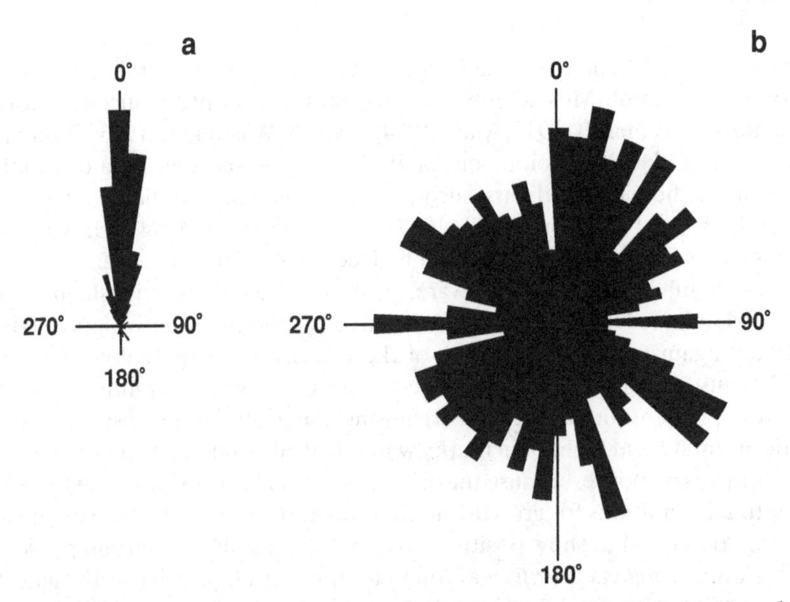

Figure 5.3. Histograms of gravitactic orientation of *Euglena gracilis* **(a)** before launch and **(b)** after some time in microgravity during a parabolic flight on a TEXUS rocket. (After Häder et al., 1990b.)

comparison between swimming in microgravity and at $1 \times g$, the sedimentation velocity can be calculated; for *Euglena*, it is 7 μm/s (Vogel et al., 1993).

To a certain extent, microgravity can be simulated on Earth by rotating the organisms on a fast-rotating horizontal clinostat (Hemmersbach & Häder, 1999; Hemmersbach-Krause et al., 1994; cf. Section 2.1). The principle is that, during rotation, the cells are exposed to forces that change so fast that the cell cannot detect the direction of acceleration. When the cells are kept close to the axis of rotation, there are only small centrifugal forces. Also, under these conditions, *Euglena* moves in random directions (Vogel et al., 1993; Häder, 1994a). However, the cells cannot be completely fooled: under simulated microgravity, more frequent sudden directional changes (phobic responses) were detected in their swimming path than under real weightlessness.

The next question is what is the lower limit or threshold for detection of the gravitational field. This problem was studied during a prolonged space experiment during the IML-2 mission on the American shuttle Columbia (Häder et al., 1995, 1996). Cells were transferred into custom-made cuvettes that had two reservoirs holding about 1.5 ml each connected by a thin viewing area of 170 μm thickness. Four red light-emitting diodes (LEDs) were incorporated in the lateral walls to irradiate the cells during the 12-day mission. The cuvettes were stored on the Shuttle 19 h prior to launch; some of them were positioned on a $1 \times g$ reference centrifuge until subjected to the experimental procedure in space, and others were stored in an incubator where they were exposed to microgravity while in orbit. On the second, sixth, seventh, and twelfth days into the mission, samples were retrieved from either storage under microgravity conditions or from the $1 \times g$ reference centrifuge and transferred into the analysis facility. This was the flight version of NIZEMI (Niedergeschwindigkeits-Zentrifugenmikroskop), a slow-rotating centrifuge microscope (Joop et al., 1989; Häder, 1996b), on which the moving organisms could be observed and tracked under defined centrifugal accelerations (cf. Section 2.5). The terrestrial version of this device had previously been used to study the responses of flagellates under hypergravity (Häder et al., 1991a,b). In space, it was used to cover the range between microgravity and $1.5 \times g$.

As expected, under microgravity, the cells moved in random directions. This behavior was not altered by a centrifugal acceleration of up to $0.08 \times g$ (Fig. 5.4). At $0.16 \times g$, the cells started to orient themselves with respect to the vector of the centrifugal force, at least after some lag phase; and, at higher accelerations, the cells were oriented as under terrestrial conditions. The dependence of the precision of orientation on acceleration follows a sigmoidal curve starting at the threshold below $0.16 \times g$ and saturating above $0.32 \times g$. This behavior was the same independent of whether the experiment was performed from low to high accelerations or from high to low – independent of the duration of the preceding time in microgravity or on the $1 \times g$ reference centrifuge. Thus, there was no adaptation to the conditions of microgravity during this 12-day mission. On a subsequent TEXUS flight, the threshold for gravitactic orientation in *Euglena* could be established with more precision to be at $0.12 \times g$ (Häder et al., 1997).

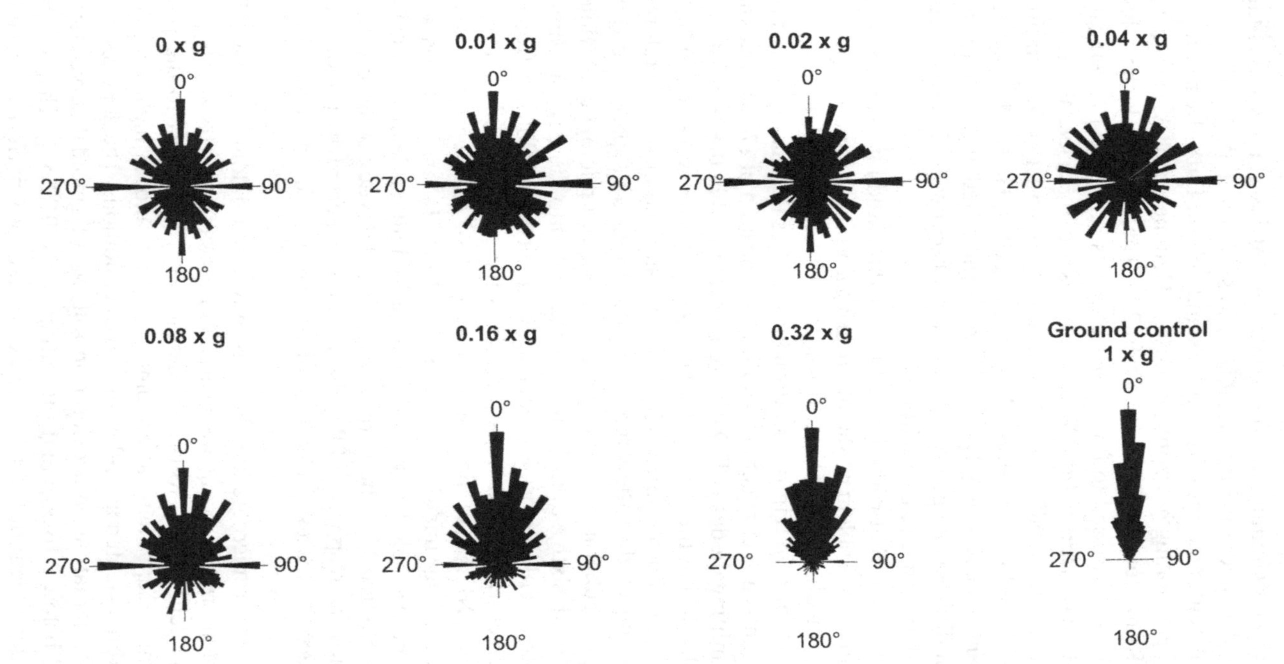

Figure 5.4. Histograms of gravitactic orientation of *Euglena gracilis* at increasing accelerations obtained on a slow-rotating centrifuge microscope (NIZEMI) on board the shuttle Columbia during the IML-2 mission. The threshold for graviperception is found between 0.08 and 0.16 × *g*. (After Häder et al., 1995.)

5.2.2 Passive orientation vs. active sensing

One of the most intensive discussions concerning gravitactic behavior has been on the mechanism of orientation. Although some researchers hold that the phenomenon is based on a pure passive physical orientation in the water column (Mogami et al., 2001), others claim that gravitaxis is based on an active physiological process – including a subcellular gravireceptor, signal amplification, and physiological control of the flagellar movement – which results in a realignment of the long cell axis with the gravivector after a deviation of the cell path from the vertical (cf. Chapter 9).

Brinkmann (1968), as many other authors before him (cf. Chapters 1 and 9), suggested that the rear end of the cell is heavier than the anterior pole, thus bringing the cell body into a vertical position with the front end and the emerging flagellum pointing upward. Because this assumed buoy effect (Fig. 5.5) would be based on an uneven distribution of the cell mass, the orientational movement is not gravitaxis in the strict sense – as indicated in the title of the aforementioned paper. Brinkmann observed that, at a concentration of 10^6 cells/ml in a sealed 36-cm-long tube, the population settled after a 15-h incubation period in the dark, when the tubes were wide. However, in narrower tubes (with diameters of the order of the free path length for *Euglena*), the cells accumulated at the top. He concluded that gravitaxis does not occur in *Euglena* and that settling of the cells

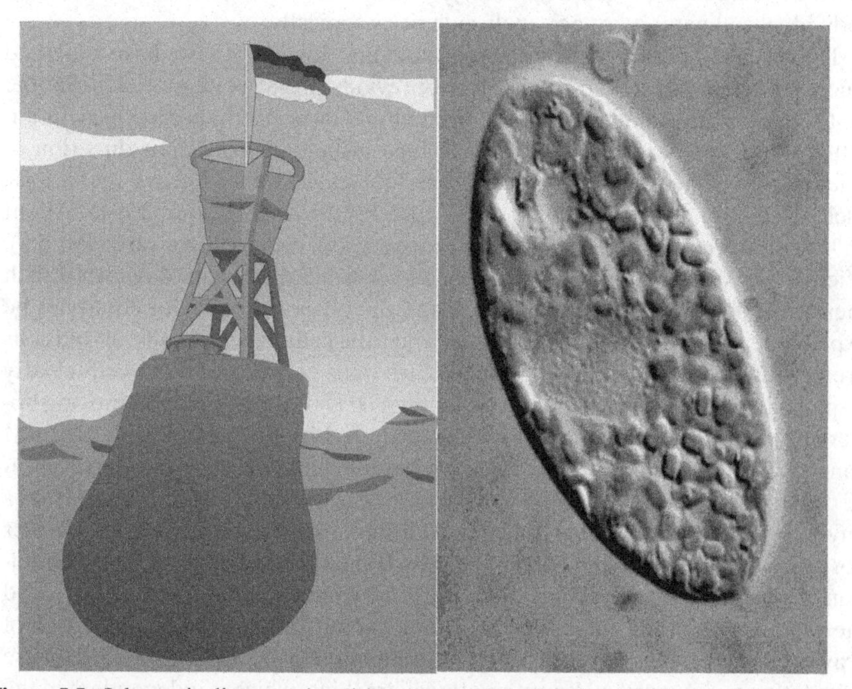

Figure 5.5. Schematic diagram visualising the concept of the buoy effect explaining gravitactic orientation by a physical principle based on an uneven mass distribution in the cell.

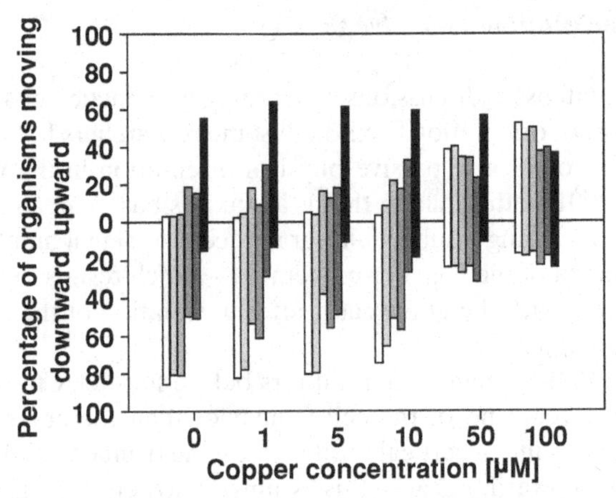

Figure 5.6. Reversal of gravitactic orientation in young cells from downward (positive) to upward (negative) by the addition of heavy metal ions. (After Stallwitz & Häder, 1994.)

is caused by sedimentation, whereas upward movement results from a hydrodynamic restriction imposed by the diameter of the capillary. This conclusion is not conclusive, because this behavior may be induced by the physiological conditions (such as decreased oxygen availability), which were not reported for the study. Individual cell paths were not analyzed microscopically.

However, a number of results contradicting this hypothesis have emerged since. One is the observation, discussed previously, that young cells during the first few days after inoculation of a new culture show positive gravitaxis. In addition, these cells can be induced to change instantaneously their direction of orientation by the addition of micromolar concentrations of heavy metal ions, such as cadmium, lead, mercury, or copper (Stallwitz & Häder, 1994). About 85% of young control cells, 4 days after inoculation, moved downward (Fig. 5.6). Microscopic observation indicated that the cells actively swam downward with their posterior end pointing upward. Thus, the observed behavior could not be explained by a passive sedimentation of immotile cells. Also, 5- to 8-day-old cells preferentially showed positive gravitaxis, whereas 11-day and older cells clearly displayed negative gravitaxis, with more than 60% of the population moving upward. This behavior was not significantly altered by the addition of 1–10 μM concentrations of copper (as sulfate) 4 h before cell track analysis. However, at 50 or 100 μM, cells of all ages switched to negative gravitaxis. This indicates that a physiological switch is activated by the addition of heavy metal ions that reverses positive to negative gravitaxis in young cells. It should be mentioned that the addition of copper at these concentrations did not change the cell form, but decreased the average swimming velocity of the cells by about 50%. However, inhibition of gravitaxis cannot be due to the observed reduction in swimming speed, because orientation was quantified by automatic cell tracking, which is independent of the cell velocity. The opposite effect (reversal from negative to positive gravitaxis) can be induced by adding NaCl (10–15 g/liter) to an older culture or by irradiating

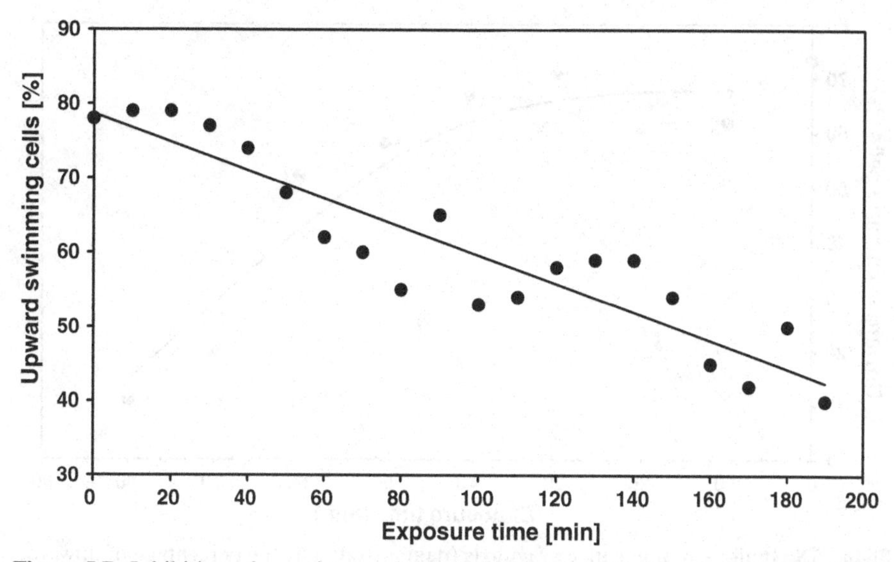

Figure 5.7. Inhibition of negative gravitaxis in *Euglena gracilis* by exposure to artificial UV radiation. (After Häder & Liu, 1990b.)

the cells with strong blue/ultraviolet (UV) radiation (Richter et al., unpublished results).

The precision of gravitactic orientation is not constant over the day. Rather, it follows a circadian rhythm (Lebert et al., 1999a), with an increasing degree of gravitaxis during the day and a pronounced lower gravitactic precision at night, when the cell population is entrained by the light/dark cycle (cf. Section 5.5). When, after some days of synchronization, the cells are subjected to constant light or darkness, the rhythm persists for several days – indicating that it follows an endogenous rhythm, synchronized by the external light/dark cycle. Another supporting observation for an active physiological mechanism for gravitactic orientation was found during the study of effects of solar and artificial UV radiation on motility and orientation of *Euglena* (Häder & Liu, 1990b). Older cells, which showed a pronounced negative gravitaxis, were exposed to either solar or artificial radiation from a UV source, with a peak at 312 nm for different periods of time and then analyzed in a vertical cuvette using an automatic computer-controlled tracking system (Häder & Lebert, 1985) to quantify gravitaxis. Although unexposed control cells showed a high degree of orientation, histograms of cell tracks recorded after increasing exposure times indicated a strong deterioration of the orientation (Fig. 5.7). It is important to note that gravitactic orientation was affected before motility of the cells was impaired, which proves that the reduced precision of gravitactic orientation is not the result of decreased motility or swimming velocity.

A similar effect was found in a number of other flagellates: In the absence of a light stimulus, the freshwater dinoflagellate *Peridinium gatunense* showed a pronounced negative gravitaxis, with more than 80% of the cells moving toward

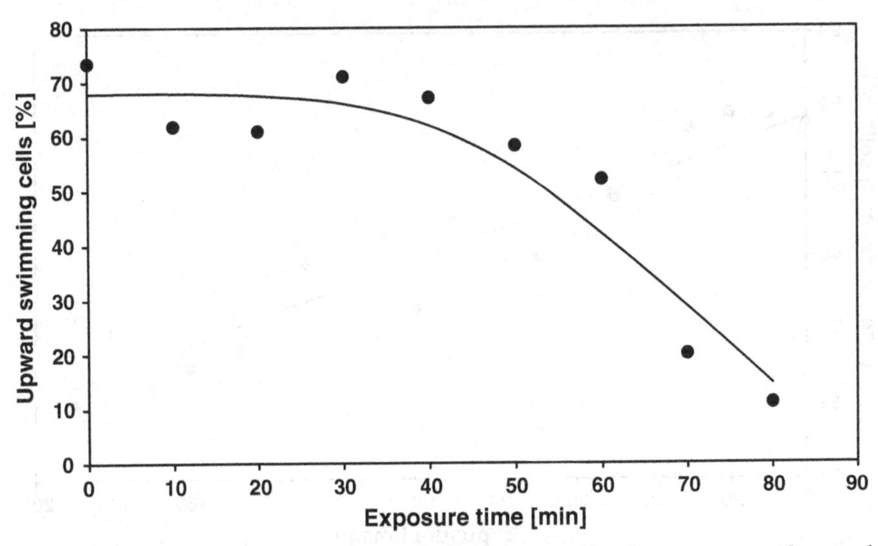

Figure 5.8. Inhibition of negative gravitaxis (demonstrated by the percentage of upward swimming cells) in *Peridinium gatunense* at 2 × g (produced on the slow-rotating microscope centrifuge NIZEMI) by exposure to artificial UV radiation. (After Häder & Liu, 1990a.)

the upper hemisphere (Häder & Liu, 1990a). Exposure to solar or artificial UV radiation significantly affected the precision of orientation and, in this case, even the direction of movement. While obvious at 1 × *g*, the effect is even stronger at 2 × *g*, produced by exposing the cells to a centrifugal force on the NIZEMI (cf. Section 2.5; Fig. 5.8). After 10 min of exposure to the UV source (at 1 × *g*), part of the population reversed their direction of gravitactic orientation to positive (during measurement on the centrifuge); after 70 min, almost all cells swam downward in the vertical cuvette. The marine dinoflagellate *Gymnodinium* also shows a pronounced negative gravitaxis. Exposure to unfiltered solar radiation affects gravitaxis within minutes (Schäfer et al., 1993). After about 5 min, the cells moved in random directions; after about 100 min, the cells were found to be positively gravitactic. Also, this behavior cannot be explained by impaired motility, because the average swimming velocity was not affected at all during the first 125 min of exposure, and, whereas the percentage of motile cells decreased with time, motile cells were found even after 130 min (and only these were tracked by the image analysis system).

The inhibition of gravitaxis by UV radiation is paralleled by the effect of UV on phototaxis. The precision of phototactic orientation decreases during exposure to solar or artificial UV radiation in green and dark-bleached *Euglena gracilis* (Häder & Häder, 1988a,b), *Chlamydomonas nivalis,* and *Astasia longa* (Häder & Häder, 1989c), in freshwater (Häder & Häder, 1989a, 1990), and a marine *Cryptomonas* species (Häder & Häder, 1991a), *Cyanophora paradoxa* (Häder & Häder, 1989b), and *Peridinium gatunense* (Häder & Liu, 1990a; Häder et al., 1990a). In *Euglena gracilis*, the photoreceptor pigments have been identified as flavins and pterins located in the PAB (Brodhun & Häder, 1990). The chromophoric groups are linked covalently to four proteins that have been purified from isolated PABs. Biochemical analysis revealed that the proteins are partially destroyed and the chromophores bleached during exposure of the cells to UV radiation (Brodhun & Häder, 1993). Although not yet proven, a similar mechanism of selective destruction of proteins in the gravity-sensing organelle could be responsible for the loss of gravitaxis after UV exposure.

Also, Creutz and Diehn (1976) assumed that *Euglena* uses an active physiological mechanism involving a sensor, a transduction chain, and a modulation of the flagellar beat pattern. However, the observed behavior was interesting: in a 2 × 2 × 3 mm cuvette, the cells swam in a transverse direction, from which the authors concluded that *Euglena* has a gravity-sensing device that modulates the motor apparatus to maintain the cell at right angles to the force of gravity. This observation is in contrast to earlier observations that the cells show negative gravitaxis (Schwarz, 1884; Jensen, 1893; Thiele, 1960).

5.2.3 Sensor for gravity perception

The first approach during a search for an organelle involved in gravity perception is to look for statoliths (Häder, 2000). Their hypothetical involvement in gravitaxis has been proposed by several authors (Loeb, 1897; Lyon, 1905; Kanda, 1918; Köhler, 1922, 1930, 1939). Amyloplasts are being discussed as potential gravisensing organelles for gravitropism in mosses (Kern et al., 2001; Jenkins et al., 1986; Kuznetsov et al., 1999), ferns (Edwards & Roux, 1998), and higher plants (Sievers & Volkmann, 1977; Sack, 1997; Kiss et al., 1996; Baluska & Hasenstein, 1997; Wunsch & Volkmann, 1993; Evans & Ishikawa, 1997). In the green alga, *Chara*, statoliths have been identified as barium sulfate-filled vacuoles (Braun & Sievers, 1993). Also, in fungi, statoliths have been identified (Kern et al., 1997; Kern & Hock, 1993). In microorganisms, the ciliate *Loxodes* contains statoliths in the form of the so-called Müller bodies (Fenchel & Finlay, 1984; cf. Section 4.2). *Euglena gracilis* also contains organelles with reserve substances (paramylon bodies); however, these are considerably smaller and lighter than amyloplasts in higher plants or statoliths in *Chara* or *Loxodes*.

There is an unambiguous test for the involvement of intracellular statoliths in graviperception. In water or in culture medium, a population of older cells moves upward. When the density of the medium is increased by the addition of high molecular polysaccharides, such as Ficoll (which is a nonionic sucrose polymer) or Percoll, the precision of gravitactic orientation decreases (cf. Fig. 9.6). These substances have the advantage that they only slightly increase the osmolarity (100 cm H_2O at 10% w/v) and only marginally affect the viscosity of the medium (10% Ficoll: 5 cP; Häder, 1999). At a density of 1.05 g/ml, the cells swim in random directions. Increasing the specific density of the medium even more results in a positive gravitaxis (Lebert & Häder, 1996; Lebert et al., 1996; Häder, 1997b). A similar effect was found in the ciliate *Paramecium*, but not in *Loxodes* (cf. Section 4.1.6). The conclusion we can draw from this result is that *Euglena*

Table 5.1. *Dependence of specific density of* Euglena gracilis *cells in relation to culture age and conditions*

Culture conditions (age)	Specific density (kg m^{-3})	Cell length (μm)	Cell width (μm)	Volume (m^3)*
Mineral medium (10 days)†	1,049	55	8.4	2.031×10^{-15}
Mineral medium (10 days)‡	1,053	55	8.4	2.031×10^{-15}
Complex medium (42 days)	1,046	52	9.6	2.509×10^{-15}
Mineral medium (60 days)	1,054	32	8	1.072×10^{-15}
Tap water (60 days)	1,053	50	8	1.676×10^{-15}
Tap water (300 days)	1,054	46	8	1.541×10^{-15}

Note: Specific density was determined using Ficoll step gradients in an isopygnic centrifugation approach (Lebert et al., 1999b).

* Volume was estimated based on the assumption that cells form a rotational ellipsoid ($V = \frac{4}{3}\pi ab^2$; a = long axis radius, b = short axis radius).

† Light band.

‡ Heavy band.

(and *Paramecium*) cells do not use intracellular statoliths for detecting the gravitational field of the Earth, because an intracellular mechanism would not be affected by a change in the extracellular density. Because this hypothesis was falsified, what are the alternatives? One option is that the whole cell operates as a statolith. If the cell content has a higher specific density than the surrounding medium, it would press on the lower membrane and could there activate a gravity-specific sensor.

The next question is: What is the specific density of the cell? This problem can be solved by isopygnic centrifugation (Porst, 1998; Häder, 1997c). A stepwise density gradient prepared with Ficoll is filled into centrifuge tubes (with increasing density from top to bottom); every second step is colored with neutral red for better visibility, and the cells are layered on top of the gradient. After centrifugation, the cells accumulate at a density of 1.045 –1.06 g/ml (cf. Fig. 8.1). The density increases with increasing age of the culture and depends on the medium in which the cells are kept (Lebert et al., 1995). Freshly inoculated cultures often split into two distinct bands, one of which represents newly divided cells and the other older, not yet divided, cells. Cell size and volume also depend on the culture conditions and cell age (Table 5.1). These parameters modify the force the cell body exerts on the lower membrane (Lebert et al., 1999b; cf. Chapter 8) and is closely correlated with the precision of gravitactic orientation.

If this reasoning holds, the next logical question is: What is the subcellular receptor for the force which the cell body exerts? Interpolating from other systems, possible candidates are mechanosensitive channels, which are triggered by pressure or shear forces. Such channels have been found in many biological systems, from bacteria and microbes to plants and vertebrates (Kloda & Martinac, 2001a,b; Buechner et al., 1990; Gustin et al., 1988; Morris, 1990; Sachs, 1991; Ding et al., 1993; Reifarth et al., 1999). Some of these channels can be selectively inhibited

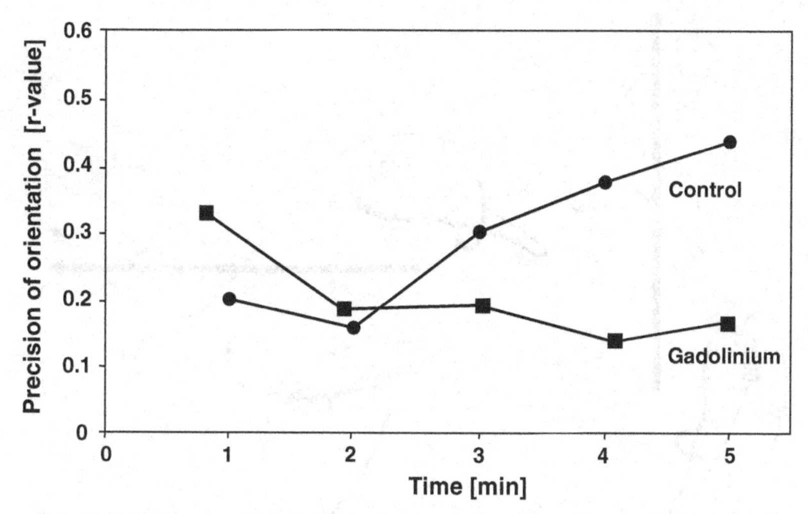

Figure 5.9. Inhibition of gravitaxis in *Euglena gracilis* by gadolinium, a specific blocker of mechanosensitive ion channels. (After Lebert & Häder, 1996.)

by the application of gadolinium ions (as chloride) (Franco et al., 1991; Hamill & McBride, 1994). Indeed, addition of this inhibitor selectively blocks gravitaxis in *Euglena gracilis* (Häder & Hemmersbach, 1997; Lebert & Häder, 1996; Lebert et al., 1997). Even at low concentrations (100 μM), the cells move in random directions (Fig. 5.9). The phototactic orientation, which is also found in *Euglena* (cf. Section 7.2.3), is not affected by the drug, indicating that the inhibitory effect is selective for gravitaxis (Häder, 1999). Inhibition by similar concentrations has been found in other systems (Lacampagne et al., 1994; Yang & Sachs, 1989). Application of cadmium has the same inhibitory effect on negative gravitaxis in *Euglena* (Häder et al., 1999; Lebert et al., 1996). This result seems to indicate that the cell body, which in water is heavier than the surrounding medium, exerts pressure on the lower membrane and activates (or deactivates) mechano- or stretch-sensitive ion channels.

Obviously, these putative gravireceptors cannot be equally distributed over the surface of the cell, because in this case some channels would always be activated no matter which direction the cell swims; thus, the cell could not discriminate the direction of the gravitational vector. Based on a number of other observations, it is believed that the mechanosensitive channels are located at the front end of the cell under the flagellum (Fig. 5.10; cf. Chapter 9). In this configuration, the channels would not be activated when the cell swims upward, but they would be activated when it deviates from the vertical (Lebert et al., 1997). Furthermore, it should be kept in mind that, during forward locomotion, the cell rotates around its long axis. Thus, a horizontally swimming cell would perceive a modulated signal at the frequency of rotation. This signal – after appropriate amplification and sensory transduction – could trigger flagellar movements that result in a course correction of the cell's path toward the vertical.

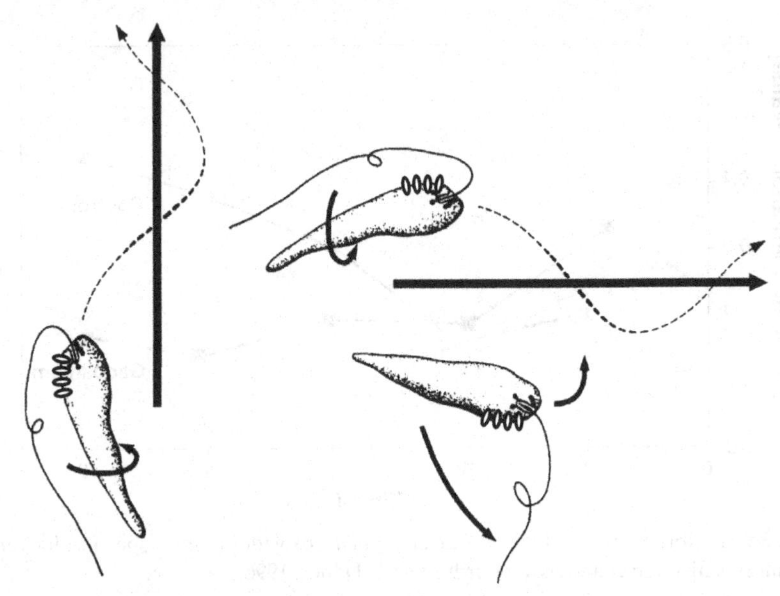

Figure 5.10. Hypothetical location of the mechanosensitive channels in *Euglena gracilis* at the front end under the flagellum. The channels are activated (and the flagellum swings out) when the cell deviates from the vertical direction and the flagellum points downward during cell rotation around the long axis.

5.2.4 Sensory transduction chain of gravitaxis

If, indeed, gating of a mechanosensitive channel is the primary event in graviperception, what is the nature of the gated ion and in which direction is it translocated? In many systems, calcium is involved in sensory and regulatory processes. To verify if this is also true for gravitaxis in *Euglena*, a number of strategies can be followed: one is to use a calcium-selective ionophore that incorporates into the cytoplasmic membrane and operates as a bypass to the gating ion channels. A23187 (calcimycin) is one such ionophore, and its application significantly impairs gravitaxis in the flagellate, even at submicromolar concentrations (Fig. 5.11). Likewise, an increase in the external calcium concentration or a depletion in the medium by the addition of EGTA [ethyleneglycol-bis(β-aminoethyl ether)-*N,N,N′,N′*-tetraacetic acid] affects the gravitactic response (Lebert et al., 1997).

Eukaryotic organisms tend to keep their intracellular calcium concentration rather low on the order of 100 nM, compared with a calcium concentration in the millimolar range in pond water. This is achieved by the activity of energy-dependent calcium-ATPase in the cytoplasmic membrane. This enzyme pumps calcium ions out of the cell using energy in the form of ATP. In contrast to the plastidic and mitochondrial ATPases, the cytoplasmic membrane ATPase can be inhibited by vanadate. For gravitactic stimulation, this would mean a decrease in the gradient between the inside and outside and thus a reduced influx

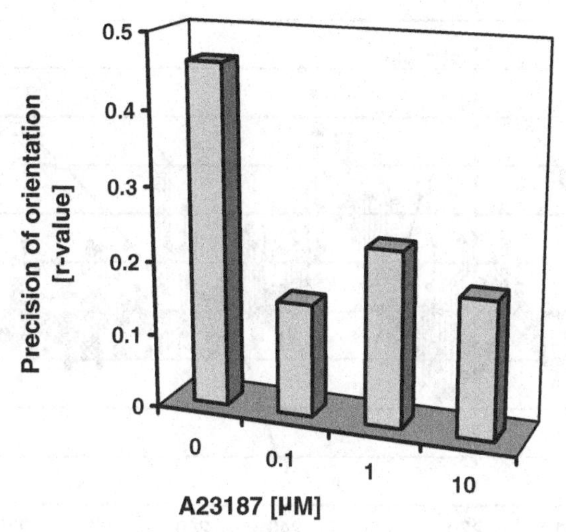

Figure 5.11. Inhibition of gravitaxis in *Euglena gracilis* by calcimycin (A23187), a calcium-specific ionophore. (After Lebert et al., 1997.)

during reorientation. In fact, the precision of gravitactic orientation is significantly affected by the addition of vanadate (Lebert et al., 1997).

All of these results seem to indicate that an influx of calcium ions into the cells during a gravitactic response is involved in the primary steps of graviperception in *Euglena*. To visualize this process, calcium-sensitive fluorescence dyes can be used (cf. Chapter 3). However, most of the dyes, such as Fura or Quin, are activated and/or fluoresce in the UV/blue region. This corresponds with the sensitivity range for phototaxis and photophobic responses in this organism (Diehn, 1969a; Checcucci et al., 1976). Therefore, the calcium fluorescence indicator – Calcium Crimson – was selected, which absorbs between 500 and 600 nm, with a maximum at 589 nm and has an emission peak at 607 nm (Eberhard & Erne, 1991). This is the calcium indicator with the longest excitation wavelength available (Haugland, 1997). To avoid overlap with the fluorescence from photosynthetic pigments, chlorophyll-free strains of *Euglena* or its colorless relative *Astasia* can be used. There are several methods available to load the dye into the cells. One is by using the acetoxy methyl (AM) ester of the substance that passes the cytoplasmic membrane. Intracellular esterases cut the ester bond, after which the dye can no longer permeate out of the cell. However, the dye can equally well enter into intracellular organelles, such as the plastids or nucleus. To overcome the problem of compartmentalization, a high molecular weight dextran conjugate of Calcium Crimson (10,000 molecular weight) was used. This can be loaded into the cell by electroporation (Richter et al., 2001b).

The calcium signal can be quantified with a sensitive spectrofluorometer. A typical experiment is shown in Figure 5.12. Cells are loaded with Calcium Crimson and allowed to orient gravitactically in a vertical cuvette. After alignment with the gravity vector, only a small fluorescence signal is detected. Subsequently, the cells are disturbed by shaking the cuvette, which results in an increase in the

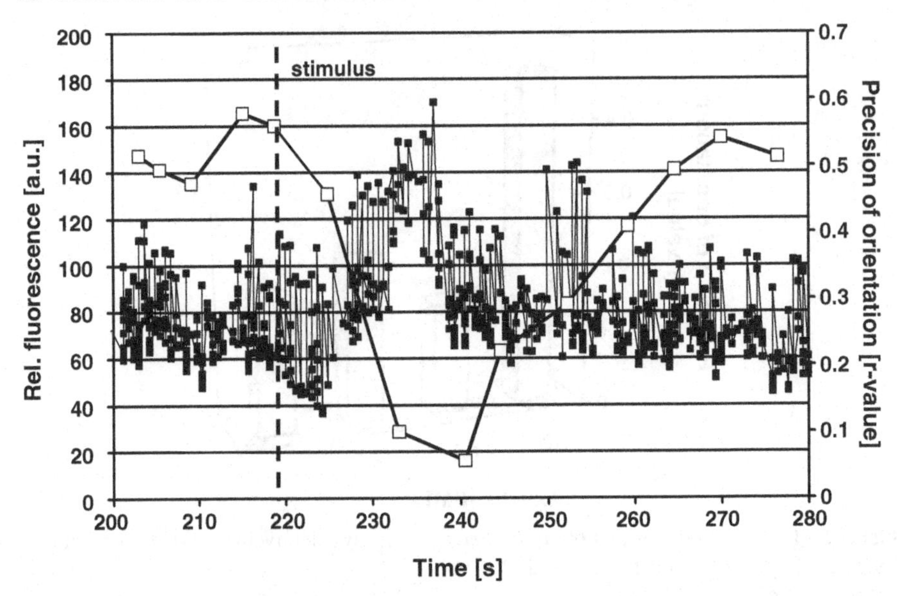

Figure 5.12. Changes in intracellular concentrations of free calcium as indicated by the fluorescent calcium probe Calcium Crimson associated with gravitactic reorientation of *Astasia longa* after mechanical disturbance of the population. **Open squares**, r-value; **closed squares**, relative fluorescence in arbitrary. (After Richter et al., 2001b.)

fluorescence intensity. The maximum is reached after 30–60 s; and, afterward, the fluorescence signal, decays. Addition of gadolinium (250 μM), which blocks the mechanosensitive channels, effectively impairs the increase in the fluorescence signal, as well as gravitactic reorientation as expected (Richter et al., 2001b).

To visualize the weak fluorescence signal, an image intensifier camera needs to be used that can pick up light signals on the order of 10^{-5} lx. The excitation beam is produced with a 20 mW laser at 538 nm. The fluorescence signal from individual cells was detected with a real-time imaging system. When the experiment described previously was repeated, reorientation of the cells was accompanied by a transient increase in calcium fluorescence with the same kinetics. Of course, a large number of cell tracks had to be analyzed because of the high variation between individual cells.

A similar experiment was performed in space on a MAXUS rocket. The experimental setup was mounted on a centrifuge that allowed accelerations between microgravity and $0.3 \times g$. Cells previously adapted to microgravity showed a substantial increase in the calcium fluorescence signal when accelerated above the threshold determined during the IML-2 mission. The image analysis system made it possible to correlate the calcium fluorescence signal with the swimming direction: Cells moving parallel to the acceleration vector showed a low signal and cells moving perpendicular to the vector a high signal (Richter et al., 2001b), which is in good agreement with the assumption that there is a substantial calcium influx only during reorientation of misaligned cells. Similar results were obtained during a recent parabolic flight experiment on board an aircraft (ESA 29th parabolic flight campaign; cf. Section 2.4.1), where the cells were exposed to changes in acceleration between $1 \times g$, $1.8 \times g$, and microgravity. A transient increase in the intracellular calcium concentration was detected from lower to higher accelerations ($1 \times g$ to $1.8 \times g$, or microgravity to $1.8 \times g$), but not in the opposite direction (Richter et al., 2002).

If, indeed, a gated calcium influx is the primary event in the gravitactic signal transduction, one should expect changes in the membrane potential to occur that might eventually trigger the flagellar reorientational beat pattern. In *Chlamydomonas*, electrical potential changes are clearly involved in phototactic reorientation of the cell (Holland et al., 1996; Govorunova et al., 2001), and the beating mode of the two flagella can be manipulated by electrical stimulation (Holland et al., 1996). Electrical potential changes can be recorded from the cell both by intracellular electrodes and by external methods, and the electrical events can be detected in a whole population (Sineshchekov & Govorunova, 1991; Sineshchekov et al., 1992). A similar approach was chosen for *Euglena*. However, despite many attempts in several laboratories, no one has ever successfully recorded electrical potentials from the cell, let alone shown the dependence of electrical potential changes on external light or gravitational stimuli. But, there is hope: there are indirect methods to measure the membrane potential using indicators that change their spectroscopic properties according to the membrane potential (Bashford et al., 1979; Armitage & Evans, 1981). The cells were labeled with Oxonol VI (Haugland, 1997), which changes both its absorption and fluorescence properties in dependence of the membrane potential. The incorporation of the dye does not seem to harm the cells or alter their gravitactic behavior.

The membrane potential is derived from the ratio of the absorption at 590 and 610 nm. Richter and coworkers (2001a) developed a computer-controlled device to determine electrical potential changes during gravitactic reorientation in *Euglena* (Fig. 5.13). Two arrays of three light-emitting diodes (AlInGaP LEDs) were switched on alternately by a computer program via the serial port. The resulting absorption changes were recorded by an array of phototransistors – the output of which was digitized and stored in the computer over time. From the raw data, the ratios could be calculated and plotted either against time or the experimental treatment of the cells. Cells were allowed to orient in a vertical cuvette. When a stable and precise gravitactic orientation was reached, the cells were gently mixed with a pipette and the potential monitoring continued. Immediately after the stimulus, there was a transient increase in the absorption ratio for 5–10 min, followed by depolarization for 60–100 s. The membrane potential returned to its initial value within 180–200 s, which closely corresponds to the complete reorientation of the population. Similar results were obtained during a reflight on a recent parabolic flight experiment (Richter et al., 2002).

There are also a number of indirect confirmations that electrical potential changes are involved in gravitactic orientation of *Euglena*. Triphenyl methyl

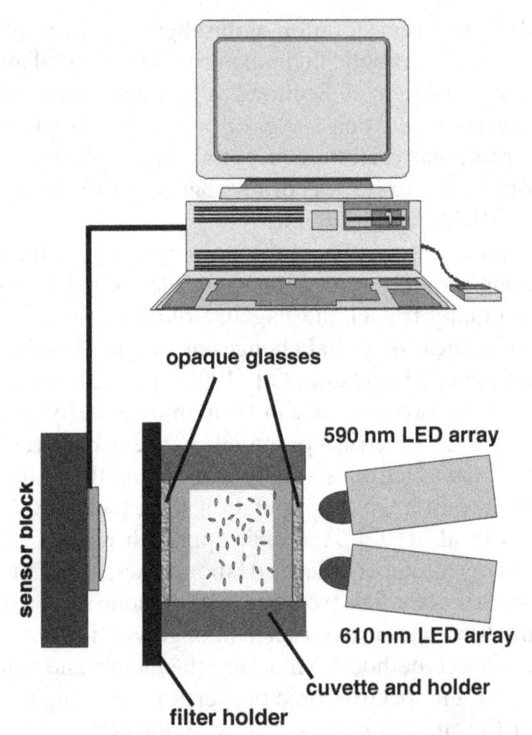

Figure 5.13. Instrument to determine changes in the membrane potential associated with gravitactic reorientation of *Euglena gracilis* after mechanical disturbance of the population using the potential-sensitive dye Oxonol VI. Two sets of LEDs alternatively irradiate the cell suspension at 590 and 610 nm, respectively, under computer control. The absorption is measured by an array of phototransistors, the signal of which is digitized and recorded by the host computer. The membrane potential can be determined from the ratio in absorption at the two wavelengths.

phosphonium ($TPMP^+$), usually applied as bromide, is a lipophilic cation that penetrates membranes and follows the existing membrane potential. Most eukaryotic cells maintain a negative electric potential inside with respect to the outer medium, so $TPMP^+$ diffuses through the cytoplasmic membrane into the cell and decreases the intracellular negative potential due to its own positive charge. Addition of this cation to a suspension of gravitactic *Euglena*, even at a concentration of 10 μM, significantly reduces the precision of orientation (Häder et al., 1998). Also, a sudden step-up in the outer potassium concentration impairs gravitaxis, but this is only transient, and the former gravitactic orientation is restored after some adaptation time (Lebert et al., 1997). It is unlikely that the primary influx of calcium through the mechanosensitive channels is solely responsible for the observed membrane potential change. Quantitative calculations indicate that the resting concentration of about 100 nM increases to about 300 nM during gravitactic reorientation. Probably biochemical mechanisms involving second messengers are responsible for signal amplification.

Calcium could interact with calcium binding proteins, such as calmodulin, which have been found in the cell body and flagella of *Chlamydomonas* (Schleicher et al., 1984; Gitelman & Witman, 1980; van Eldik et al., 1980). Also, in *Euglena*, calmodulin was detected and its amino acid sequence revealed (Toda et al., 1992). Calmodulin has been proposed to be a regulator of calcium homeostasis in the cell by regulating the membrane-bound calcium pump (Schuh et al., 2001). Calmodulin inhibitors – such as W7, trifluoperazin, and fluphenazin – inhibit gravitaxis in this organism. In addition, calmodulin controls the actin/myosin system and the cell shape (Lonergan, 1985). This is of interest for the mechanism of gravitactic orientation, because during reorientation, the cell shape changes: the cells become more elongated during course corrections and are more rounded when they are aligned with the vertical (Lebert & Häder, 1999c). It is interesting to note that changes in the form factor also follow an endogenous rhythm that parallels that of the precision of gravitaxis and also the cellular concentration of cAMP (Fig. 5.14; Lebert et al., 1999a). In *Chlamydomonas*, calmodulin inhibitors cause flagellar wave reversal (Marchese-Ragona et al., 1983). Calmodulin also controls the adenylate cyclase bound to the flagellar membrane. This is of interest because cAMP affects motor responses in this flagellate (Pasquale & Goodenough, 1988).

Earlier experiments indicated that cAMP is involved in gravitactic signal transduction in *Euglena*, but not cGMP – the concentration of which was below the detection limit of the radioimmunoassay used to quantify these nucleotides (Tahedl et al., 1998). 8-Bromo-cAMP is an analog of cAMP, which is not hydrolyzed by the phosphodiesterase. Addition of this drug to a cell suspension resulted in an increased precision of orientation (Lebert et al., 1997). Inhibition of the phosphodiesterase by caffeine or theophylline or by IBMX (3-isobutyl-1-methylxanthine) or indomethacine also increased the precision of gravitaxis (Tahedl et al., 1998). This indicates that, during reorientation, there is an increase in the intracellular cAMP concentration. To verify this assumption, a space experiment was launched on a MAXUS rocket. The cells were transferred into 2.5-ml syringes and placed on a centrifuge inside the rocket. Each syringe was connected to a second one holding ethanol as a chemical fixative. The connecting tube was blocked by a small rubber ball until a hydraulic piston pressed the fixative into the cell suspension. A total of 112 syringe pairs were mounted on two oppositely rotating centrifuges, so that cells could be fixed at various points of time after the onset of microgravity and centrifugal acceleration, respectively. Different accelerations could be achieved by placing the syringes at different radii. Also, several independent parallel experiments were performed to increase the statistical significance. Some experiments were done in the presence of known inhibitors of gravitactic orientation (Tahedl et al., 1998). The cAMP concentration increased to almost double the value found in control cells, which was about 0.05 fmol/cell, within 2 s after the onset of centrifugal acceleration, provided the force was above the threshold for gravitaxis ($\geq 0.12 \times g$; Fig. 5.15). When the cells were fixed 40 s after the onset of acceleration, the cAMP concentration had already dropped to the control value (Fig. 5.15). After transition from acceleration to microgravity, no change in cAMP was recorded (Häder et al., 1999; Streb et al., 2001).

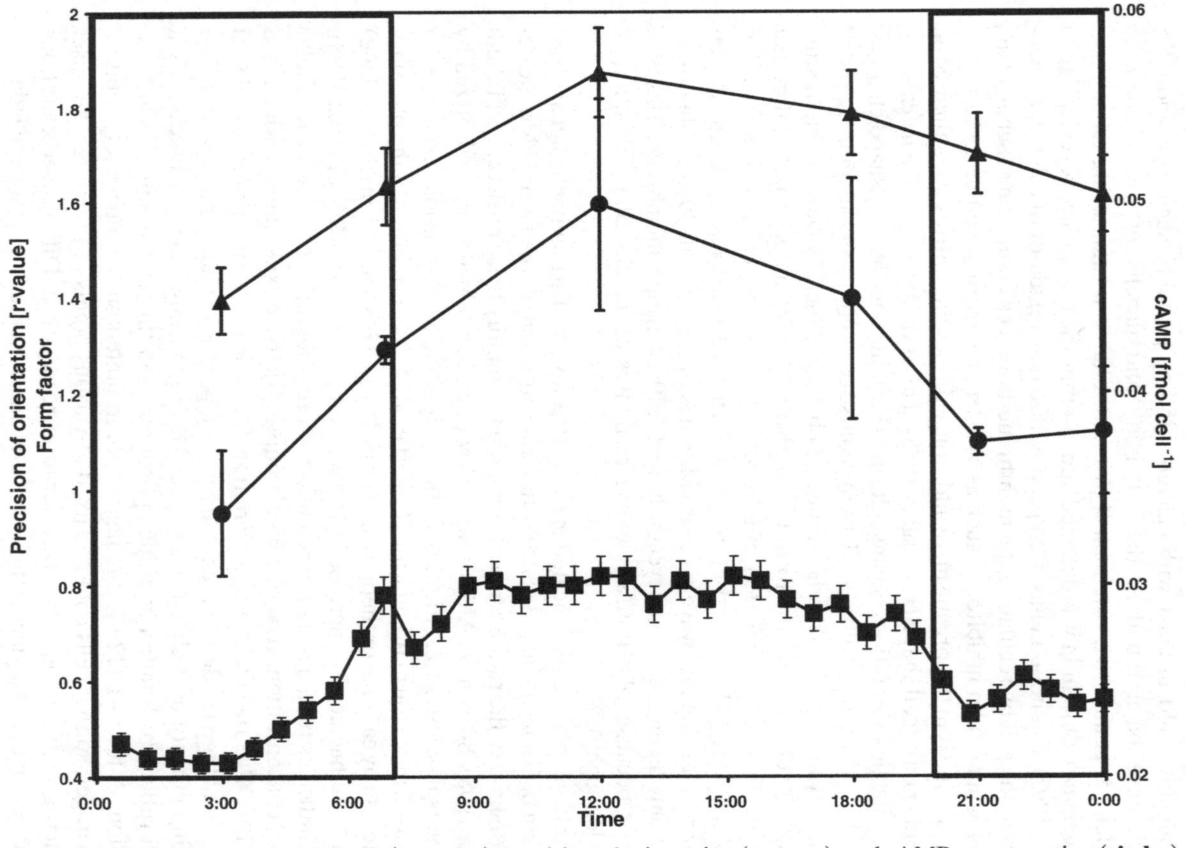

Figure 5.14. Changes in the cell shape (**triangles**), precision of orientation (**squares**), and cAMP concentration (**circles**) in a synchronized culture of *Euglena gracilis* over the day. (After Lebert et al., 1999a.)

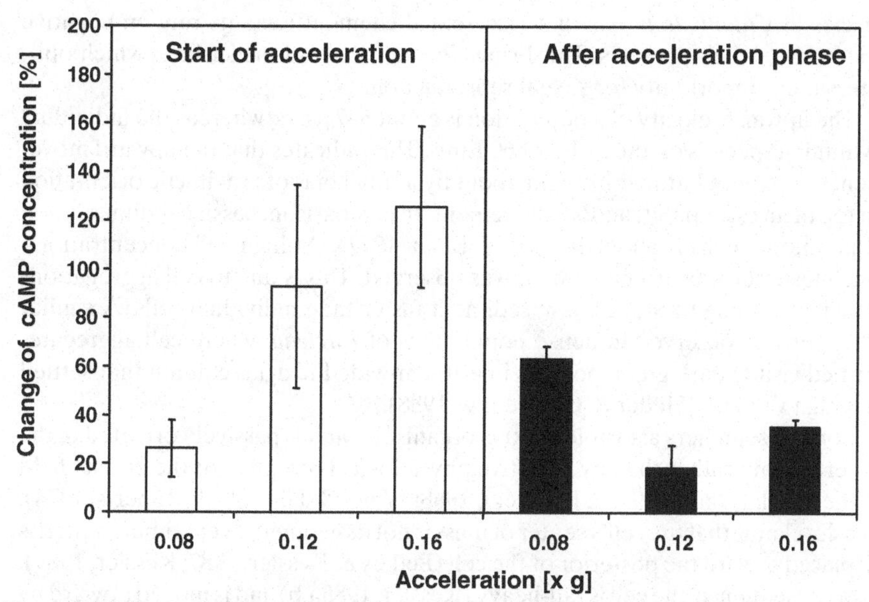

Figure 5.15. Intracellular concentration of cAMP in *Euglena gracilis*, 2 s after the onset of acceleration on a centrifuge during a TEXUS rocket flight (**left panel**) and 40 s after the acceleration phase (**right panel**). (After Tahedl et al., 1998.)

When the inhibitor of mechanosensitive channels, gadolinium, was added at 1 mM, no increase in the cAMP concentration could be detected. This result indicates that the gating of mechanosensitive channels occurs before cAMP synthesis in the sensory transduction chain of gravitaxis in *Euglena*. Thus, it supports the initial hypothesis that cAMP operates as a second messenger, whereas activation of the channels is the primary event in graviperception. In caffeine-treated cells, the level of cAMP was about three times higher than in control cells, but no further significant increase could be induced by accelerating the cells after adaptation to microgravity. Inhibition of the phosphodiesterase by IBMX or theophylline had the same effect.

5.3 Gravitaxis in *Chlamydomonas*

Chlamydomonad flagellates respond to a multitude of environmental stimuli, including light, temperature, gravity, and chemicals (Desroche, 1912; Kuwada, 1916; Pfeffer, 1897). Pronounced upward swimming was observed in *Chlamydomonas* long ago (Schwarz, 1884; Aderhold, 1888; Verworn, 1889b). In the absence of other stimuli, such as light, several species of this large genus were found to display negative gravitaxis, like *C. reinhardtii, C. moewusii*, and *C. nivalis* (Bean, 1977; Kessler, 1992). This behavior brings the population toward the water surface, which is of vital importance for photosynthetic organisms. Gravitaxis also improves the reproductive success of flagellates by spatial concentration. Those species of *Chlamydomonas* that lack sexual chemoattractants rely on negative gravitaxis of their gametes. This brings the cells toward the surface, which optimizes their opportunity for sexual reproduction.

The upward velocity of a population is about 6–7 μm/s, whereas the individual swimming speed is on the order of 60 μm/s. This indicates that the upward movement of the population is brought about by a small bias of gravitactic orientation on top of an essentially random movement. In comparison, passive sedimentation of nonmotile cells is about 1–3 μm/s (Bean, 1977). At high cell concentrations, much faster downward movement was observed. This is due to cell aggregations which, according to Stokes' law, sediment faster than individual cells. A similar behavior was observed in dense populations of *Euglena*, where cell aggregates formed visible dark green pockets in a 10-cm-wide Plexiglas column that settled at a high velocity (Häder & Griebenow, 1988).

Some researchers also hold for this organism that it is passively oriented in the water column rather than by an active physiological receptor for the gravity field of the Earth (Haupt, 1962a; Kuznicki, 1968; Winet & Jahn, 1974; Roberts, 1974). Kessler claims that the cell's center of mass is not its geometric center but, rather, is displaced toward the posterior of the cell (Pedley & Kessler, 1987; Kessler, 1989). As a consequence, the cell is tail-heavy (Kessler, 1985a,b) and is moved upward by its two flagella inserted at its anterior pole. However, microvideographic analysis did not show any sign of passive reorientation of freely suspended cells during momentary cessation of forward locomotion (Bean, 1977). Cells also did not turn or tumble during periods of passive sedimentation. *Chlamydomonas* has a far smaller cell mass than ciliates or *Euglena*, and it was claimed that this is not sufficient to trigger mechanosensitive channels (Sineshchekov et al., 2000). This passive orientation results in an interesting behavior in a flowing liquid. In the shear flow of the ambient fluid, a cell is subjected to a viscous torque, which turns the cell to an angle from the vertical that the authors dubbed gyrotaxis (Pedley & Kessler, 1990). If a cell suspension slowly flows downward in a cylindrical tube, the velocity is higher in the center than at the outer wall, resulting in a higher torque at the sides of the cells pointing toward the center of the tube. Therefore, the cells swim toward the center, where they head upward in a focused stream (Timm & Okubo, 1994).

This mechanism may also be responsible for the formation of bioconvection patterns found in several microorganisms (Fig. 5.16), including *Euglena, Tetrahymena, Chlamydomonas, Paramecium*, and *Polytomella* (Pedley et al., 1988; Bees & Hill, 1999, 2000b; Ghorai & Hill, 2000a,b; Childress et al., 1975; Levandowsky et al., 1975). The formation of these patterns strongly depends on the cell density, geometry of the container, depth of the water column, capacity for gravitactic orientation, and the absence or presence of other environmental stimuli. When the population is irradiated horizontally, the geometry of the pattern markedly changes (Kessler, 1986).

Bean was the first author in modern times to reconsider the possibility of an active gravireceptor linked to a sensory transduction chain (Bean, 1975, 1977). Studying different strains – including different mating types as well as phototaxis and other mutants – the author observed that gravitaxis is not discernable at high

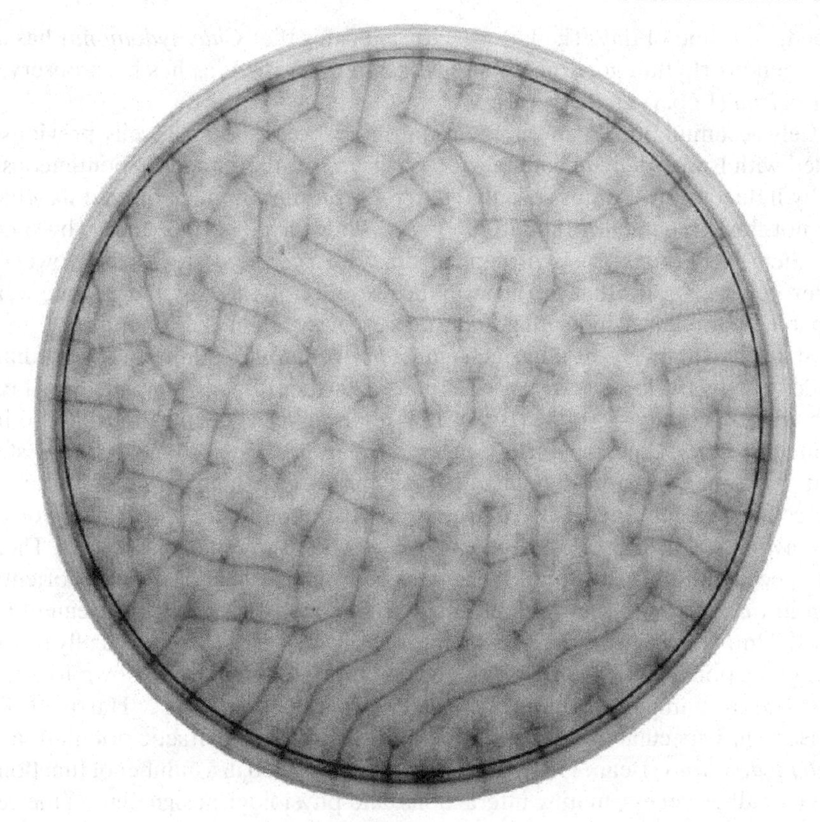

Figure 5.16. Bioconvection pattern in a shallow layer of a *Euglena gracilis* suspension.

cell concentrations, when the free-swimming path is below 200 μm. The free-swimming path L can be calculated from an equation derived from kinetic theory

$$L - \frac{1}{2\pi d^2 n} \tag{5.1}$$

where d is cell diameter and n is cell concentration. For *Chlamydomonas* with an average diameter of 8 μm, the mean free-swimming path is about 350 μm at a density of 10^7 cells/ml. At 3×10^7 cells/ml, the mean free path length decreases to about 120 μm, which is not sufficient for the expression of gravitaxis as shown by microvideographic observation.

Therefore, gravitaxis is limited in small diameter tubes, whereas its precision increases with capillary diameter. This indicates that the cells reorient themselves by long gradual turns. This was found by a population method, where the cells were placed in a vertical 100-μl capillary tube 116 mm long. After about 20 or 30 min, the capillary was broken in several parts and the number of cells in each part evaluated using a Coulter counter. The kinetics with which the cells accumulated in the top portion of the capillary depended on the initial cell density, percentage of nonmotile cells, adhesiveness of the cells to the glass walls, swimming speed, and time of day. The latter factor indicates that *Chlamydomonas* has an endogenous rhythm governing its gravitactic orientation, as has been observed in *Euglena* (Lebert et al., 1999a).

Cell accumulation is an energy-requiring process, because cells previously killed with formaldehyde, iodine, etc., or which lost their flagella spontaneously or by induction or mutant cells that carry the *pfl* mutation (paralyzed flagella), did not show gravitactic orientation. Also, cells deprived of their energy by keeping them in darkness for a prolonged time did not show gravitaxis. However, heterotrophic strains that can grow in the presence of, for example, acetate were capable of gravitaxis, indicating that the presence of light is not required.

Stavis had shown that phototaxis in *Chlamydomonas* is inhibited by sodium azide (Stavis, 1974a,b). Later, it was found that the drug also impairs gravitaxis at a concentration $>500\ \mu M$ (Bean, 1977). At higher concentrations, it also inhibited motility, especially after longer incubation times. Stavis had suggested that azide may cause a depolarization of the electrochemical gradient across the cellular membrane. If this is true, electrical phenomena might be involved in gravitactic signal transduction as has later been shown for phototaxis. There are specific inhibitors for phototaxis, such as copper, which block photoorientation in *Chlamydomonas*, but not gravitaxis (Bean & Yussen, 1979; Bean et al., 1978). Unfortunately, no specific inhibitor has been found that specifically blocks graviperception, but not phototactic orientation. All inhibitors known to affect gravitaxis also impair the motor behavior, such as nickel (Bean & Harris, 1978).

Bean has speculated on the organelles involved in gravitactic orientation in *Chlamydomonas* (Bean, 1984). The axoneme is involved in a number of functions, such as adhesiveness, mating interactions, and physiological signaling to the cell body. Many of these responses involve calcium ions (Solter & Gibor, 1977; Claes, 1980; Goodenough et al., 1982). Due to the asymmetrical position on the cell, the cytoplasm of the cell body could be regarded as a statocyst relative to the motor apparatus. Bean also hypothesized that the membrane could be involved in gravity perception. Pressure of the cell body may perturb ion partitioning or other membrane-associated processes or it could operate on membrane-bound enzymes. Calmodulin-mediated processes may also be involved. Although these suggestions sound rather modern and have been proven later on for other organisms such as *Euglena*, the conclusion for *Chlamydomonas* is that more than 100 years of study have resulted in many interesting observations, but have failed to generate a definitive explanation for the gravitactic orientation in this and other flagellates.

5.4 Other flagellates

Gravitactic behavior is not restricted to photosynthetic flagellates and has been observed, for example, in saprophytic flagellates (Pringsheim, 1922). A number of early studies appeared around the turn of the nineteenth to the twentieth centuries describing the physical, chemical, and behavioral character of gravitaxis. In addition, different models were developed to explain the observed behavior. This literature has been covered in a number of reviews (Davenport, 1908; Kuznicki, 1968; Haupt, 1962b; Hemmersbach et al., 1999b). Most flagellates show negative gravitaxis (Bean, 1975; Jahn & Votta, 1972; Jennings, 1906; Kuznicki, 1968; Massart, 1891; Roberts, 1974). This behavior is interesting because most species often have a higher density than the surrounding medium and have to spend metabolic energy to move upward in the water column (Bean, 1977). Other organisms show positive gravitaxis, and also transverse (horizontal or inclined) swimming has been reported (Fornshell, 1980; Fenchel & Finlay, 1984; Bean, 1984; Hemmersbach et al., 1996a). Depending on the physical or physiological conditions, the same species or even individual can switch from positive to negative responses (Moore, 1903; Fox, 1925). In some species, such as the heterotrophic *Astasia*, a close relative of the green *Euglena*, both responses can occur simultaneously in the same population, so that a fraction of the cells moves upward and the other downward, but all are precisely oriented (Lebert & Häder, 1997a). This was also observed in a white mutant strain of *Euglena*.

Many dinoflagellates undergo daily vertical migrations in the water column (Yentsch et al., 1964; Tyler & Seliger, 1978, 1981; Taylor et al., 1966). Many species move to lower layers at night and return to the surface during daytime (Estrada et al., 1987; Holmes et al., 1967). In darkness, the dinoflagellate *Prorocentrum micans* shows a pronounced negative gravitaxis in a vertical cuvette (Eggersdorfer & Häder, 1991a). The precision of orientation is modulated by an endogenous rhythm entrained by a light/dark cycle. The highest precision is found around noon and the lowest in the early morning and late evening (Fig. 5.17). Together with phototaxis, this behavior controls the position of the cells in the water column.

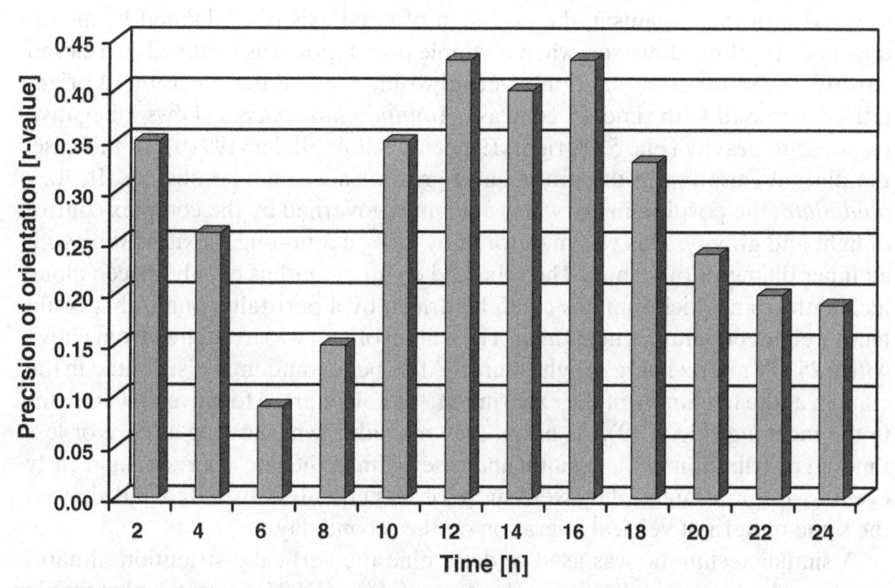

Figure 5.17. Precision of gravitactic orientation in *Prorocentrum micans* controlled by an endogenous rhythm entrained by a light/dark cycle (light from 6 a.m. to 10 p.m.). (After Eggersdorfer & Häder, 1991a.)

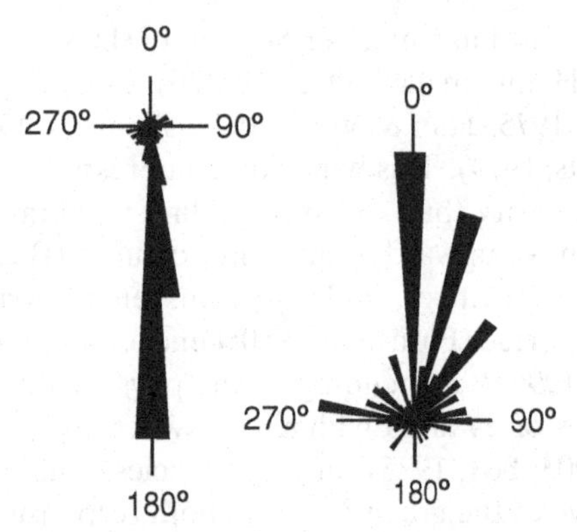

Figure 5.18. Positive gravitactic orientation in *Peridinium faeroense* (**left**) and negative gravitaxis in *Amphidinum caterea* (**right**). (After Eggersdorfer & Häder, 1991b.)

Exposure to solar or artificial UV radiation changed the sign of gravitaxis from negative to positive within a few minutes of exposure (Sebastian et al., 1994). A similar behavior was found in the dinoflagellate *Gymnodinium* in which UV exposure first decreased the precision of gravitactic orientation and subsequently reversed its sign (Tirlapur et al., 1993). In contrast, negative gravitaxis in the green flagellate *Dunaliella bardawil* was not affected by UV exposure, whereas phototaxis was affected (Jimenez et al., 1996).

The dinoflagellate *Peridinium faeroense* shows positive gravitaxis (Eggersdorfer & Häder, 1991b). Older cultures orient with high precision (Fig. 5.18, left). Also in this organism, the precision of gravitaxis is modulated by an endogenous rhythm. However, when a sample population was confined to a closed cuvette, the sense of orientation inversed within 1 h and the precision of orientation increased with time. In contrast, *Amphidinium caterea* shows a negative response to gravity (Fig. 5.18, right; Eggersdorfer & Häder, 1991b). In this case, the diurnal variation in the precision of gravitaxis was not so obvious. In *Amphidinium*, the position in the water column is governed by the complex control of light and gravity. This was monitored by using a 3-m-long Plexiglas tube with an inner diameter of 70 mm. The tube had eighteen outlets evenly spaced along its length from which samples could be drawn by a peristaltic pump that could handle eighteen samples in parallel. The water column was irradiated from above with a 250 W quartz halogen light source. After being randomly distributed in the column at the beginning of the experiment, the cells started to move upward later in the morning (Fig. 5.19). At noon, they moved down, showing a more or less random distribution in the column and rose again in the late afternoon and early evening. Later at night, they were more or less equally distributed and showed the same pattern of vertical migration on the second day.

A similar technique was used to determine the vertical distribution of natural phytoplankton populations in the ocean (Häder, 1995). Twenty submersible pumps were lowered into the water column at equidistant intervals. Samples of 1 l each were drawn simultaneously from all pumps at predefined time intervals.

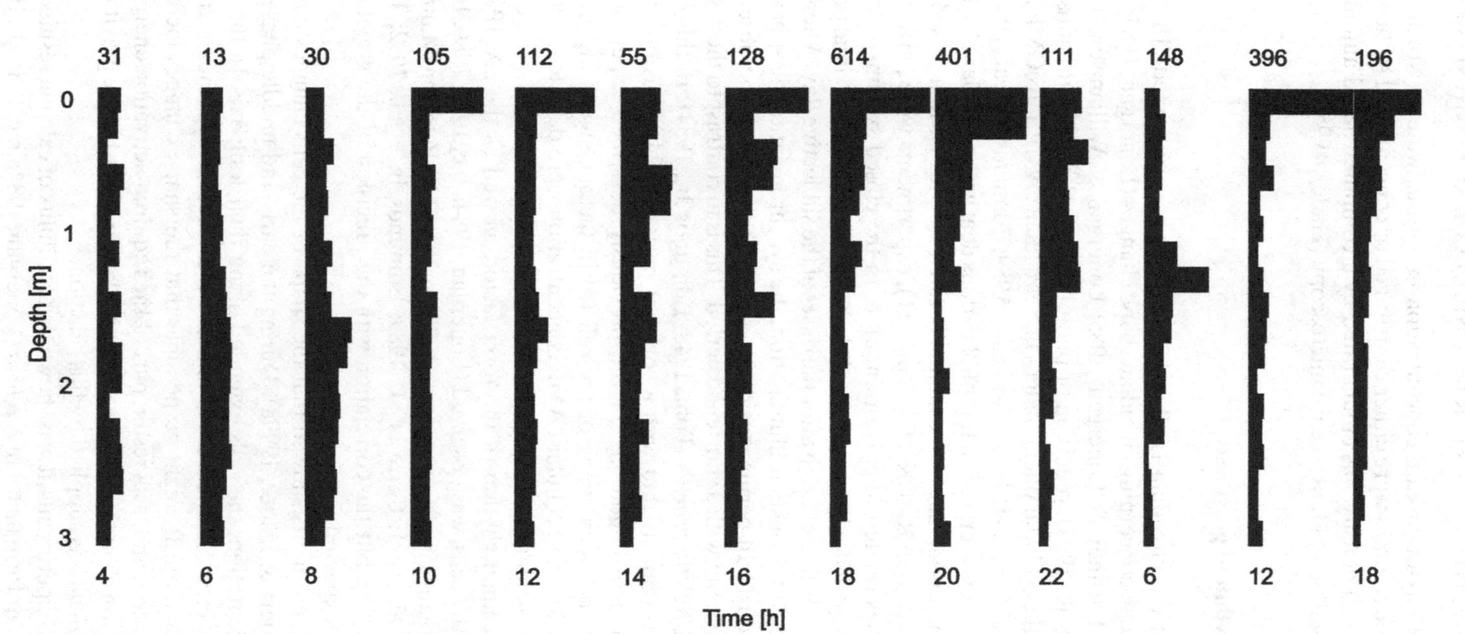

Figure 5.19. Vertical migration of *Amphidinium caterea* in a 3-m Plexiglas tube (inner diameter 70 mm) during two subsequent days after being randomly distributed at the beginning of the experiment. The cell population was irradiated from above with a 250-W quartz halogen light source. The numbers above the columns indicate the values for the χ^2-test. (After Eggersdorfer & Häder, 1991b.)

Depending on the phytoplankton concentration in various waters (North Sea, Mediterranean), the samples were either centrifuged or concentrated by tangential filtration. The cell density was enumerated by a computer-based automatic image analysis based on epifluorescence microscopy (Häder, 1995).

5.5 Circadian rhythm of gravitaxis

Our knowledge of the circadian rhythm of gravitaxis is very limited. There is significant knowledge concerning circadian shape changes (Lonergan, 1984a,b,c, 1985, 1986, 1990; Lachney & Lonergan, 1985; Lonergan & Williamson, 1988; Petersen-Mahrt et al., 1994), precision of phototaxis (Pohl, 1948; Edmunds Jr., 1984), and many other cellular phenomena – like circadian cAMP and cGMP concentration changes (Tong et al., 1991; Carre et al., 1989; Edmunds Jr. et al., 1992) in single, motile cells. However, only a few publications deal with circadian rhythms of gravity-related phenomena. One example is *Euglena gracilis*. In a closed, artificial ecosystem, (AQUARACK, cf. Chapter 11), experiments were performed addressing questions related to graviorientation. In the closed system, a strong circadian rhythm of gravitaxis was observed (Lebert et al., 1999a). Maximal precision of orientation was detected 5 h after the onset of the subjective day. When the culture was subjected to constant illumination, the synchronization disappeared almost instantaneously. In permanent darkness, the rhythm could be observed for more than 5 days, but with a shortened period. This is in contrast to the results on phototaxis of *Euglena gracilis* (Pohl, 1948; Edmunds Jr., 1984). In this case, a free running circadian rhythm with a constant period of 24 h was observed under all experimental conditions. In the previously described experiments, strong correlations between shape of the cell (form factor), swimming speed, gravitactic orientation, and cellular cAMP concentration were detected. All parameters showed an identical rhythm within the same period. As previously described, the circadian rhythm of the form factor, as well as the cAMP and cGMP content of the cells, was observed (Lonergan, 1984a,b,c, 1985, 1986, 1990; Lachney & Lonergan, 1985; Lonergan & Williamson, 1988; Petersen-Mahrt et al., 1994; Tong et al., 1991; Carre et al., 1989; Edmunds Jr. et al., 1992; Tong & Edmunds Jr., 1993), but the correlation with the precision of the gravitactic orientation was unexpected.

Single-cell measurements showed that the shape of the cells changes during reorientation (Lebert & Häder, 1999a). During the early and middle phases of the gravitactic orientation, the cells extended along their long axis. In the late phase, the form factor returned to its initial value. Several hypotheses could explain this observation. It might be possible that the shape changes increase the membrane tension, and, as a result, part of the required activation energy of mechanosensitive ion channels – which are most likely involved in the gravitactic orientation of *Euglena* – is supplied by this reaction.

In independent experiments, it was shown that the influx of calcium results in a change of the cellular form factor in *Euglena* (Lonergan, 1984a,b,c, 1985a,b, 1986, 1990; Lonergan & Williamson, 1988). Consequently, an alternative hypothesis is based on the activation of an adenylate cyclase by a calcium influx. This would increase the cAMP concentration inside the cell. The cAMP could activate kinases, which in turn could influence the dynamics of the cytoskeleton and by this means the cell form by phosphorylation. If this hypothesis holds, cAMP could modulate the gravitactic sensitivity.

The outer hair cells, which are involved in the hearing of vertebrates, react to changes of the membrane potential with an extension in their long axis, while the volume remains constant (Geleoc et al., 1999). This shape change is based on the activity of membrane potential-controlled motor molecules. The change of the form factor in *Euglena*, controlled by analogous motor molecules, could result in a shift of the center of gravity of the cells. This shift would apply a torque by the heavier rear end of the cell and by this means rotate the cell passively into the 'correct' orientation. Following this line of reasoning, gravitaxis in *Euglena gracilis* would be the result of the combination of an active (signal perception, membrane potential, and resulting form factor change) and a passive phenomenon (a torque applied by a spatial distance between the center of gravity and the center of the cell).

6

Other Organisms

This chapter summarizes our knowledge from gravitational biological experiments performed with "other" organisms, which means other than ciliates and flagellates. Amoeba, cellular and acellular slime molds, swimming reproductive stages – such as zoospores and sperm cells – and bacteria have been exposed to altered gravitational stimulation to analyze the impact on behavior and, in few cases, on biochemical processes. In all examples given, a clear hypothesis on the mechanism of graviperception is still missing and should be a task for the future.

6.1 Amoeba

Amoeboid cells are characterized by their actin- and myosin-driven (amoeboid) movement along surfaces (for a review, see Hausmann & Hülsmann, 1996). A weak tendency for negative gravitaxis in *Amoeba* has been stated (Klopocka, 1983). Cultivation of *Amoeba proteus* at 40 × *g* for 36 days did not induce detectable changes in cell form or function (Montgomery et al., 1965). Cultivation of *Pelomyxa carolinesis* in microgravity on Biosatellite II for 2 days indicated a slightly increased division rate (Ekberg et al., 1971), whereas another experiment stated no effect on growth rate and morphology (Abel et al., 1971). The mechanism of graviperception of amoeba needs to be investigated.

6.2 Slime molds

6.2.1 Dictyostelium

The cellular slime mold *Dictyostelium discoideum* is characterized by a life cycle alternating between a multicellular pseudoplasmodium (slug) stage and a unicellular amoeboid stage. A slug (0.5–2.0 mm long, 0.1 mm diameter) contains between 10^3 and 10^5 ameboid cells. Both developmental stages of the organism respond to environmental stimuli to find optimal living conditions for growth and spore discharge. Besides phototactic, chemotactic, and thermotactic orientation (for a review, see Fisher et al., 1984), pseudoplasmodia (but not amebae) of *Dictyostelium* show negative gravitaxis (Häder & Hansel, 1991).

6.2.2 Physarum

The single-celled, multinucleated (millions of nuclei!) organism *Physarum polycephalum* (Myxomycetes, acellular slime mold) can react to environmental stimuli, such as light, chemicals, and gravity (Sauer, 1982; Wolke et al., 1987). The slowly migrating plasmodium can reach several square meters in area and a thickness of 1–2 mm (Fig. 6.1). It consists of a network of tube-like strands and a progressing front zone. Cross-sections of the strands reveal a gel-like ectoplasm and a sol-like endoplasm. The actomyosin-driven rhythmic contraction of the ectoplasm induces a streaming of the endoplasm, which periodically alters its direction in the minute range. Due to a more pronounced streaming in one direction, which might be determined by external stimuli, a slow movement of the organism in the range of 1 cm/h is induced. The streaming also guarantees transport and distribution of nutrients and metabolites within this giant cell. The slime mold has been used as a model system to study actomyosin-driven movements in single cells (Wohlfahrt-Bottermann, 1979). Furthermore, it shows gravitaxis (Wolke et al., 1987), initiating studies in gravitational biology with this system. Early experiments on Biosatellite Kosmos-1129 revealed a reduced growth of the myxomycete, but maintenance of its migration capability after exposure to microgravity (Tairbekov et al., 1981). Due to the fact that the velocity of the cytoplasmic streaming and the contraction rhythm of the protoplasmic strands determine the motor response of the system, these parameters should also reveal early steps in the gravity signal transduction chain that finally result in gravitaxis. By using a light microscope equipped with a photodiode system, both phenomena were tested under variable accelerations, ranging from single 180°-horizontal turns and clinorotation to the conditions of microgravity on several missions (Spacelab D1, IML-1, and IML-2).

Exposing *Physarum* to the conditions of simulated or real microgravity induced a decrease in the contraction period and thus an increase in contraction frequency of protoplasmic strands of about 10% (Block et al., 1986a,b). The maximum response occurs between 20 and 30 min after exposure to microgravity. This is followed by a backregulation to initial values within 30 min accompanied with strong oscillations of the period length (Fig. 6.2). The streaming velocity increased by about 40% on the clinostat and more pronouncedly by about 120% in microgravity, compared with the $1 \times g$ behavior of the corresponding strand (Block et al., 1986a,b). Exposing *Physarum* simultaneously to microgravity and a light stimulus (white light) induced a mutual suppression of the opposing responses; however, exposing the cell subsequently to microgravity and light changed the

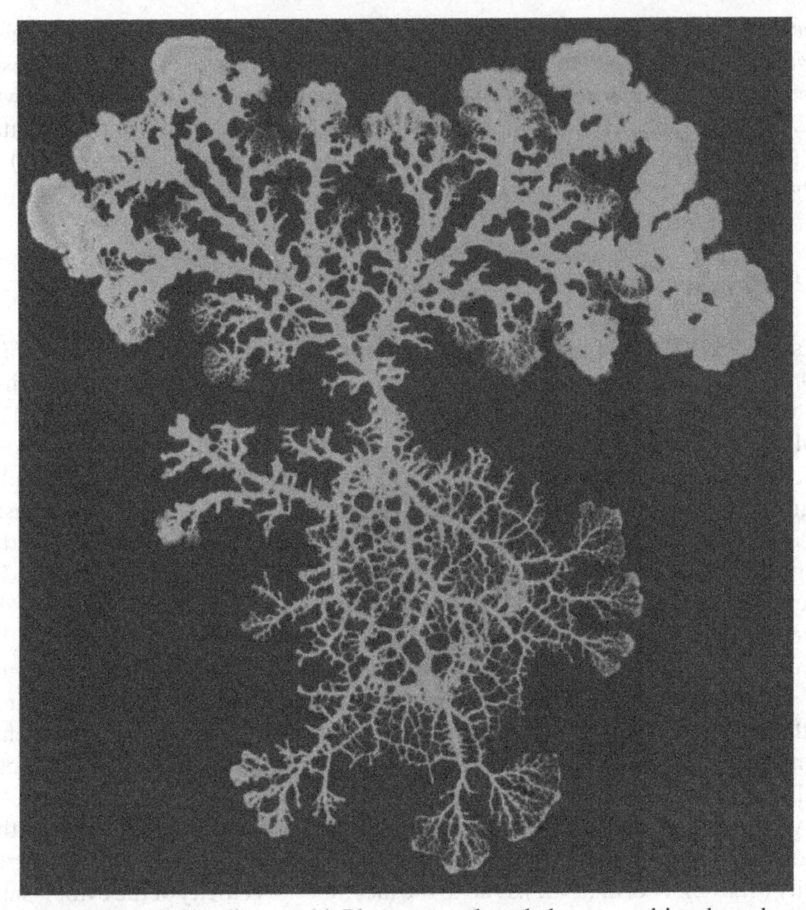

Figure 6.1. The cellular slime mold *Physarum polycephalum*, a multinucleated organism that may reach a cell size of several meters. This vertically oriented plasmodium is clearly moving upward following the differentiated front zone. (Courtesy of I. Block, DLR, Cologne, Germany.)

time course of the individual responses. These mutual influences indicate a common sequence in the processing of the gravity and light stimuli. Furthermore, simulated weightlessness shortened the time course of mitosis (Sobick & Briegleb, 1983). However, this short-term process (few minutes) is most probably not related to the observed graviresponse. Nevertheless, the nucleoli – which show a striking eccentric location in the interphase nuclei – might be somehow involved in graviperception (Sobick et al., 1983).

To determine the threshold of the graviresponse, the slime molds' oscillating contractions were measured on the centrifuge microscope NIZEMI (Niedergeschwindigkeits-Zentrifugenmikroskop; cf. Section 2.5) under varying acceleration steps between microgravity and $1.5 \times g$, each lasting for 15 min. The lowest acceleration inducing a significant change in the contraction period (increase) was $0.1 \times g$, though individual differences were observed ($0.2 \times g$ and

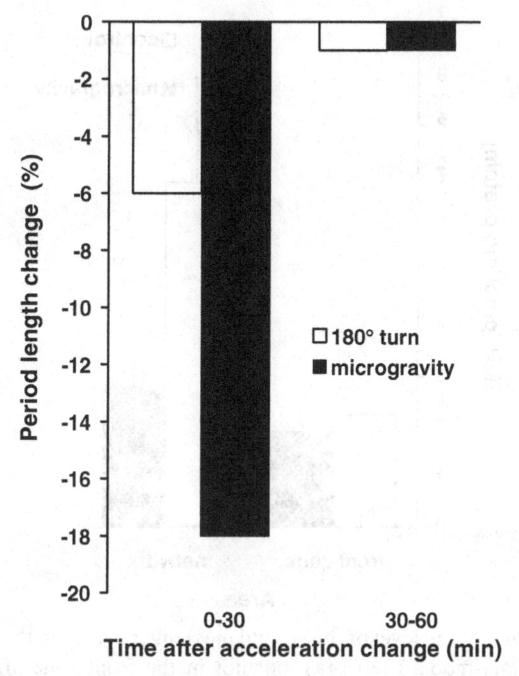

Figure 6.2. Periodic changes of the contraction frequency of cytoplasmic strands of *Physarum* plasmodia in response to gravistimulation, compared with undisturbed 1 × *g*. Analysis by means of a photodiode revealed time-dependent responses to a 180° horizontal turn on ground (**white bars**) and microgravity (**black bars**) during the IML-1 mission. Both stimulations result in a short-term decrease in periodic contraction, though the response in microgravity is more pronounced. (Courtesy of I. Block, DLR, Cologne, Germany.)

0.4 × *g*; Block et al., 1996). Thus, the threshold was in the range determined for the onset of gravitaxis in free swimming cells [*Paramecium*: 0.3 × *g* (Hemmersbach et al., 1996a,b); *Euglena*: 0.16 × *g*, 0.12 × *g* (Häder et al., 1996; Häder, 1997b)].

To test whether gravistimulation in *Physarum* is coupled to changes in the level of cyclic nucleotides, the organism was exposed for 3 days to microgravity (Space shuttle mission STS-69) and frozen at −196°C under these conditions. Plasmodia were either processed as a whole or front area and network separately. As in *Paramecium* (Hemmersbach et al., 2002), a decrease in cAMP was measured, compared with the 1 × *g* controls on ground. Interestingly, the decrease was restricted to the rear areas of the plasmodium (i.e., where the motive force of the actomyosin-based ameboid movement is generated). Furthermore, turning *Physarum* upside down increases cAMP, as well as its graviresponses (Fig. 6.3). The results indicate an involvement of cAMP in the gravity signal transduction chain in *Physarum* (Block et al., 1998, 1999). Exposing *Physarum* to a medium with a density exceeding its own of 1.076 g/cm^3 neither affects the magnitude nor the time course of the rhythmic cell contractions that are responsible for gravitaxis (Block et al., 1999). The gravireceptor candidates in *Physarum* remain

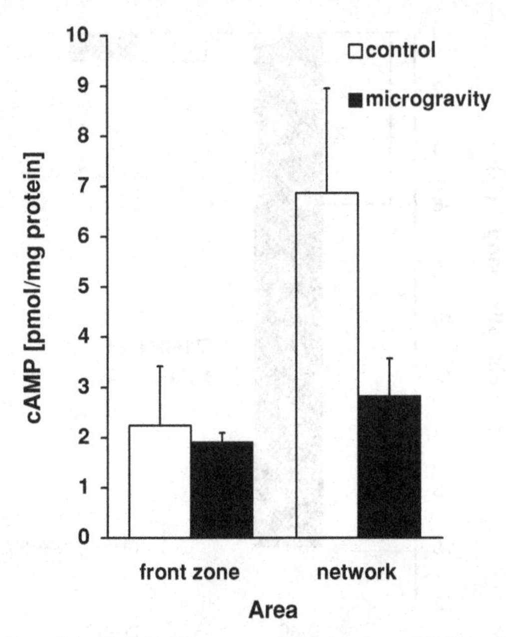

Figure 6.3. Reduction of the level of the second messenger cAMP in the force-generating area of *Physarum* plasmodia (network), but not in the front zone after cultivation in microgravity (STS-69) for 3 days. (Courtesy of I. Block, DLR, Cologne, Germany.)

speculative: nuclei and mitochondria have been proposed, both occurring in high numbers in this giant cell (Block et al., 1998).

6.3 Reproductive unicellular stages

6.3.1 Fungal zoospores

Accumulation of zoospores below the surface of the medium has been stated for different species. As in the case of protists, a discussion arose whether this behavior is guided by oxygen (aerotaxis) or by gravity (Cameron & Carlile, 1977). Capillary-tube experiments with *Phytophthora* under different conditions with respect to the gas compositions within the medium, light, darkness, and gravity (horizontally or vertically positioned tubes, 180° turns of the tubes) suggested that the behavior of the zoospores is guided by gravity. A detailed analysis of orientation, swimming velocities, sedimentation velocities, and density of the cells is necessary to discriminate between the active or passive character of gravitaxis in zoospores. Cameron and Carlile (1980) proposed that the chemotaxis of zoospores is regulated by changes in the membrane potential, which in turn determines the flagellar activity. Whether a comparable mechanism might be valid for gravitaxis has yet to be shown.

6.3.2 Sperm cells

Behavioral and biochemical studies with sperm cells in microgravity revealed an increase in sperm velocity and motility correlated with an accelerated phosphorylation of axonemal proteins (Engelmann et al., 1992; Tash & Bracho, 1999). In contrast, hypergravity (as low as $1.3 \times g$) did not only reduce the percentage of motile sperm cells and their straight-line velocity, but also led to a rapid decline of the phosphorylation status of the proteins FP160 and FP130 – both of which are involved in flagellar activation. In addition, hypergravity reduced the ratio of sperm–egg binding and fertilization (sea urchin) by about 50% (Tash et al., 2001). Another study using hind limb suspension as a method to simulate microgravity revealed a significant inhibition in spermatogenesis in rats (Tash et al., 2002). It is important to study the long-term effect of microgravity on spermatogenesis, sperm function, and fertilization, because they have a direct impact on reproduction, for example, of fish that will be an essential food source during space travel (Tash et al., 2002).

6.4 Bacteria

In the beginning of space experimentation, prokaryotes have been exposed to space conditions (radiation and microgravity) to investigate whether they can survive under these circumstances. The experiments over the years – starting on Vostok and Biosatellite II in 1967 – showed that microgravity and radiation influence basic cell functions (Gmünder & Cogoli, 1988). Effects on growth rate, sporulation, phage productivity, and resistance to radiation have been reported. Various studies in space of different bacterial strains showed that the growth rates increased (for reviews, see Mattoni et al., 1971; Kordium et al., 1978; Mennigmann & Lange, 1986; Gmünder & Cogoli, 1988), though the reason remained speculative. The fast growth and high density of the cultures were taken as explanation for the reduced sporulation of *Bacillus subtilis* in space (Mennigmann & Lange, 1986). On Spacelab D-1, *Escherichia coli* was taken as a model system to study the mechanism of genetic transfer. Conjugation (transfer of part of the chromosome from a donor cell to a recipient cell), transduction (transfer of DNA between bacteria by bacteriophages), and transformation (transfer of genetic material between bacteria; genes are transformed by incorporating DNA of a genetically different bacterium) have been investigated. Although transformation and transduction were unchanged, compared with the $1 \times g$ controls, conjugation showed a 40% enhancement. Interestingly, the length of transferred DNA during conjugation increased up to 4 times, leading to the assumption that, due to the lack of sedimentation, the process of conjugation might be less disturbed in microgravity (Ciferri et al., 1986). Furthermore, bacteria showed an increased resistance toward antibiotics in space (Tixador et al., 1985; Lapchine et al., 1987). Possible reasons were discussed: either an increased bacterial proliferation in microgravity and consequently an insufficient amount of antibiotics, or changes at membrane level thus altered transport of the antibiotics. To contribute to the underlying mechanism, suspension cultures of *Escherichia coli* were studied during seven U.S. shuttle missions between 1991 and 1994 (STS-37, STS-43, STS-50, STS-54, STS-57, STS-60, and STS-62). Data indicate a shortening of the lag phase, an increase in the duration of the exponential growth phase, and an approximate doubling of the final cell population density. According to the model derived by Klaus and coworkers (Klaus, 2002; Klaus et al., 1997b), it seems likely that a "cumulative effect of gravity may have a significant impact on suspended cells via their fluid environment, where an immediate, direct influence of gravity may otherwise be deemed negligible."

Some bacterial species have magnetosomes, that means magnetite-containing inclusions enclosed in a membrane and arranged as a chain along the long axis of the cell (cf. Section 7.5; Gorby et al., 1988). It was postulated that, by means of magnetosomes, bacteria are able to orient to and migrate along geomagnetic lines. As a consequence, they move downward toward the sediment and thus to optimal habitats (Blakemore, 1975). Alternatively, it has been presumed that magnetosomes primarily enhance the cell's density and the cellular response to gravity, whereas magnetotaxis is a secondary effect. To discriminate the effect of gravity from the magnetotactic response, *Magnetospirillum magnetotacticum* was exposed to microgravity in combination with magnetic forces well above and below the strength of the Earth's magnetic field of 0.1 mT. In contrast to $1 \times g$ conditions, bacteria showed no attraction to the north or south pole of a magnet even during an exposure time of 120 days aboard MIR. The inability of magnetotaxis was accompanied with a loss of magnetosomes. It is therefore suggested that magnetosomes enhance the gravitactic response under normal gravity conditions (Urban, 2000).

7

Responses to Other Stimuli

Microorganisms respond to a multitude of external stimuli in their habitat to select a suitable niche for survival and reproduction. Light and gravity are probably the most important cues for most motile microorganisms. Several types of light-induced behavior can be distinguished in microorganisms, including phototaxis, photophobic responses, and photokinesis. Chemical gradients, such as oxygen or carbon dioxide, are sensed by many organisms. Bacteria recognize and follow gradients of attractants (e.g., nutrients such as sugars) or avoid sources of toxins (e.g., phenol). Heterotrophic eukaryotic microorganisms also use chemical gradients to find their food. Pheromones are produced and emitted to attract gametes of the opposite sex. Some prokaryotic and eukaryotic organisms are capable of sensing extremely small thermal gradients very close to the physical limits. Almost all motile organisms, from bacteria to vertebrates, recognize and use the magnetic field of the Earth. Responses to electrical fields are not easy to explain, because these stimuli are not expected in nature. The responses to multiple stimuli may be additive or connected in a complex network of signal transduction chains. In other cases, responses to certain stimuli may override those to others.

7.1 Introduction

Microorganisms respond to a host of stimuli in their environment to search for and stay at favorable habitats optimal for their growth and survival, as well as reproduction. Their responses to these environmental clues may vary depending on their developmental stage. For example, vegetative cells of photosynthetic organisms may aim at optimal light conditions at intermediate depths in the water column to satisfy their energy needs, whereas gametes of the same species may aim for the surface to facilitate meeting gametes of the opposite sex and spores – again of the same species – may be guided to the bottom for attachment.

The responses to the multitude of stimuli may be additive, resulting in a vectorial addition of the resulting response paths. In other cases, the response to one stimulus may override that to others. For instance, often, the reaction to strong light overrides graviorientation to avoid exposure to detrimental radiation at the water surface. In some cases, an intricate web of synergistic and antagonistic responses is governed by multiple stimuli.

In many eukaryotic microorganisms, the responses to light and gravity are the most prominent reactions to external clues in their habitat (Häder, 1998). In addition, a number of unicellular and multicellular microorganisms has been found, which results in a complicated pattern of responses to chemical gradients, including oxygen and carbon dioxide (Moir, 1996). Heterotrophic prokaryotes and eukaryotes are guided by chemicals to their food (attractants) or away from hazardous toxic concentrations (repellents; Ueda et al., 2001). Pheromones are produced to attract gametes of the opposite sex chemotactically (Kuhlmann et al., 1997b). Mainly gliding organisms have been studied, which show an astonishing sensitivity in responding to thermal gradients. Microorganisms, as well as invertebrates and vertebrates – including mammals – have been found to be guided by the magnetic field lines in their environment (Nemec et al., 2001; Emura et al., 2001; Deutschlander et al., 1999; Vainshtein et al., 1998; Kirschvink, 1997). This response is used for long-range navigation (e.g., by migratory birds), but can also be of use to guide organisms to the surface or bottom of the water column (Blakemore, 1982). Many motile microorganisms have sensors for mechanical stimuli to avoid obstacles in their swimming path or to avoid predators (Iwatsuki et al., 1996; Sikora et al., 1992). It is interesting to note that a number of microorganisms are guided by electric fields (Korohoda et al., 2000; Morris et al., 1992; Gruler & Nuccitelli, 1986), which are not routinely expected to be standard elements in their natural habitat.

In this chapter, we give an overview of the responses of microorganisms to other stimuli. Comparison with the responses to gravity is important to understand the interaction between the various sensory transduction chains. Furthermore, methods and techniques used to identify the structural elements and the molecular mechanisms involved in the responses to other stimuli may be of interest for further experimental approaches to the understanding of graviresponses.

7.2 Photoorientation

Light is one of the major clues for microorganisms to orient in their habitat. This behavior is obvious for photosynthetic organisms, but also heterotrophic organisms utilize this stimulus for habitat selection. A number of different responses can be distinguished:

7.2.1 Photokinesis

Photokinesis is a clear photomovement reaction found in many prokaryotic and eukaryotic organisms (Iwatsuki, 1992; Zhenan & Shouyu, 1983; Nultsch, 1975). It is defined as the dependence of the steady-state movement velocity on the ambient irradiance (Häder, 1979; Diehn et al., 1977). This behavior is independent of the direction of light. Photokinesis is defined as positive when the velocity at a given light intensity is higher than in darkness, and it is called negative when the speed of movement is lower than in the dark control (Haupt, 1975). The organisms can be immotile in darkness or become motionless at high fluence rates. In many photosynthetic prokaryotes, this behavior has been found to be controlled by energetic aspects. For example, in cyanobacteria, higher irradiances (within physiological limits) result in a higher rate of photophosphorylation, which is used to increase the activity of the motor apparatus (Nultsch, 1975b). This can be proven by applying uncouplers, which block the photosynthetic ATP production while the photosynthetic electron transport chain is not affected (Nultsch, 1973). Photokinesis is also found in many eukaryotic microorganisms (Cohn, 1993; Iwatsuki, 1992; Posudin et al., 1988). Here, the mechanism of response is not understood, but it may be independent of the photosynthetic apparatus, because heterotrophic organisms – such as the colorless *Astasia longa* – also show photokinetic behavior (Mikolajczyk & Walne, 1990; Häder & Häder, 1989c). In the flagellate *Euglena*, photokinesis saturates at about 300 lx in white light (Wolken & Shin, 1958). Mast (1911) observed a 10- to 15-min lag period before a new steady state of velocity is reached, indicating an involvement of metabolic processes. In this flagellate, photokinesis seems to be due to an increased beating frequency of the trailing flagellum, as indicated by laser Doppler velocimetry (Cantatore et al., 1989). The photoreceptor for this reaction is not yet clear. Although some findings indicate the involvement of the photosynthetic pigments (chl *b* and/or β-carotene; Wolken & Shin, 1958; Nultsch, 1975; Ascoli, 1975), others stated a strong effect of blue light (Diehn, 1973).

7.2.2 Photophobic responses

Another distinctly different class of responses mediated by light are the photophobic reactions. When Engelmann watched photosynthetic bacteria under the microscope, he observed them reverse their swimming direction when they left the irradiated field and entered a dark field as if frightened (Engelmann, 1882, 1883). The older literature is summarized in a number of reviews (Haupt, 1959; Feinleib & Curry, 1971; Nultsch & Häder, 1979, 1988). Cells can be concentrated in a light field based on this mechanism: no response occurs when the cells enter a light field from the surrounding dark area, but each time they try to leave it, they reverse their swimming direction (light trap). However, there is a zero threshold in the irradiance above which it is detected (Clayton, 1959), which varies widely between different organisms by orders of magnitude.

Figure 7.1. When a photographic negative is projected into a homogeneous suspension of cyanobacteria (*Phormidium uncinatum*) in an agar layer, the filaments accumulate in optimal light intensities using step-up and step-down photophobic responses. The resulting "photographic positive" can be stabilized by drying the agar.

In cyanobacteria, this behavior is also linked to the photosynthetic electron transport chain. The redox state of an electron carrier (most probably plastoquinone) in the linear electron transport connecting photosystems II and I seems to govern the reversal of gliding motility in these filamentous organisms (Häder, 1987b). At higher fluence rates, some cyanobacteria of the genus *Phormidium* have been found to show the opposite response: when they enter a high-irradiance light field, they reverse the direction of movement, but do not respond to the transition from light to darkness. To discriminate these two responses, the second behavior is termed "step-up photophobic response," as opposed to a "step-down photophobic response" mediated by a decrease in light intensity. Organisms do not only respond to a change from light (at a given irradiance) to darkness, but also to small changes in the irradiance as low as 4%. This percentage difference is called difference threshold and holds for a wide range of intensities (Weber-Fechner law).

This seemingly primitive behavior is very effective in guiding populations of microorganisms to suitable habitats. This can be seen in nature: when one removes a leaf laying on the surface of a wet area, its form may be outlined by accumulated organisms. The precision of this response can be demonstrated impressively by shining a photographic negative into a homogeneous suspension of organisms (e.g., in an agar layer). After a while, the cells accumulate in areas of suitable irradiances avoiding too bright or too dark fields and form a remarkably detailed photographic positive (Fig. 7.1), which can be "frozen" by drying the agar layer.

The response can also be induced by a temporal change in light intensity: organisms undergo a phobic response when the experimenter increases or decreases the light intensity, provided that the change in intensity occurs fast enough; a slow change is not detected. The same organisms may respond with a step-down phobic reaction at low fluence rates and with a step-up response at higher irradiances (Häder, 1987c; Doughty, 1993).

The actual phenomenon of the response depends on the organism under consideration. Although cyanobacteria reverse their direction of movement, flagellates may stop and tumble for a short while and then head off in a seemingly

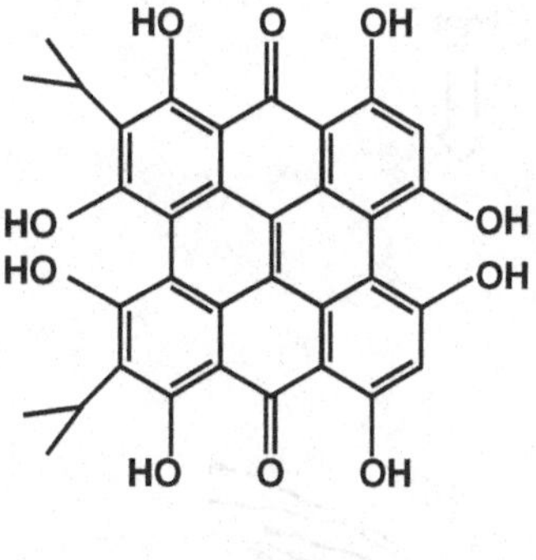

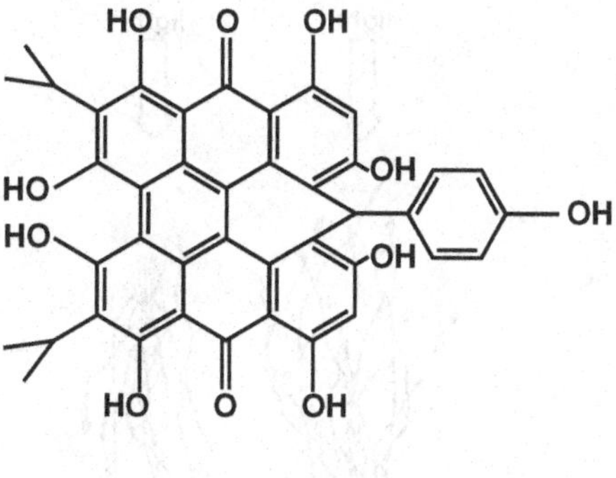

stentorin blepharismin

Figure 7.2. Chemical structures of stentorin and blepharismin, the putative photoreceptor molecules in the ciliates *Stentor* and *Blepharisma*, respectively.

random direction (Diehn et al., 1975). Ciliates often stop when they move into a bright light field, back up a few cell lengths, often turning in an arc-like fashion and then resume forward movement (Song et al., 1980). If this brings them again into the (too bright) light field, the response is repeated until they steer into preferred darker areas, a behavior that has been studied in detail in the colored ciliates *Stentor* and *Blepharisma*.

The photoreceptors for this step-up photophobic reaction are believed to be stentorin and blepharismin, respectively, derivatives of the hypericin family (Fig. 7.2). These substances are photosensitizer: upon excitation by light, they form long-lived triplet states that can transfer their excitation energy to, for example, oxygen, which in its ground state is the triplet form. This excitation results in the short-lived, but very aggressive, singlet oxygen ($^1O_2$), which is known to destroy vital cellular components – such as proteins, DNA, lipids – and to disrupt organelles, such as membranes and chloroplasts. It is an unsolved puzzle why these ciliates produce these dangerous molecules in the first place and then use them as photoreceptors to guide them away from detrimental radiation.

In *Euglena*, both step-down and step-up photophobic responses also occur at different irradiances, separated by an indifferent irradiance range with no evident reactions (Diehn, 1969a,b, 1973). Based on experiments with numerous inhibitors, ionophores, ion channel blockers, and various pH and ion concentration studies, Doughty and Diehn developed a model for the step-down photophobic response (Doughty & Diehn, 1979, 1982, 1983, 1984; Doughty et al., 1980; Barghigiani et al., 1979a; Walne et al., 1984). The model is based on the change in irradiance detected by the paraxonemal body (PAB). This is assumed to modulate the activity of a flagellar Na^+/K^+ exchange pump. The resulting increase of the sodium concentration in the intraflagellar space is hypothesized to open sodium-dependent calcium channels. The increasing intraflagellar calcium concentration eventually mediates a change in the beating pattern of the flagellum. The action spectrum for this photophobic response is characteristic of a flavin photoreceptor (Barghigiani et al., 1979b; Diehn, 1969a).

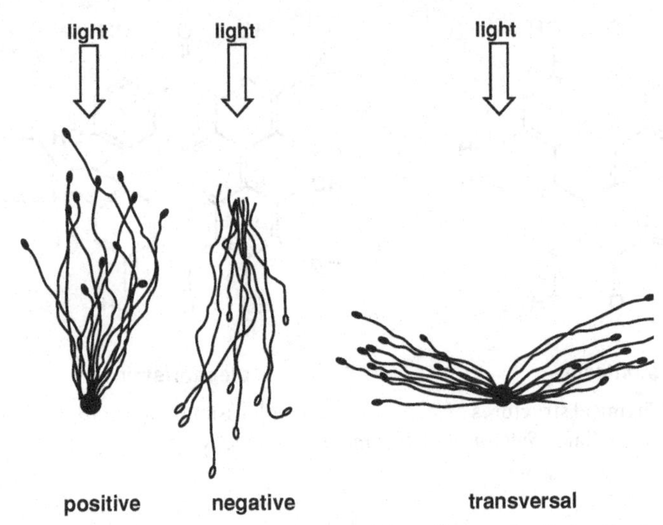

Figure 7.3. Tracks of organisms showing positive phototaxis (**left**, *Euglena gracilis*), negative phototaxis (**center**, *Euglena gracilis*), and diaphototaxis (**right**, amebae of *Dictyostelium discoideum*).

7.2.3 Phototaxis

Phototaxis is an orientation mechanism guided by the direction of light. Organisms may be swimming toward the light source, which is called positive phototaxis (Fig. 7.3, left). This behavior is often found at low fluence rates. The same organism may show negative phototaxis (swim away from the light source) at higher fluence rates (Fig. 7.3, center). In some cases, a movement at a certain angle with respect to the incoming light rays (e.g. perpendicular) has been found, which is called transversal phototaxis or diaphototaxis (Fig. 7.3, right). Phototaxis in unicellular flagellates has been known for more than 100 years (Pfeffer, 1881; France, 1908, 1909; Mast, 1914; Buder, 1919). In many cases, spectral sensitivity is highest in the ultraviolet (UV)-blue-green spectral region (300–550 nm). The older literature is reviewed by Bowne and Bowne (1967, 1968).

Although phototactic orientation may be conceptionally easy to understand, it involves more than a photoreceptor to detect the presence of light or its intensity. It requires an apparatus to determine the vectorial component of light, as we will see later in the discussion of a few case examples. Some ciliates have developed complicated cellular structures for this purpose, which resemble eyes in invertebrates with a lens and a photosensitive layer of pigments (Fig. 7.4), even though they are unicellular organisms (Kuhlmann et al., 1997a; Omodeo, 1975).

Haupt (1966) introduced the concept of a "one-instant mechanism," where two (or more) photoreceptor(s) detect the light facing different directions. A comparison of the readings provides a cue for the cell where the light is coming

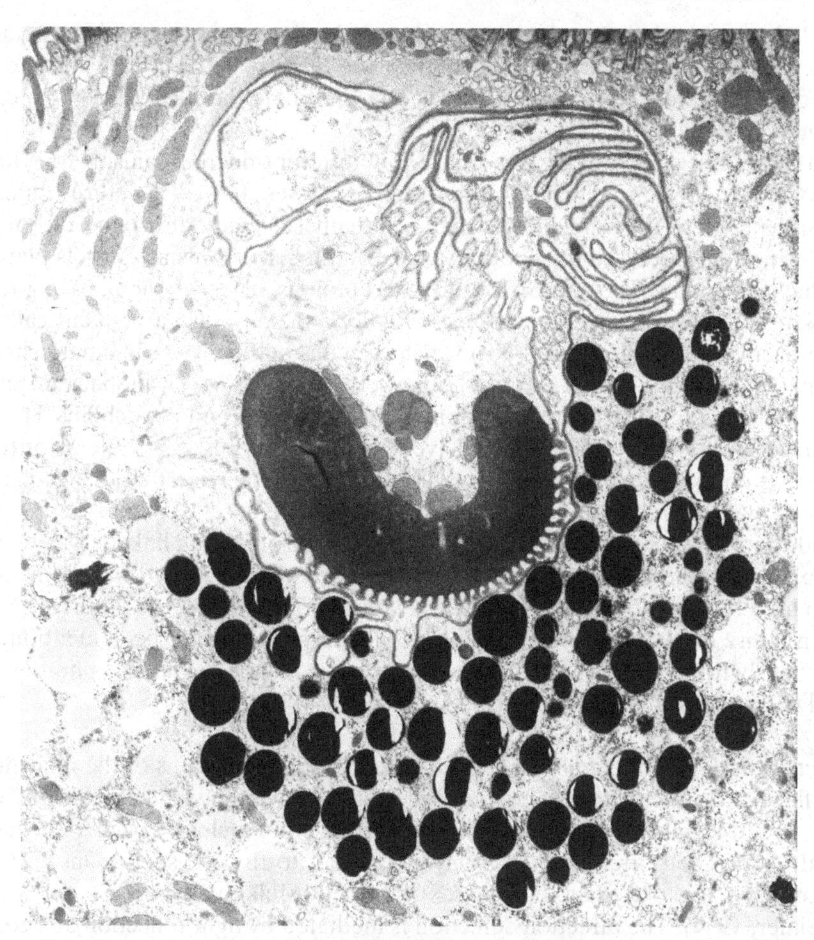

Figure 7.4. Photoreceptive structures in the ciliate *Ophryoglena catenula* with a crystalline lens and a layer of pigment granules. The convex side of the organelle displays ridges that are covered by the plasma membrane and two alveolar membranes. (Courtesy of H. Kuhlmann, Münster, Germany, with permission.)

from. The alternative model is the "two-instant mechanism." In this concept, one photoreceptor monitors the irradiance over time. When the cell, as many flagellates and ciliates do, rotates during forward locomotion, the photoreceptor scans the horizon and a comparison over time indicates the direction of the light source (Omodeo, 1975; Boscov, 1974; Boscov & Feinleib, 1979; Feinleib, 1975). Buder (1919) devised a conclusive experiment to answer the question whether phototaxis is a result of a directional movement in respect to the light source or an orientation in the spatial gradient of light intensity. A parallel light beam was passed through a biconvex lens to make it converge in the focal point. After the focal point, it diverged. Phototactically responding cells were observed moving toward the light source through the focal point. They continued to swim toward the light source also after they passed the focal point, even though they experienced a decrease in the irradiance. This behavior indicates that the organisms did not orient with respect to the changing irradiance, but to the vectorial component of light.

The photoreceptor molecules utilized by different microorganisms for phototaxis vary substantially and fall in different classes. It is as if nature has experimented with several concepts and eventually selected a few photoreceptor molecules that are used for photoresponses in higher organisms, such as plants or higher animals. The number of photoreceptors is especially high among the prokaryotes. Some bacteria, such as *Ectothiorhodospira*, use a pigment called photoactive yellow protein, (PYP) which contains a 4-hydroxycinnamate chromophore (Hendriks et al., 2002; Hellingwerf et al., 1996). Cyanobacteria use the photosynthetic pigments chlorophyll *a*, carotenoids, and phycobilins. However, recently, some cyanobacteria have been found to possess pigments of the phytochrome type, which were previously thought to be restricted to algae and higher plants. Mainly archaea use a class of pigments closely related to the rhodopsins. This chromophore is derived from β-carotene and linked to a seven-helix, membrane-bound protein by a Schiff's base (Fig. 7.5). Rhodopsins are also used by a number of eukaryotic algal microorganisms, whereas others have been found to be using flavins and/or pterins. Hypericins have been mentioned previously for ciliates, where they mediate phobic responses but also phototaxis and in some cases photokinesis (Matsuoka et al., 1999).

Phototaxis in Chlamydomonas. The photoreceptive apparatus of the unicellular flagellate *Chlamydomonas* is located in and above the stigma – an array of carotenoid-stained lipid droplets that are arranged in a closely packed, hexagonal pattern and often stacked in layers parallel to the cell surface inside the chloroplast, and placed on the equator of the cell (Melkonian & Robenek, 1984; Kreimer, 1994). The direction detection is mediated by light reflection and constructive interference (Foster & Smyth, 1980; Kreimer & Melkonian, 1990). The layers reflect the light, and the reflected waves interfere with the incoming waves, provided that $\lambda/2$ coincides with the distance between the layers (Fig. 7.6a,d). With light impinging perpendicular to the cell perimeter, the resulting maximum appears at the plasmalemma of the cell, where an array of large protein particles has been found, assumed to be the site, or associated with the photoreceptor pigments (Melkonian & Robenek, 1984; Foster & Smyth, 1980). Figure 7.6b shows a differential interference image of *Tetraselmis chuii*. The efficiency of reflection and interference improves with the number of layers. Light impinging from the rear of the stack is effectively shielded by the chloroplast located toward the cell center (Yoshimura, 1994). The reflective properties of the stigma can be impressively visualized by epifluorescence (Fig. 7.6c).

The cells rotate counterclockwise during forward locomotion, and the perceptive apparatus scans the environment for the changing irradiance. The photoreceptor perceives a modulated light signal, which ultimately controls the flagellar beat frequency and direction until the cell is aligned with the incoming light rays. In *Chlamydomonas*, the stigma is positioned on the cell equator 20°–40° out of the

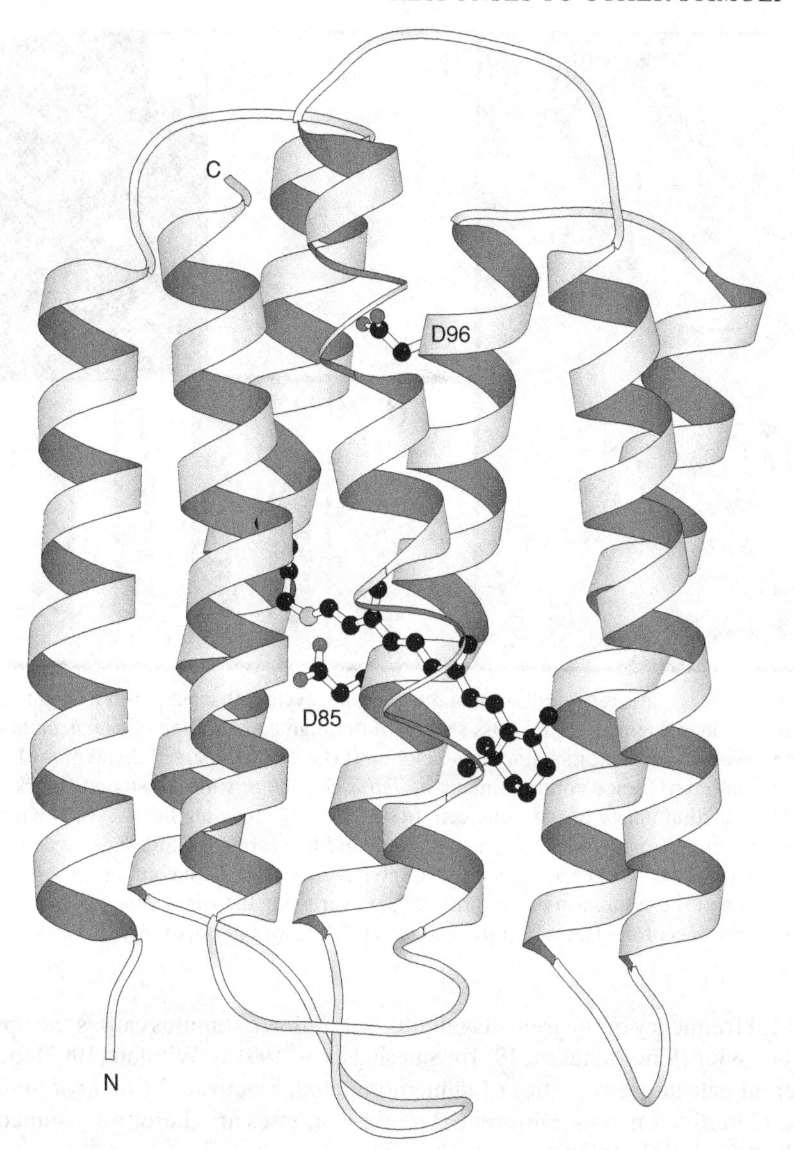

Figure 7.5. Chemical structure of bacteriorhodopsin, a 7-helix membrane-spanning protein with the retinal chromophore bound via a Schiff's base. (Modified after Häder, 1999b.)

beating plane of the two flagella. Thus, maximal light perception occurs a fraction of a second earlier than the reorientation of the flagella, a time interval that agrees well with the time needed for signal transduction from the photoreceptor to the flagella. The stigma is connected to the microtubular rootlet that arises from the basal body of the *cis*-flagellum and extends in the direction of the distal end under the plasmalemma (Melkonian & Robenek, 1984). Step-up light stimuli induce a transient increase in the beat frequency of the *cis*-flagellum and a decrease of

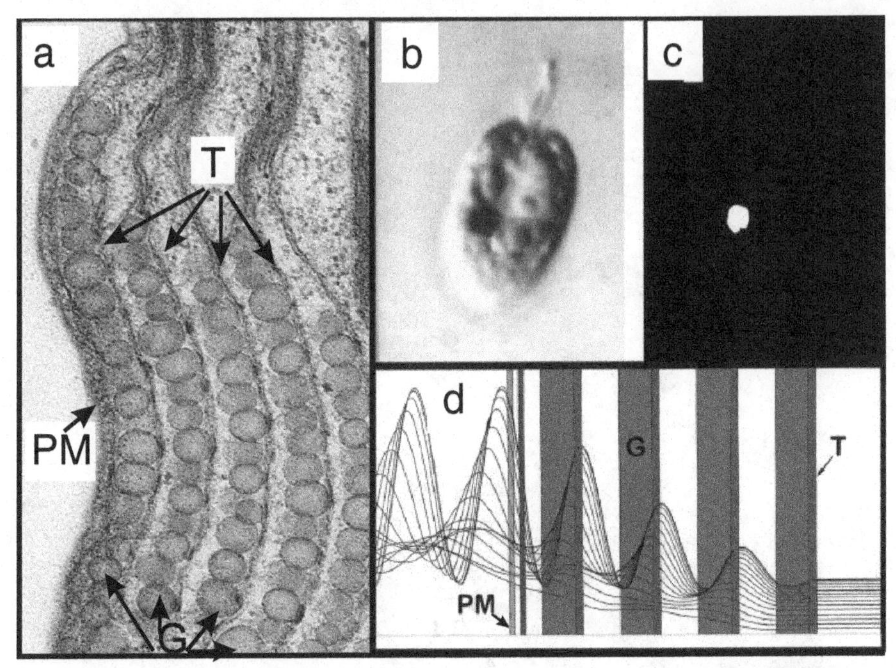

Figure 7.6. Schematic representation of the quarter wavelength mechanism for light direction detection in flagellates. (**a**) Cross-section of the stigma of *Hafniomonas reticulata* with several layers of carotenoid-stained lipid globuli (G) with interspersed thylakoids (T). (**b**) Differential interference contrast image of *Tetraselmis chuii* with the stigma (**black dot**) and (**c**) reflection image of the same cell. (**d**) A laterally incoming light wave is partially reflected at interfaces between layers with high and low refractive indices, spaced at $\lambda/2$. The reflected wave undergoes constructive interference with the incoming wave forming a maximum at the cytoplasmic membrane (PM) overlaying the stigma, the postulated site for the photoreceptor. (Reprinted from Kreimer, 2001, with permission.)

the beat frequency of the *trans*-flagellum. A step-down stimulus causes the opposite behavior (Sineshchekov, 1991b; Sineshchekov, 1991a). Witman (1993) found different calcium sensitivities of demembranated, reactivated *Chlamydomonas cis*- and *trans*-axonemes. Differential motor responses are therefore assumed to be the basis for phototaxis.

Maximal wavelength sensitivity is determined by the distance between the reflective layers and changes with the angle of the impinging light (Land, 1972; Hegemann & Harz, 1998). Identification of the photoreceptor by means of action spectra is difficult, because the action of the photoreceptor may be distorted by shading pigments and adaptation phenomena. Therefore, it has been suggested to measure action spectra for phototaxis at low irradiances (threshold action spectra; Foster, 2001). These action spectra have maxima between 460 and 560 nm, and are interpreted to be due to the action of rhodopsins. Action spectra for phobic responses are thought to be less distorted and peak between 490 and 520 nm (Schletz, 1976; Uhl & Hegemann, 1990).

The nature of the photoreceptor in *Chlamydomonas* was experimentally shown by reconstituting blind (retinal-deficient) cells. Both phototaxis and phobic responses were restored in these mutant cells by the addition of all-*trans* retinal. Mono-*cis* isomers were much less effective (Nakanishi & Crouch, 1995; Spudich et al., 1995). Judging from a number of reconstitution experiments with different chromophores suggests that the *Chlamydomonas* photoreceptor contains an all-*trans*, 6-*S*-*trans* retinal chromophore similar to bacterial rhodopsins. In contrast, animal rhodopsins contain 11-*cis* retinal or the derivatives 11-*cis*-3-hydroxy, 11-*cis*-4-hydroxy, or 11-*cis* 3,4-dehydroretinal. The chromophore undergoes a 13-*trans*- to *cis*-isomerization upon excitation, probably resulting in a conformational change of the protein via the 13-methyl group. The chromophore is easily accessible by hydroxylamine, which bleaches the chromophore and inhibits phototaxis. After washing out the hydroxylamine, retinal restores the chromophore and the photoresponse reappears (Hegemann et al., 1988). It is interesting to note that the β-ionon ring is not essential for photoperception, and a chromophore with only three conjugated double bonds plus methyl groups also restores photosensitivity (Nakanishi & Crouch, 1995; Spudich et al., 1995).

In light-grown cells, small amounts of 13-*cis* and 11-*cis* isomers were found besides all-*trans*-retinal (Hegemann et al., 1991; Derguini et al., 1991). Kreimer and coworkers (1991) isolated all-*trans* and 11-*cis* retinal in the chlorophycean alga *Spermatozopsis similis*, confirming the results on *Chlamydomonas*.

Using ^{3}H-retinal at a concentration just sufficient to fully reconstitute phototactic activity, a 30-kDa protein was found as the only labeled retinal protein in the membrane fraction (Beckmann & Hegemann, 1991). By treating cytoplasmic membranes with detergent, the photoreceptor retinal was exchanged for ^{3}H-retinal and the labeled opsin visualized by fluorography (Deininger et al., 1995). The photoreceptor was dubbed chlamyopsin, and polyclonal antibodies were used to localize the receptor in fixed and permeabilized cells (Deininger et al., 1995). It was found in a sharply localized spot of 1 μm in diameter at a position where, in living cells, the eyespot is seen. This was also observed in retinal-deficient cells and in stigma preparations of the green alga *Spermatozopsis* (Calenberg et al., 1998). In the latter organism, the stigma has a light-regulated GTPase activity that is suppressed by addition of the antibodies against chlamyopsin (Calenberg et al., 1998; Schlicher et al., 1995; Kreimer, 2001).

The complete chlamyopsin gene (*cop*) was sequenced from an EMBL3 clone (Fuhrmann, 1996; Fuhrmann et al., 1999). In this gene, introns are especially variable in size, ranging between 63 and 955 bp. The coding regions of the gene show strong codon biases, with preferentially G or C at the third position (Ikemura, 1985; Fuhrmann et al., 1999). The protein is about 65% identical to a similar one found in the colonial alga *Volvox*. Both proteins are highly charged, and transmembrane segments are not obvious; there are only 2–4 segments that are long enough to define transmembrane helices. Algal rhodopsins show sequence homology to the corresponding animal proteins, but not to archaean rhodopsins.

Algal rhodopsins are suspected to be either closely attached to an ion channel with calcium conductance or constitute the channel themselves (Harz et al., 1992). Lysin-containing sections with interspersed hydrophobic sections resemble corresponding motifs of voltage-gated or cGMP-gated ion channels (Hegemann & Deininger, 2001).

Light activation results in changes of the electrical membrane potential in *Chlamydomonas* (Harz et al., 1992; Sineshchekov, 1991a,b; Sineshchekov & Govorunova, 1999). In this organism, intracellular recording of the potential changes by micropipettes is difficult due to its small size. The light-induced potential changes can be monitored either by a population method, which is based on the asymmetric localization of the signal sources within the cell (Sineshchekov et al., 1992) or by whole-cell recording by means of suction pipettes (Litvin et al., 1978; Harz et al., 1992). The latter measurements were performed with cell wall-deficient strains. The first electrical event after a light flash is an inward current across the portion of the plasma membrane overlaying the stigma. This current has been named "photoreceptor current" (PC) (Litvin et al., 1978; Harz & Hegemann, 1991). There is a strict correlation between the irradiance and the peak amplitude. There is also a correlation between the delay time (between stimulus and PC) and the irradiance that is as low as 5 μs for high irradiances. The action spectrum for the induction of the PC correlates well with the action spectra of photophobic responses and phototaxis.

The PC is actually the sum of two processes: a fast component – which saturates only at extremely high irradiance levels – and a slow component (Sineshchekov, 1991b). The maximal amplitude amounts only to 10% of the fast component. The fast potential depends solely on the photoconversion rate of the photoreceptor and is the result of a localized calcium influx (Litvin et al., 1978; Harz & Hegemann, 1991). The late PC is driven by the transport of about 10^7 elementary charges across the membrane. Because this is achieved by the absorption of approximately 10^3 photons, there is an amplification of about 10,000 (Fuhrmann et al., 1999). This amplification could be due to the activation of GTP: in animal vision, one excited rhodopsin can activate up to 500 G proteins, which in turn activate thousands of phosphodiesterase molecules. In *Spermatozopsis*, convincing evidence for the involvement of G proteins in photoperception was presented (Calenberg et al., 1998; Schlicher et al., 1995; Linden & Kreimer, 1995). Light-dependent GTPase activity in isolated eyespot apparatuses was found, with an action spectrum similar to that of rhodopsin absorption.

When the light stimulus exceeds a certain threshold, an "all-or-none" response is triggered. The PC becomes superimposed by a spikelike response (Litvin et al., 1978; Harz & Hegemann, 1991), which is related to the flagella membrane current and was therefore named "flagellar current" (FC). The FC is due to a massive calcium influx and is assumed to be driven by voltage-gated calcium channels in the flagellar membrane (Beck & Uhl, 1994; Harz & Hegemann, 1991). The FC controls the flagellar beating pattern and reverses the normal breast stroke to the so-called undulation, which results in backward swimming of the cell (Holland et al., 1997). The "fast flagellar current" (F_f) is followed by a "slow flagellar current" (F_s), which seems to be important for the kinetics of backward swimming (Holland et al., 1997; Harz et al., 1992). Subsequent to the FC, a potassium efflux polarizes the membrane again and restores light responsiveness (Braun & Hegemann, 1999; Nonnengässer et al., 1996; Govorunova et al., 1997).

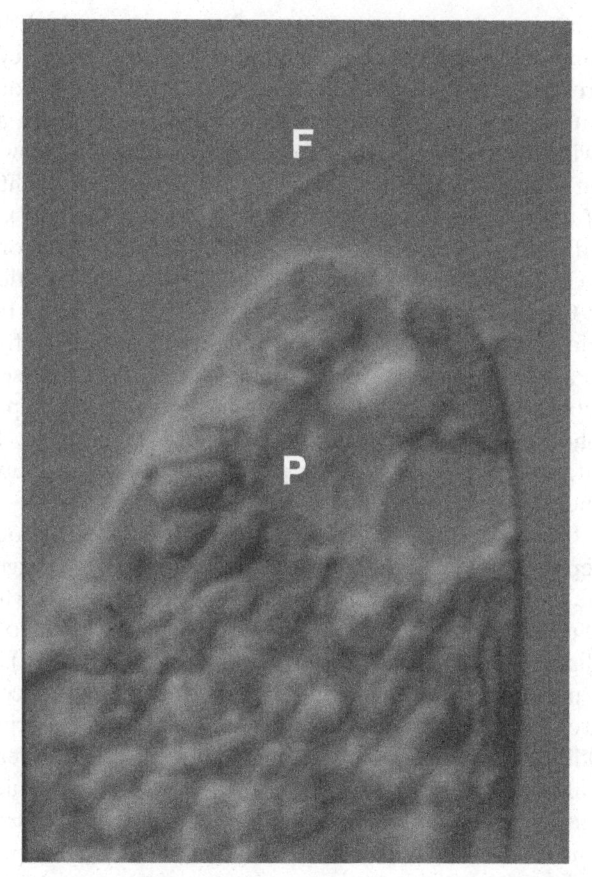

Figure 7.7. Front end of *Euglena gracilis* showing the two flagella (F), each arising from a basal body at the bottom of the reservoir and the paraflagellar body (P).

Phototaxis in Euglena. The unicellular flagellate *Euglena*, even though characterized by chlorophyll *a* and *b*, is not a green alga but belongs to the group of Euglenophytes. The cells are powered by a trailing flagellum that emerges from a basal body inside an invagination at the front of the cell (reservoir). A second flagellum also arises from a basal body next to the former, but it does not leave the reservoir, and its tip is glued to the emerging flagellum.

In addition to photokinesis, and step-up and step-down photophobic responses, the cells show positive phototaxis at low fluence rates below 10 W m^{-2} and a negative one at higher irradiances. The photoreceptor in *Euglena* is assumed to be located in the PAB (formerly called paraflagellar body), a swelling at the emerging flagellum inside the reservoir where the nonemerging flagellum is attached to the longer one (Fig. 7.7). Whether or not the paraxonemal rod, which runs along the whole length of the flagellum, is involved in light-mediated responses is still an open question (Walne & Arnott, 1967; Piccinni et al., 1975; Piccinni & Omodeo, 1975; Piccinni & Mammi, 1978).

An earlier hypothesis assumes that light direction detection is based on the periodic shading of the PAB by the stigma that is located in the cytoplasm adjacent to the reservoir (Buder, 1919). However, this "shading hypothesis" is not compatible with a number of recent findings; for example, when a population is irradiated with two perpendicularly oriented light sources at low irradiances, the population splits in two: half of the cells move toward one light source, and the other half toward the second light source (Häder et al., 1987). In contrast, at higher irradiances, the cells swim on the resultant of the vector addition of both beams. Also, stigmaless mutants of *Euglena* show phototactic orientation (Häder, 1993; Checcucci et al., 1976; Gössel, 1957; Ferrara et al., 1975). A final argument against the validity of the shading hypothesis is derived from inhibitor studies: If the assumption holds that phototaxis in *Euglena* is based on repetitive photophobic responses, then inhibitors of the photophobic responses should also impair phototaxis. This could not be verified (Häder et al., 1987). Some stigmaless mutants did not show positive phototaxis, but oriented away from the light source (negative phototaxis).

Light direction detection in *Euglena* is based on a dichroic orientation of the photoreceptor pigments (Häder, 1987d). This is already suggested by the paracrystalline structure of the PAB (Piccinni & Mammi, 1978). The cells show a pronounced polarotaxis and swim at a fixed angle with respect to the e-vector of polarized light (Bound & Tollin, 1967; Creutz & Diehn, 1976). During one rotation, two positions with a maximal absorption probability occur in lateral light. The role of the stigma could be to suppress one of the two maxima. Negative phototaxis additionally requires the screening of the PAB by the rear end of the cell with its chloroplasts. White, chloroplast-free mutants consequently cannot easily distinguish light coming from the front or rear, and the population splits up in two components showing positive and negative phototaxis, respectively, over a wide range of irradiances.

A number of action spectra have been determined for phototaxis (and photophobic responses) in *Euglena* (Fig. 7.8). They are distinctly different from those measured for *Chlamydomonas* and have been interpreted to represent the action of flavins and pterins. Extensive fluorometric analysis has shown that the pterins, absorbing in the UV-A range of the spectrum, with a peak near 360 nm, operate as antenna pigments. They emit at about 450 nm in the blue wavelength band, one of the peaks for flavin absorption. The fluorescence emission of the flavins is detected at 520 nm (Lebert & Häder, 1997a; Häder & Lebert, 1998). Therefore, the proper photoreceptor is assumed to be represented by the flavins.

Alternatively, it was proposed that also in *Euglena*, a rhodopsin receptor is active (Gualtieri et al., 1989). Recently, this controversy was solved by the genetic analysis of the photoreceptor. Iseki and coworkers (2002) have identified two genes involved in phototactic perception. The first codes for a protein subunit with a molecular weight of 105 kDa and the second one with 90 kDa. Both protein subunits coded by these genes are tandem repeats of a flavin-binding motif, followed by a portion of the molecule with adenylyl cyclase activity (Fig. 7.9). Therefore, these two protein subunits have been called PAC α and PAC β; PAC stands for photoactivated adenylyl cyclase. Polymerase chain reaction with primers for

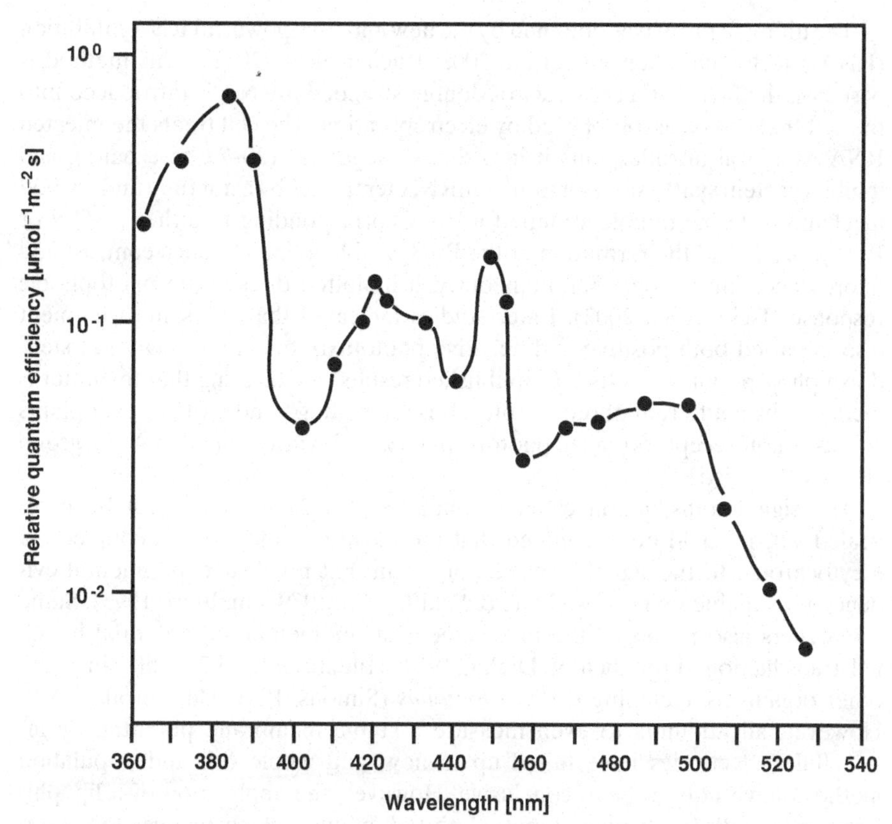

Figure 7.8. Action spectrum for phototaxis in *Euglena gracilis*. (After Häder & Reinecke, 1991.)

the two genes indicated that both PAC α and PAC β were present in a number of phototaxis mutants, indicating that they are not photoreceptor mutants, but obviously downstream mutants (unpublished data). Neither of the two proteins was found in *Astasia*, a colorless, nonphotosynthetic close relative of *Euglena*, which lacks the PAB and consequently does not show phototaxis. Polyclonal antibodies raised against these two proteins bind at the site of the PAB in the cell, indicating that they represent the photoreceptor.

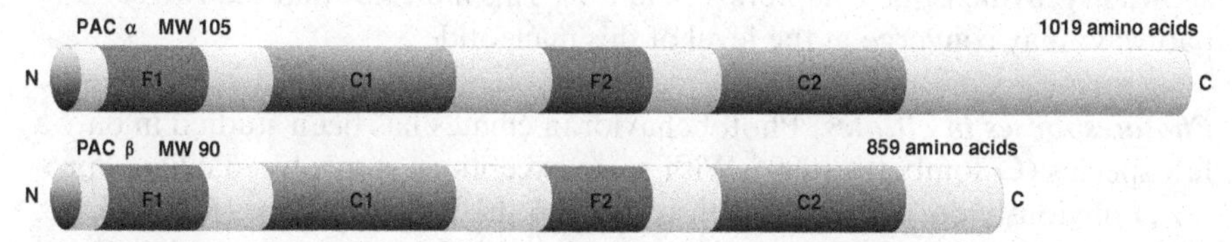

Figure 7.9. Structure of the PAC α and PAC β genes, which code for the two subunits of the photoreceptor protein with tandem repeats of a flavin-binding domain and an adenylyl cyclase. N, N-terminus, C, C-terminus of the protein. MW, molecular weight. (Modified after Iseki et al., 2002.)

The ultimate proof was obtained by the new and very powerful RNA inhibition (RNAi) technique (Martens et al., 2002; Bucher et al., 2002). This method is based on the fact that a construct of double-stranded mRNA is introduced into the cell that can be accomplished by electroporation. The cell treats the injected RNA as a viral intruder, cuts it into shorter segments (20–23 base pairs), and inhibits protein synthesis of all similar mRNA templates by a not fully understood mechanism. Using double-stranded mRNA corresponding to either PAC α or PAC β prevented the formation of the PAB, visible in interference contrast and fluorescence microscopy. Simultaneously, it inhibited the step-up photophobic response (Iseki et al., 2002). Later studies indicated that the same treatment also impaired both positive and negative phototaxis, but surprisingly not step-down photophobic responses (unpublished results) – indicating that the latter is mediated by a different photoreceptor. It is interesting to note that higher plants also use photoreceptors (called phototropins) with flavin as a chromophoric group (Briggs & Christie, 2002).

The signal transduction chain of phototaxis in *Euglena* has not been revealed yet. It could be speculated that the flavin chromophore is coupled via a cytochrome to the signal transduction chain, but no clear experimental evidence is available up to now (Fong & Schiff, 1978, 1979; Gualtieri, 1993). Some researchers also proposed the involvement of the membrane potential in signal transduction (Froehlich & Diehn, 1974; Hildebrandt, 1974), like in many other organisms, including *Chlamydomonas* (Simons, 1981; Harz et al., 1992). However, all attempts to even measure a stable membrane potential by intracellular electrodes have failed up to now, and whole cell and population methods have not yet been conclusive. However, the application of a lipophilic cation, $TPMP^+$ (triphenyl methyl phosphonium), which reduces the membrane potential by following the existing gradient, shifted the threshold irradiance between positive and negative phototaxis to lower irradiances, indicating that the membrane potential is involved in some aspects of phototaxis. The involvement of membrane potential changes in *Euglena* gravitaxis is described in Chapter 5.

The recent results on the genetic analysis described here indicate a different linkage to the subsequent steps in the sensory transduction chain. The adenylyl cyclase activity, which has been shown to be light-induced by the flavin chromophores on the same molecule, suggests that cAMP is involved in the signaling cascade. This is of specific interest since the involvement of cAMP has been shown in gravitaxis (cf. Chapters 4, 5, and 9). This indicates that the two sensory pathways may converge at the level of this nucleotide.

Photoresponses in ciliates. Photobehavior in ciliates has been studied in only a few species (Colombetti, 1990). With some exceptions, noncolored ciliated protozoa obviously do not orient with respect to the light direction (Matsuoka & Nakaoka, 1988). Symbiont (e.g., *Chlorella*)-bearing forms (e.g., *Paramecium* and *Euplotes*) are sensitive to light in the form of kinetic or phobic, but not tactic, responses (Reisser & Häder, 1984). True phototaxis in ciliates seems to be coupled to morphological features: bright coloration due to pigmented granules [considered to be the site of the photoreceptor (Lenci et al., 2001; Song, 1985; Song & Poff, 1989)], eyespot structures, or even watch-glass organelles (organelle of Lieberkühn; Kuhlmann, 1998). True phototaxis has been stated for the ciliate *Chlamydodon.* Direction and precision of the cells depend on their feeding status, which is coupled to the occurrence of an eyespot structure. If the cells are well-fed, they have numerous bluish food vacuoles due to the uptake of cyanobacteria. In this physiological state, they show a precise negative phototaxis if illuminated in unilateral constant white light. A reversal to positive phototaxis occurs in slightly underfed cells, which are characterized by an accumulation of several hundred orange vesicles at their anterior left side (= eyespot). It is proposed that the eyespot takes part in the photoorientation of *Chlamydodon* (Kuhlmann & Hemmersbach-Krause, 1993b). A similar structure has been stated for *Nassula citrea*, whose precision of negative phototaxis depends on its feeding status. The phototactic response is most pronounced in slightly starved cells, which are characterized by a pigment spot consisting of an accumulation of almost 2,000 orange vesicles (Kuhlmann & Hemmersbach-Krause, 1993a). From the center of the eyespot, a membranelle row extends to the oral opening, which might have a "sensory role" (Tucker, 1971). The identification of the primary pigment for photoreception and the corresponding signal transduction chain are still lacking.

Intracellular microelectrode recordings in the ciliates *Stentor* and *Blepharisma* suggest a light reception via the stentorin (blepharismin) photochemistry inducing a photoreceptor potential that determines the light-dependent photoresponses (Fabczak et al., 1993c,d). The mechanism of the photoreceptor function remains to be clarified. cGMP is considered as a possible mediator in the light signal transduction chain (Fabczak et al., 1993a,b).

7.2.4 Other light-induced responses

The differentiation between the three classes of responses described here (photokinesis, photophobic responses, and phototaxis) may be somewhat formalistic. Some responses may not be properly described by the definitions given previously, and some responses are not fully understood to be classified. In addition, some researchers do not follow these definitions, but rather use their own. Researchers of bacterial responses often loosely describe light-induced reactions as phototaxis, even though they are clearly photophobic (Krah et al., 1994). Also, the methodology used to study these light-induced responses contributed to confusion. The so-called phototaxigraph (Lindes et al., 1965) does not really measure what the name promises it does, but rather measures photoaccumulations in a light field, which can be brought about by phobic responses, by photokinesis (e.g., when cells become immotile in the light field), by light-induced attachment of the cells to the glass walls (which is observed in a number of organisms), and also by phototaxis induced by light scattered from organisms already inside the light field.

One obvious and well defined light-induced response is phototropism, which is observed in sessile organisms. It occurs in higher plants, where shoots bend toward the light (positive phototropism) and roots often away from the light (negative phototropism). Leaves are often oriented perpendicular to the incoming light rays (transversal or diaphototropism). Some sessile microorganisms also show phototropic behavior, for example, the mold *Phycomyces* shows a remarkably sensitive phototropism of its sporangiophores, which bend toward light sources at irradiances as low as 10^{-9} W m^{-2} (Galland & Lipson, 1984). For many of these responses, flavin-containing photoreceptors are being discussed (Christie et al., 1998). As an aside: in some plants, organs such as fruit stems have been found to bend toward dark areas, which is different from bending away from the light direction (Strong & Ray, 1975). This response, which has been termed "skototropism," is advantageous for a plant if, for example, seeds are stored in crevices of rocks or walls where they can later germinate.

7.3 Orientation in chemical gradients

The direction to a light source can be determined from a distance, at least with the use of a "one-instant mechanism." In contrast to light, the source of a chemical cannot be detected; an organism does not "know" where a food source is located. Rather, all chemotactic orientation is based on a trial-and-error mechanism of some kind following a chemical gradient, which eventually will guide the organism to the source. The organisms move in an arbitrarily chosen direction and sense an increase or decrease in the concentration of a chemical as they follow their path. In the case of an attractant (e.g., a pheromone or a nutrient such as a sugar), sensing an increase in the concentration enforces moving in this direction, while sensing a decrease results in a reversal of movement or a turn until a "correct" direction is found. Repellent gradients are answered by the opposite behavior.

Chemotactic orientation has been found in many taxonomically very divergent groups from bacteria (Berg, 2000; Aizawa et al., 2000) to eukaryotic microorganisms (Almagor et al., 1981; Antipa et al., 1983; Doughty & Dodd, 1976) via fungi (Allen & Newhook, 1973; Akitaya et al., 1984; Andre et al., 1989), algae (Boland et al., 1982, 1983, 1989), mosses (Akerman, 1910; Godziemba-Czyz, 1973), and higher plants (Blasiak et al., 2001) to lower and higher animals (Garbers, 1988; Ward, 1978; del Portillo & Dimock, 1982; Bareis et al., 1982) – including men (Gallin & Snyderman, 1983).

Chemotaxis is clearly best investigated and understood in eubacteria (Manson et al., 1998). These prokaryotic organisms swim in their aqueous habitat powered by one to many flagella. The flagellum is a passive propeller rotated by a basal motor, which is driven by protons or in some cases other ions (Imae & Atsumi, 1989). The structure and the forty individual gene products of this remarkable rotary motor are well known (Fig. 7.10). The basal apparatus consists of two to four (in some cases five) rings that are either attached to the cytoplasmic membrane, the cell wall, or the rod – the central rotating axis – and build from four different

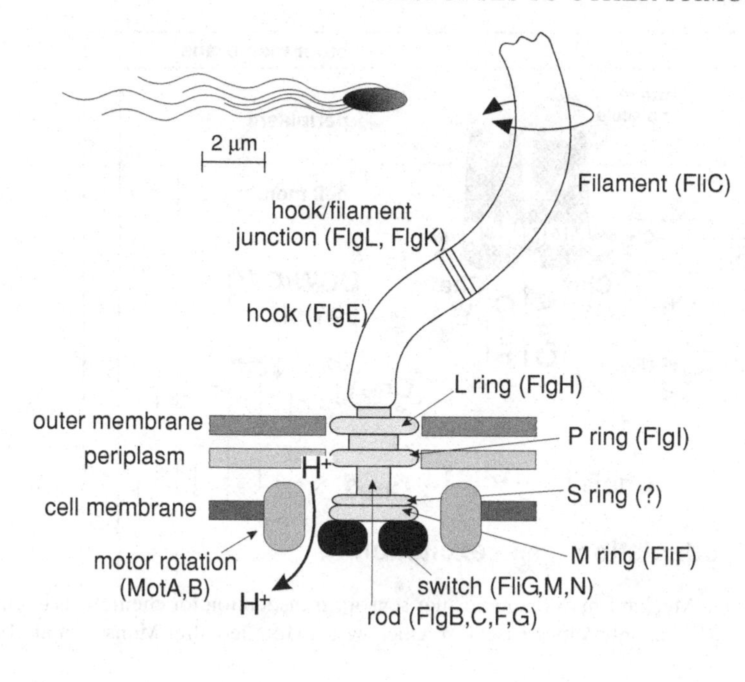

Figure 7.10. Structure of the bacterial flagellum with its basis consisting of some 40 gene products. (Modified after Manson et al., 1998.)

proteins (FlgB, FlgC, FlgG, and FlgF). The force for rotation is generated between the stator complex, constructed from the motA and motB proteins anchored to the S ring, and the M ring, which forms the rotor and is linked with the rod. The P and L rings serve as bushing and are anchored to the outer membrane (MacNab & Aizawa, 1984). The rod is connected by a flexible hook (consisting of a single protein species, FlgE) to the long filament, also produced from a single protein (FliC; MacNab, 1990). The motor can rotate clockwise or counterclockwise (Silverman & Simon, 1974), and the direction of movement is controlled by a number of switching elements at the basal body. During counterclockwise rotation, bacterial cells – such as *Escherichia coli* – swim forward on a smooth track at 20–60 μm/s, which corresponds to up to thirty body lengths. This is interrupted periodically (at random intervals of about 1 s) by short chaotic tumbles lasting about 0.1 s, in which the cell turns randomly into a new direction. During tumbles, the flagella rotate clockwise. It is interesting that, during forward swimming, the individual rotating flagella – which are inserted peritrichously distributed over the cell surface – form a bundle, in which all flagella operate in phase. Bundle formation is explained by hydrodynamic interaction (MacNab, 1977).

This swimming behavior, consisting of tumbles and runs, results in a three-dimensional random walk. In a chemical gradient of an attractant (e.g., ribose),

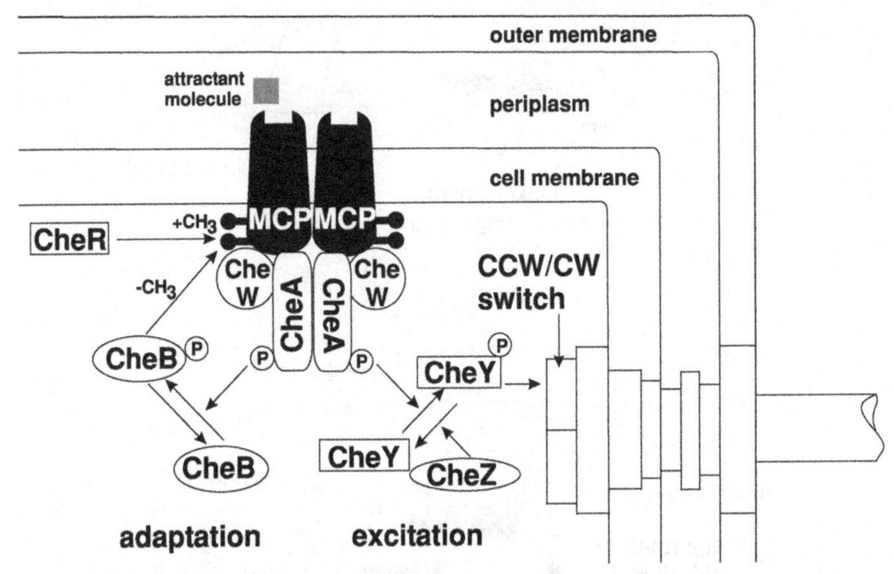

Figure 7.11. Mechanism of the molecular sensory transduction for chemotaxis in enterobacteria. CCW, counter clock wise; CW, clock wise. (Modified after Manson et al., 1998.)

the runs are extended and tumbles suppressed when the cell happens to swim toward the source of the attractant (i.e., experiences increasing concentrations of the chemical over time). Swimming away from the attractant source, resulting in decreasing concentrations, causes the cells to increase the frequency of tumbles until a favorable direction is found by accident. Repellents induce the opposite behavior. From this analysis, we can conclude that the cells do not "know" the direction to the source of the chemotaxis-inducing substance, but rather orient by a bias to the random walk.

Bacteria have a limited number (five known in *E. coli*) of different chemoreceptors for various attractant and repellent chemicals. These chemoreceptors are 2-helix protein dimers that span the inner membrane underneath the periplasmic space (Manson et al., 1998) and form a ternary complex with two CheA and two CheW proteins on the cytoplasmic side (Fig. 7.11). Binding an attractant molecule initiates a conformational change of the chemoreceptor, which results in methylation of the cytoplasmic part of the molecules (Manson et al., 1998). There are four or five glutamate residues that can each bind a methyl group. Therefore, the chemoreceptors are also called methyl-accepting chemotaxis proteins (MCPs). CheA is a histidine protein kinase. The linked phosphate is transferred to mobile CheY in the cytoplasm, which interacts with the switching elements on the motor apparatus. Docking of an attractant molecule suppresses the autokinase activity of CheA. Therefore, the level of phosphorylated CheY decreases, resulting in less frequent tumbling of the cell.

Adaptation is accomplished by CheR (methyltransferase) and CheB (methylesterase). CheR methylates the cytoplasmic domain of the MCPs. CheB is phosphorylated by CheA and in this state demethylates the MCPs. This mechanism warrants an intermediate level of MCP methylation and thus controls the run/tumble frequency.

Protists also provide suitable model systems to study chemoresponses and the corresponding signal transduction chains. The ciliate *Paramecium* reacts by moving toward or away from the source of the chemical stimulus. Chemical attractants are, for example, substances produced by bacteria (food), cAMP, and amino acids, whereas repellents are extremes of pH and salt (for review, see van Houten & Preston, 1987; van Houten, 1990). The general model for a sensory transduction pathway starts with binding of a ligand to a receptor in the cell membrane, followed by transduction probably by the involvement of second messengers and finally the motor response. In analogy with the electromotoric coupling in ciliates (cf. Chapter 4), it seems likely that similar events are involved in the behavioral responses to chemicals, as kinetic (speed control) and turning based on ciliary reversal are essential. So far, three different pathways for chemosensory signal transduction of attractants in *Paramecium tetraurelia* have been identified. All of them hyperpolarize cells and thus induce fast and smooth swimming, resulting in an accumulation of cells (van Houten, 1990). Yang and coworkers (1997) were able to show that L-glutamate initiates a signal transduction chain involving cAMP production. Stimulation of cells with L-glutamate increased cAMP within 30 ms, whereas cGMP levels remained unchanged. Interestingly, other attractants, which also hyperpolarize the cells, did not induce a change in cAMP. It can be assumed that hyperpolarization might be used differently within a cell: some might activate the adenylyl cyclase (or inhibit the phosphodiesterase), others, not. How the L-glutamate binding sites and adenylyl cyclase or phosphodiesterase are coupled has to be clarified.

A number of flagellates are also known to respond chemotactically. Aerotaxis (also called oxytaxis) – i.e., orientation in an oxygen concentration gradient – has been observed in *Euglena gracilis* (Colombetti & Diehn, 1978; Porterfield, 1997). The receptor is assumed to be a cytochrome *c* oxidase (Miller & Diehn, 1978). Also, ciliates respond to oxygen: For the microaerophilic ciliate *Loxodes* (cf. Section 4.2), oxygen is toxic and consequently the cells avoid being exposed to high concentrations of oxygen by negative chemotaxis (Fenchel & Finlay, 1990; Finlay et al., 1986). The cytochrome oxidase seems to be the oxygen sensor also in *Loxodes*; because the organism respires oxygen, its respiration is sensitive to KCN, and the cells show faster and random swimming at 10^{-4}–10^{-6} M KCN (Finlay & Fenchel, 1986). Oxytaxis is also found in the ciliate *Tetrahymena* (Shvirst et al., 1984). Mammalian cells also possess receptors for oxygen concentration. In these cells, a family of hypoxia-inducible transcription factors has been discovered, which are activated by a decrease in the concentration of molecular oxygen and that regulate the expression of downstream target genes that mediate adaptation and survival of cells and the whole organism (Bruick & McKnight, 2002). The heterotrophic flagellate *Astasia longa* has been found to respond to gradients of dissolved carbon dioxide (Borgers & Kitching, 1956). Gametes of several algal groups utilize pheromones for sexual attraction (Boland et al., 1983, 1989). Due to the fast responses of protists toward chemicals, their sensory characteristics have been used as sublethal endpoints in ecological risk assessment.

Chemotaxis plays a key role in cellular slime molds, such as *Dictyostelium discoideum*. These simple eukaryotic organisms undergo a life cycle between unicellular amebae and multicellular pseudoplasmodia (Bozzaro & Ponte, 1995). The amebae feed on bacteria that grow on decaying litter on the forest floor. The amebae detect their food by chemotaxis to folate excreted by the bacteria. When food is exhausted, the multicellular life stage is initiated. One or several cells start signaling by excreting an attractant molecule, acrasin (Bonner et al., 1969), which has been identified as cAMP. This molecule diffuses outward and induces neighboring cells to move toward the primary source. Simultaneously, the receiving cells break down the cAMP by a phosphodiesterase (Yamasaki & Hayashi, 1982) and produce new cAMP on the distant side by an adenosine cyclase (Fontana & Devreotes, 1984). This relay system produces a concentric wave pattern by which up to hundreds of thousands unicellular amebae aggregate into a multicellular pseudoplasmodium.

7.4 Orientation in thermal gradients

Orientation in thermal gradients is frequent among animals; but, also, many microorganisms orient with respect to temperature (Haupt, 1962c). As with light, we can discriminate between phobic and tactic responses. We can further discriminate between responses to conducted versus radiative heat, but only the first case is of importance for motile microorganisms. One word of caution is in order in the evaluation of thermotactic responses. Often, the observed reaction is only indirectly due to thermal gradients and rather mechanistically induced by, for example, resulting convection currents of the medium (Clayton, 1959). To establish a true orientation with respect to the temperature gradient, the movement of dead cells or inorganic particles can be compared with that of active motile cells (Mendelssohn, 1902). Metzner (1920) microscopically observed phobic responses of the organisms, and Reimers (1928) chose the dimensions of the observation vessel such that convection currents were largely subdued. Another experimental complication is the altered solubility of gases, such as oxygen and carbon dioxide, at different temperatures. Many organisms are known to respond chemotactically to gaseous gradients (e.g., aerotaxis; see Section 7.3).

Metzner (1920) described a clear thermotaxis in spirilla, whereas other bacteria have been found not to be thermotactic (Reimers, 1928). The response is a typical phobic one: decreasing temperatures result in a reversal of movement. This behavior can be used to collect the bacteria such as *Spirillum volutans* in a heat trap analogously to the light trap experiments described previously. Increasing temperatures did not cause a phobic response, but resulted in higher swimming speeds (thermokinesis; Metzner, 1920). In contrast, *Spirillum serpens* showed negative thermotaxis (i.e., it reversed its direction of movement on increasing temperatures). Recently, the eubacterium *Thermotoga maritima* was found to show thermotaxis (Gluch et al., 1995). The bacterium is motile at temperatures between 50° and 105°C. The response is similar to chemotactic reactions. Also, the average swimming speed is controlled by the temperature (thermokinesis), with a maximum of about 60 μm/s at 85°C. As other eubacteria, this organism shows runs and tumbles. With increasing temperatures, the rate of tumbles increases and with decreasing temperatures it decreases. Thus, the cells tend to swim more along a straight line when they experience increasing temperatures, effectively heading for higher temperatures and fleeing cooler areas.

Positive thermotaxis is also found in the cyanobacterium *Oscillatoria brevis* (Mendelssohn, 1902), whereas Reimers (1928) could not detect thermotaxis in other *Oscillatoria* and *Spirulina* species. Several, not identified diatoms were also observed to show thermotaxis, by gliding more smoothly in a temperature gradient, reversing less often, and stopping for shorter periods of time. These reactions resemble those found for light-induced behavior (Nultsch, 1975). Some *Nitzschia* and *Navicula* species are positive thermotactic and move to a temperature optimum of 28°–35°C (Reimers, 1928).

Among the flagellates, *Haematococcus* and *Volvox* show negative thermotaxis moving to an optimum of 5°–10°C. In contrast, *Euglena* and *Phacus* have been found to show positive thermotaxis (Franze, 1893; Günther, 1928). *Carteria multifilis* and some other flagellates are positive thermotactic at low temperatures and negative at higher temperatures. This behavior guides the organisms to a temperature optimum (Reimers, 1928). Metzner (1929) could show in the dinoflagellates *Peridinium* and *Ceratium* that thermotaxis is more pronounced the steeper the temperature gradient.

Most ciliates avoid extreme temperatures and accumulate at the cultivation temperature. *Paramecium* reacts to shifts in temperature: Heating and cooling affect the frequency of the ciliary beat and swimming direction. Due to our current knowledge, thermosensation by heat-detecting molecules seem unlikely. Alternatively, it is proposed that the ambient temperature directly affects ion conductances and hence the membrane potential, intraciliary calcium concentrations, and enzymatic reactions (Nakaoka et al., 1987). While local thermal stimulation of the anterior cell pole induces a depolarizing membrane potential response, stimulation at the posterior part results in a hyperpolarizing response in *Paramecium*. Consequently, two kinds of thermoreceptor channels have been suggested in analogy to the two kinds of mechanoreceptor channels. Voltage-clamp experiments revealed that the thermoreceptor currents depend on different ion channels than the ones for mechanoreceptor currents. Nevertheless, the ionic pores for the channels share common properties (Tominaga & Naitoh, 1994). With respect to the question of how the sensory response is amplified, it seems likely that changes in G-protein activity are involved in cold sensory transduction, though the mechanisms for how the cooling stimulus is transmitted and the activity of the ion channels is regulated remain speculative (Nakaoka et al., 1997).

Myxomycetes have been found to be surprisingly sensitive in their thermotactic responses. Two organisms have been studied to some extent: *Fuligo varians* and *Dictyostelium discoideum*. *Fuligo* was found to be positive thermotactic at temperatures below 31° to 36°C and negative above (Stahl, 1884; Wortmann, 1887; Clifford, 1897). The reaction is mediated by a mechanism in which the side of a plasmodium that is closer to the temperature optimum continues moving. This indicates that the organisms sense a temperature gradient in a "one-instant mechanism." Pseudoplasmodia of the cellular slime mold *Dictyostelium* turn in a temperature gradient, provided it exceeds 0.05°C/cm (Raper, 1940; Bonner, 1959). Because the width of the pseudoplasmodium is about 0.1 mm, the organism senses a gradient across its body of below 0.0001°C, which is very close to the physical limit caused by thermal fluctuations in the medium. Later on, this phenomenon was studied in more detail: The cells show a positive thermotaxis, when they are transferred to a midpoint temperature on the gradient that is higher than their current growth temperature (i.e., they move to higher temperatures); but, above 28°C, they become immotile. In contrast, when the cells are transferred from their growth temperature to a lower midpoint temperature, they migrate to lower temperature (negative thermotaxis) until they also become immotile at 13°C. Thus, in any case, they move away from an optimal temperature to either higher or lower temperatures – a behavior that is not easily understood from an ecological point of view.

It is interesting to note that these responses are due to conducted heat. The equally sensitive phototaxis of these organisms is independent and cannot be used to explain thermotaxis. In addition to thermotaxis, thermokinetic effects have been considered that may superficially look like oriented movements as shown for chemokinesis (Clayton, 1959).

7.5 Guidance by the Earth's magnetic field

A wide variety of organisms are known to orient their movement with respect to the magnetic field of the Earth. Migratory birds use it often in conjunction with optical clues (Walcott et al., 1988; Able & Able, 1990, 1993; Wiltschko et al., 1993, 2000; Wiltschko & Wiltschko, 1996). There is evidence that the magnetoreceptor is located in the eye (Ritz et al., 2000), whereas in newts the receptor seems to be extraocular (Deutschlander et al., 1999). Bull sperms orient themselves in static magnetic fields (Emura et al., 2001), and rodents can detect the geomagnetic field (Olcese et al., 1988; Thomas et al., 1986; Trzeciak et al., 1993; Burda et al., 1990). Other mammals – such as dolphins – have been found to possess magnetic material in their heads (Zoeger et al., 1981), and even humans have magnetic bones in their sinuses (Baker et al., 1983). Magnetite seems to constitute the vertebrate magnetoreceptor (Diebel et al., 2000). Most of the work on vertebrates has been reviewed by Kirschvink (1989, 1997) and Gould (1984). Among invertebrates, lobsters (Lohmann et al., 1995), honeybees (Walker & Bitterman, 1989a,b,c, Hsu & Li, 1994), and ants were found to orient with respect to magnetic fields (Camlitepe & Stradling, 1995; Anderson, 1993).

Higher plants respond to the Earth's magnetic field (Schwarzacher & Audus, 1973). Pollen tubes grow oriented by magnetic fields (Sperber et al., 1981), and the rhythmic leaflet movements in *Desmodium gyrans* are affected by static magnetic fields (Sharma et al., 2000). Roots of wheat and corn respond to very low magnetic fields, which modulate the gravitropic curvature (Bogatina et al., 1986; Sineshchekov & Sineshchekov, 1988; Kato, 1990). Even algae have been found to possess magnetite and show magnetotaxis (de Araujo et al., 1986), and some cyanobacteria respond to magnetic fields (Rai et al., 1998).

With respect to the question whether magnetic fields influence cellular behavior, *Paramecium* was exposed to different kinds of magnetic fields. Different effects with respect to swimming behavior and proliferation rate have been stated (for review, see Hemmersbach et al., 1997). *Paramecium* was found to be guided by static electromagnetic fields (Brown, 1962; Kogan & Tikhonova, 1965; Roberts, 1970). Under the influence of a low-frequency electromagnetic field (50 Hz and 2 mT), a decrease in the linearity of the swimming paths (Hemmersbach et al., 1997) and a decrease in 5′-methoxytryptamin have been found, the latter one being discussed with respect to its role in cancer promotion in humans. It is postulated that the magnetic field induces a decrease in calcium and an inhibition of specific enzymes (Wilczek, 2001).

Many bacteria have been described to respond to the magnetic field of the Earth (Bazylinski et al., 1988; Blakemore, 1975; Balkwill et al., 1980; Keim et al., 2001). Most prokaryote studies show that they produce minute magnets inside their bodies (Vainshtein et al., 2002; Mann et al., 1987; Matsunaga & Sakaguchi, 2000; Meldrum et al., 1993) consisting of magnetite (Fe_3O_4) crystals often arranged in rows (Farina et al., 1994; Baeuerlein & Schueler, 1995) and enclosed in a membrane vesicle (Gorby et al., 1988; Fig. 7.12). The magnetite can amount up to 2–4% of their dry weight (Blakemore, 1982). The potential role of magnetosomes with respect to gravitaxis is considered in Section 6.4. From a puristic point of view, the orientation of bacteria in the magnetic field of the Earth is not a true taxis, since the cells are passively oriented along the field lines like a compass needle, and even dead bacteria are oriented by the Earth's magnetic field. Taxis in the true sense requires a receptor and a physiological response by the organism.

Although for migratory birds and other animals the advantage of guidance by the magnetic field seems obvious, the interesting question is what ecological significance has an orientation by the magnetic field lines for bacteria; why should they want to swim North or South? Magnetotactic bacteria isolated from the Northern Hemisphere consistently swim toward the geomagnetic North Pole. But the magnetic polarity of their single-domain magnetite particles can be reversed by a single magnetic pulse of high-field strength (300–600 G, 1–2 μs; Blakemore, 1982). Maybe the bacteria are not "interested" in going North; due to inclination of the magnetic field lines, magnetotactic bacteria, which swim to the North, in fact move downward in the water column toward the sediment (Blakemore, 1982). The fact that in the Southern Hemisphere magnetotactic bacteria have been found to head South supports this hypothesis (Blakemore et al., 1980; Kirschvink, 1980). It is interesting to note that magnetotaxis has been "invented" several times in the evolution of bacteria, since the phenomenon is found in a number of phylogenetically distant bacteria (Delong et al., 1993; Burgess et al., 1993).

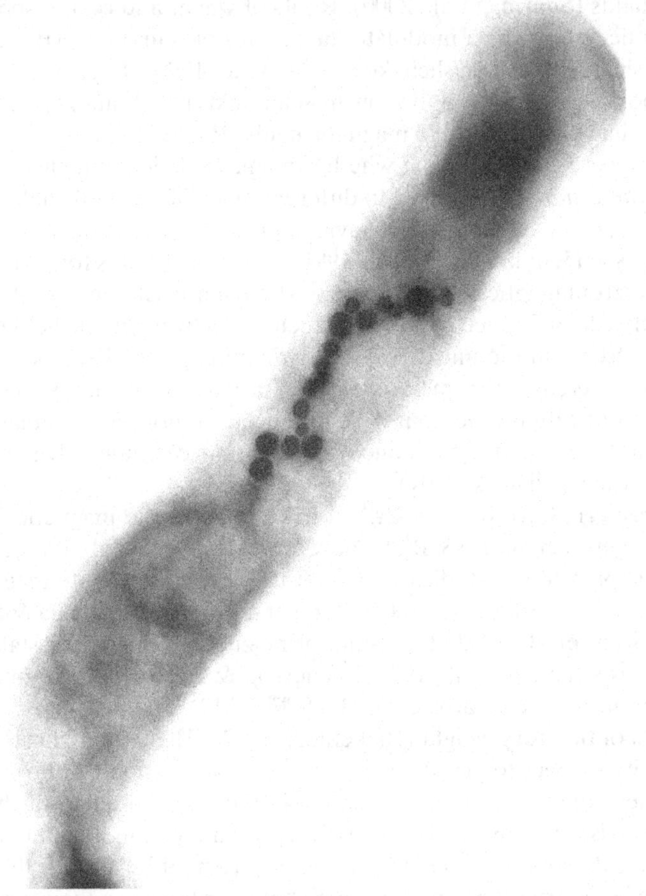

Figure 7.12. Magnetite particles in *Rhodopseudomonas palustris* VKM B-1620, shown by transmission electron microscopy. (Courtesy of E. Ariskina and M. Vainshtein, Pushchino, Russia.)

7.6 Galvanotaxis

Galvanotaxis is an artificially induced behavior in organisms, which demonstrates their voltage sensitivity. Even though electric signals may not be standard clues in natural habitats, a number of organisms are known to show galvanotaxis, ranging from bacteria (Shi et al., 1996) to protists (Jahn & Votta, 1972; Votta & Jahn, 1972a,b; Gebauer, 1930) to slime molds (Anderson, 1951; Ueda et al., 1990) and vertebrate cells (Cooper & Schliwa, 1986; Gruler & Nuccitelli, 1991). The older literature has been reviewed by Umrath (1959). Exposing the ciliate *Paramecium* in a direct current (DC) field establishes a bipolar voltage gradient across the cells and induces the ciliate to swim to the cathode (for review, see Machemer 1988b). A vertically arranged DC field enhances the gravikinetic response (gravity-dependent speed regulation to compensate sedimentation; cf. Section 1.2) of downward swimming cells three-fold, whereas the swimming velocity of upward swimming cells – which normally increases – decreases under these circumstances. It was postulated that the augmentation of the gravikinetic response by the DC field is induced by the contraction of a calcium-sensitive filament system, representing part of the gravisensory transduction chain (Machemer-Röhnisch et al., 1996).

In flagellates, galvanotaxis can be anodic or cathodic (Verworn, 1889a). In the flagellate *Chlamydomonas reinhardtii*, the precision of galvanotactic orientation depends on the external calcium concentration (Dolle & Nultsch, 1987): The higher the calcium concentration (optimal 10^{-4} M), the lower the voltage necessary to induce the response. Application of the calcium channel blockers diltiazem and nifedipine inhibits galvanotaxis in *Chlamydomonas* very specifically. Verapamie is also effective, but simultaneously causes shortening or detachment of the flagella, thus reducing motility. These results indicate that galvanotaxis may be linked or converge with phototaxis, which also involves ion gating through calcium channels (see Section 7.2.3). It is interesting to note that, also in human keratinocytes, calcium channels are involved in the galvanotactic orientation, and channel blockers impair the response (Trollinger et al., 2002).

7.7 Interaction between different stimuli and responses

Orientation of microorganisms and higher plants and animals is controlled by multiple external stimuli. The integration of the signal transduction chains and the concerted action to simultaneous stimulation is only known to a small extent and also only in a few case studies.

Both higher plants and a number of fungi respond to light. However, bending toward a lateral light source simultaneously induces gravitropism, which induces the organ to bend in the opposite direction (Nick & Schäfer, 1988; Spurny, 1974; Grolig et al., 2000). The resulting response is called photogravitropism, and the bending angle is the vectorial addition of the antagonistic movements (Neumann & Iino, 1997; Galland et al., 2002; Ensminger et al., 1990). This is not only found in shoots, but also in roots (Vitha et al., 2000). In addition, in higher plants, there are interactions between thermotropism (sensing of a temperature gradient) and gravitropism (Fortin & Poff, 1991), as well as gravitropism with touch sensitivity (Steinitz et al., 1992) and gravitropism with hydrotropism (bending in a water gradient; Takahashi & Scott, 1991).

In flagellates, an interaction between light and temperature responses has been analyzed (Kreuels et al., 1984). Phototaxis is modulated by gravitaxis in flagellates as in the corresponding responses in higher plants (Kessler et al., 1992). When *Euglena* swims in a vertical cuvette, where it is exposed to the gravity field of the Earth and simultaneously to a laterally impinging light field, the cells orient on the resultant. The angle from the vertical clearly depends on the strength of the light source. As a consequence, exclusive phototactic orientation can only be studied under weightlessness, as has been done on a ballistic rocket flight with the flagellate *Euglena* (Kühnel-Kratz et al., 1993). In the ciliate *Loxodes*, a clear relationship between oxygen, light, and degree and direction of gravitaxis has been found. The oxygen tension modifies the sign of gravitaxis and thus the swimming direction; therefore, gravity and chemosensory transduction in this ciliate have to share common steps.

Likewise, a vectorial addition of the responses to stimuli has been found, such as simultaneous acceleration. When the cells are centrifuged horizontally, they are exposed to the vertical gravity vector of the Earth and centrifugal force. Also, in this case, the deviation angle depends on the respective accelerations (Häder et al., 1991a,b; Lebert & Häder, 1997b). In addition to external factors, the endogenous rhythm can modulate the direction and strength of responses to light and gravity stimuli (Kreuels et al., 1984; Lebert et al., 1999a; Ohata et al., 1997; Byrne et al., 1992; Eggersdorfer & Häder, 1991b; cf. Section 5.5).

Observation of *Paramecium biaurelia* in its culture medium with different oxygen concentrations revealed that, at low oxygen (0.1 mg/liter), the cells show a precise negative gravitaxis. At increasing oxygen, the orientation becomes less precise; and, at >1.7 mg/liter oxygen, a random distribution of the cells was measured (Hemmersbach-Krause et al., 1991a). Similar experiments in defined inorganic experimental solutions instead of original culture medium did not show this relationship between oxygen saturation and the graviresponses (Machemer et al., 1993). As *Paramecium* did not alter its gravitaxis after treatment in NaN_2 (Taneda, 1987) or the uncoupling agent CCCP – which in mitochondria blocks ATP synthesis, leaving electron transport unaffected (Hemmersbach-Krause & Häder, 1990) – and gave no hint with respect to the oxygen receptor. Further experiments should show how oxygen and graviperception interact in this organism.

Energetics

In this chapter, we discuss energetic considerations regarding the mechanisms thought to underlie the orientation of single cells in the gravitational field. Although it is still under discussion, whether gravireactions are the result of a physical or physiological mechanisms or a combination of both, it is undisputed that the basis must be an interaction of gravity with a mass. This chapter compares potential mechanisms for gravidetection with physical and energetic limits set by nature. Several models consider mechanosensitive channels as an important part of gravity perception in single cells. The weak energy supplied by the gravity–mass interaction will be shown to be sufficient to at least potentially allow to activate such channels. As a model, the hearing system of the inner ear will be compared with the conditions in single cells.

Past and recent discussions were centered around the question of whether the reorientational movements in single cells are the result of a pure physical or a pure physiological mechanism or a combination of both (cf. Chapter 9). Independent of such considerations, the most basic event of the related movement reactions (gravitaxis, gravikinesis) will be an interaction between a cellular entity and gravity. This is the gravity stimulus perception. Depending on the model, we are looking at the following steps, including a receptor (typically a protein, but not necessarily a membrane protein), a receptor-signaling-state change (i.e., the point where a stimulus such as light or gravity is transformed into a chemical, biochemical, or electrical signal), and a signal transduction chain (that may or may not include an amplification). The energy inherited from a stimulus usually cannot directly fuel the response or, in other words – energetically speaking – we have a low-energy stimulus and a huge response with respect to energy, a signal integration (to calculate a sum response that integrates synergistic and antagonistic stimuli; e.g., light and gravity; cf. Section 1.3), and last, but not least, an effector. In a single cell system, we have it all in one – the response is normally immediate and can be pinpointed down to a single organelle: the cilium or flagellum. In a physical model, the reaction (the reorientation) will stop when the sum of the exerting forces is zero. This chapter will concentrate on the first step: the stimulus perception and possible ways to use the inherent spatial information (in other words: where is up?). Some basic principles apply to such a stimulus perception (Björkman, 1992):

1. Gravity is a body force that applies a force to a mass.
2. For a receptor-signaling-state change, work has to be done.
3. To do some work, mass has to be displaced.
4. The stimulus signal must be distinguishable from the background noise.

These principles, which include a substantial amount of physics, were formulated for perception of gravity by plants, but we feel that they should be as applicable for unicellular organisms. Thus, in the following, we will work our way through the questions implicated by the principles mentioned previously and see how the results compare with the requirements.

8.1 Gravity is a small power that applies a force to a mass

The first question to address is the effective mass of single cells or organelles. We use the term "effective mass" in contrast to passive mass, which is the actual mass of the system. In an aqueous medium (like water or cytoplasm), the mass of the water replaced by the cell or organelle must be subtracted from the passive mass to estimate the effective mass. However, first of all, the mass of a cell or an organelle has to be determined. For the determination of the mass, two pieces of information are necessary: the volume and the specific density. Specific densities (ρ[g ml^{-1}]) can be determined by isopygnic centrifugation. Cells or organelles are centrifuged in a medium of increasing density (e.g., a sucrose, Percoll, or Ficoll gradient). During centrifugation, objects will sediment (or float up when positioned at higher densities) to a position where the specific density of the cell and the medium are equal. However, time is a crucial factor in this kind of experiments. Sucrose will have a strong impact on the osmolarity of the medium and will reduce the water content of the cells and artificially increase the specific density, while the high molecular weight substances Percoll and Ficoll do not change the osmolarity significantly. In the experiments discussed next, the position of cells (which is used for the determination of the density) changed after an initial period of 30 min, where it was stable to higher densities in the next hour of centrifugation. One example for isopygnic density determination is shown in Figure 8.1. In this case, a *Euglena* cell suspension was placed on a Ficoll step gradient. Every other step was stained using neutral red to enhance visibility. Cells sedimented to the border between a specific density of 1.045 and 1.05 g ml^{-1}. The specific density, as well as the size of *Euglena*, depends on the age and the culture conditions. A variation between 1.046 and 1.054 g ml^{-1} was observed (cf. Table 5.1; Lebert et al., 1999b), and estimated volumes differed by a factor of 2.5 (1×10^{-12} liter to 2.5×10^{-12} liters). Interestingly, in younger cell cultures (i.e., freshly inoculated, <5 d),

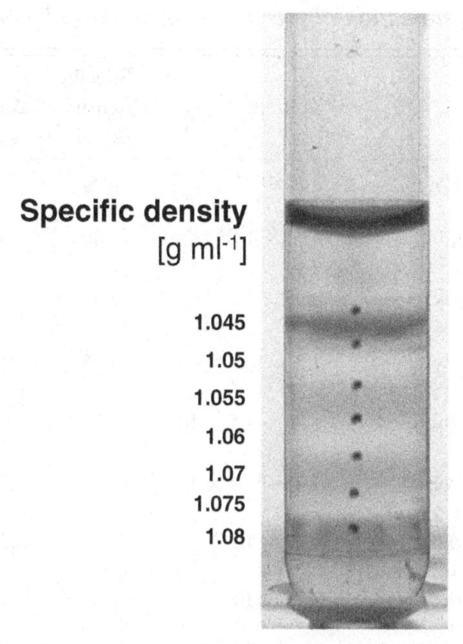

Figure 8.1. Result of an isopygnic centrifugation of a 10-d-old *Euglena gracilis* culture. A Ficoll step gradient was used. Every other step was stained by neutral red to increase visibility.

two bands were observed. It seems as if a substantial proportion of the older cells (heavier), which were used as an inoculum, had not divided and therefore remained in their original physiological state. One special case was a culture kept isolated for more than 600 d. The cells had a specific density of 1.011 g cm^{-3} and did not show any gravitaxis. After inoculating aliquots into fresh medium (mineral or complex), the specific density rose to normal values and gravitaxis recovered.

The specific density of *Euglena* is in the same range as reported for other unicellular organisms. Values of 1.04 g cm^{-3} were reported for *Paramecium caudatum* and *P. tetraurelia* (Köhler, 1922; Fetter, 1926; Taneda, 1987; Taneda et al., 1987; Kuroda & Kamiya, 1989; Nagel & Machemer, 2000b), 1.035 g cm^{-3} for *Tetrahymena* (Machemer-Röhnisch et al., 1999), and 1.03 g cm^{-3} for *Loxodes* (Hemmersbach et al., 1998).

To estimate the maximal force which can **directly** interact with a cellular entity (i.e., membranes), Equation (8.1) (Björkman, 1988) can be applied. The force can be estimated by multiplying the volume (*V*) with the acceleration (*g*) and the specific density difference (Table 8.1).

$$F = V g \Delta\rho \tag{8.1}$$

F = force, V = volume, g = acceleration ($g_n = g_{\text{Earth}} = 9.81$ m s^{-2}), and $\Delta\rho$ = specific density difference.

The estimated forces in the examples vary between 0.017 pN (*Arabidopsis* statolith), 0.018 pN (*Chara* rhizoid barium sulfate vesicle), and 11.8 nN (*Bursaria*

Table 8.1. *Force determination of specific organelles or cell bodies*

Organism	Entity	Volume (m^{-3})	Specific density ($kg\ m^{-3}$)	$\Delta\rho$ ($kg\ m^{-3}$)	Force (pN)*
Bursaria truncatella†	Cell body	3×10^{-11}	1,040	40	11,772
Paramecium caudatum‡	Cell body	3.27×10^{-13}	1,040	40	128
Paramecium tetraurelia§	Cell body	6.93×10^{-14}	1,040	40	27.2
Tetrahymena pyriformis‡	Cell body	2.2×10^{-14}	1,035	35	7.89
Euglena gracilis¶	Cell body	$1.28 - 2.5 \times 10^{-15}$	1,050	50	0.49–1.23
Loxodes striatus#	Müller vesicle (barium sulfate)	$1.4 - 3.35 \times 10^{-17}$	4,400	3,370	0.47–1.1
*Chara globularis***	Rhizoid barium sulfate vesicle	5.23×10^{-19}	4,400	3,370	0.14
*Arabidopsis thaliana***	Statoliths	3.35×10^{-17}	1,440	410	0.13

*1 At $1 \times g_n = 9.81\ m\ s^{-2}$.
†Krause, 1999.
‡Machemer-Röhnisch et al., 1999.
§Hemmersbach et al., 1998; Nagel & Machemer, 2000b.
¶Lebert & Häder, 1999c.
#Hemmersbach et al., 1998.
**Sievers & Volkmann, 1979; Volkmann et al., 1999. *Chara*: Assuming a sphere of 1 μm diameter. *Arabidopsis*: statolith size estimated from microscopic images within the publication.

cell body). The question is: Is that force sufficient to activate specialized ion channels as predicted by the models for gravitaxis or gravikinesis in single cells. To trigger an ion channel, energy is required. The lower limit of the activation energy seems to be set by the thermal noise (2×10^{-21} J at 20°C). The thermal noise ("Brownian motion") is based on the thermally driven motion of the solvent (in the normal case water) molecules. Any activation energy lower than this limit will yield an overwhelming number of "false" signals. However, it seems that mechanically stimulated ion channels in the inner ear are activated by energies that are very close or even below the thermal noise level (Holton & Hudspeth, 1986). We will discuss later how such a system may be able to discriminate between signal and noise (see principle 4).

So, following up the line of argumentation discussed previously, it seems that for interaction between gravity and cellular entity, every size will work. Is that really so? And, the answer is: No! To activate a receptor, work has to be done (principle 2). This work is done by the displacement of a mass. Although we will look at the displacement problem later on in more detail, displacement always includes sedimentation at least for our purposes. For gravity perception in a physiological model, the primary interaction [but not the amplification step(s)] must not be dependent on cellular energy sources. In other words, there cannot be a physiological response before the cell receives the information that something (reorientation or swimming speed adjustment) has to be done. Furthermore, it seems obvious that the cellular unit interacting with gravity cannot be larger than the organism (i.e., the cell). Due to the fact that, to do some work (which is the basis for receptor activation), two masses have to be moved with respect to each other, cells in the statenchyma of higher plants cannot be viewed as a cooperating unit because the position of a cell in a tissue is fixed with respect to the neighbor cells. Is there a lower limit? And, the answer is: Most likely yes. Two issues have to be considered: firstly, Brownian motion and secondly, motion at low Reynolds numbers. Brownian motion is based on the random bombardment of a particle by molecules surrounding it [Equation (8.2)].

$$x^2 = 2Dt \tag{8.2}$$

x = absolute distance, D = diffusion constant (depends on size of particle and viscosity of solvent), and t = time.

With a given diffusion constant, a particle will move in double the time only the square root of double the distance. The diffusion constant can be described by the Stoke-Einstein equation [Eq. (8.3)]. The formula states that the diffusion constant mainly depends (at a given temperature) on the size of the particle and the viscosity of the solvent.

$$D = \frac{kT}{6\pi r \eta} \tag{8.3}$$

D = diffusion constant, k = Boltzmann constant (1.38054×10^{-23} J T^{-1}), T = absolute temperature, r = radius of spherical particle, and η = viscosity of the solvent.

Let us consider some particles of different sizes in water [η = 0.001 P (N s m^{-2})]. [*Note*: the SI unit is named P after the French physicist Poiseuille (1 P = 0.1 Pa s). In the literature, commonly cP (centipoise; dyn s cm^{-2}) is used.] For a particle of 2-nm radius (a typical protein), the diffusion constant D would be 6×10^{-10} m^2 s^{-1}. Applying Equation (8.2) tells us that such a particle would move ca. 15 μm/s. Increasing the particle radius by the factor of 50 (e.g., a vesicle) will decrease D to 2×10^{-11} m^2 s^{-1}, and such an object would move 2 μm/s. A unicellular organism like *Euglena* would have a D of 2.8×10^{-14} m^2 s^{-1} and will move driven by Brownian motion of 168 nm/s (assuming a spherical cell body with a radius of 7.8 μm). For *Paramecium caudatum* (radius = 42.7 μm), D would be 5.14×10^{-15} m^2 s^{-1} and a diffusional movement of 71 nm/s can be expected.

The diffusion velocities are then given by expression 8.4:

$$v = \sqrt{\frac{kT}{m}} \tag{8.4}$$

v = velocity, k = Boltzmann constant (1.38054×10^{-23} J T^{-1}), T = absolute temperature, and m = mass.

What diffusion rates are to be expected? For $T = 300$ K, an ATP molecule (551.1 g mol^{-1}) has a speed in any given direction of ~67 m s^{-1} – an average kinetic energy of ~2×10^{-21} J/molecule, which is kind of a baseline to be considered as a lower value for activation of a receptor. Such a molecule would be hit all the times by molecules with that energy, and an activation energy lower than that energy threshold would result in many false events triggered. The bigger the particles are, the slower they move. A protein with a mass 50,000 Da would have an average speed of 7 m s^{-1}, a 200 nm vesicle with 6.25 GDa could walk with a speed of 600 μm s^{-1}, and a person (10^{28} Da) will move only 10 nm s^{-1}. *Euglena* as a typical single cell has a mass of approximately 1.27×10^{15} Da (or ~2.1×10^{-9} g) and would have an instantaneous velocity of 1.4 μm s^{-1}. [*Note*: the values given are rough estimates (e.g., all calculations are based on the assumption that the cell is a sphere, which it is not) only to get a feeling for the physical properties of a single cell.] Thus, real values might be drastically different. However, why are the speed values for *Euglena*, as well as the other examples, so different if we compare diffusion rates calculated on the basis of Equation (8.2) and instantaneous rates calculated according to Equation (8.4)? The reason for that was discovered more than 100 years ago by an English physicist, Osborne Reynolds. The Reynolds number is defined as the ratio between the inertia force over the viscous force [Eq. (8.5)]. The Reynolds number is dimensionless.

$$\mathbf{Re} = \frac{\boldsymbol{D}\,\boldsymbol{v}\rho_{\mathrm{m}}}{\eta} = \frac{\boldsymbol{D}\,\boldsymbol{G}}{\eta} \tag{8.5}$$

D = characteristic length (in our cases cell diameter), v = velocity, ρ_m = specific density of the medium, η = viscosity of medium, and G = mass velocity (i.e., mass flow).

The Reynolds number was developed for describing the behavior of fluids in pipes. Typical values for unicellular organisms are 0.026 (cell width: 50 μm, velocity: 605 μm s^{-1}; Hemmersbach-Krause et al., 1993b) for *Paramecium caudatum* or 0.0003 for *Euglena gracilis* (cell width: 8 μm, velocity: 50 μm s^{-1}), while for a person swimming in water it is ~10,000. Essentially, the Reynolds number states that the smaller you are, the more the viscosity of the medium you are in, matters due to solvent–cell interactions. Figure 8.2 shows the computed results for a particle radius, the ratio of particle area over particle volume, and the Reynolds number. It is obvious that a strong inverse relationship exists. The consequence of this "life at low Reynolds numbers" is: there is nothing like inertia for the organisms. When a typical motile bacterium like *Escherichia coli* stops its flagella rotation, it will move on for a further 0.3 μs and a distance of 0.1 Å (Purcell, 1977). The low values given indicate that the cells are swimming in molasses, as an analogy to human experiences.

Coming back to our initial question of whether there is a lower limit to the size of a cell body or a cell organelle where we do not see any sedimentation (to fulfill principle 3 that masses have to be displaced). At this point, diffusion rate and sedimenting rate must be at least equal. To estimate the

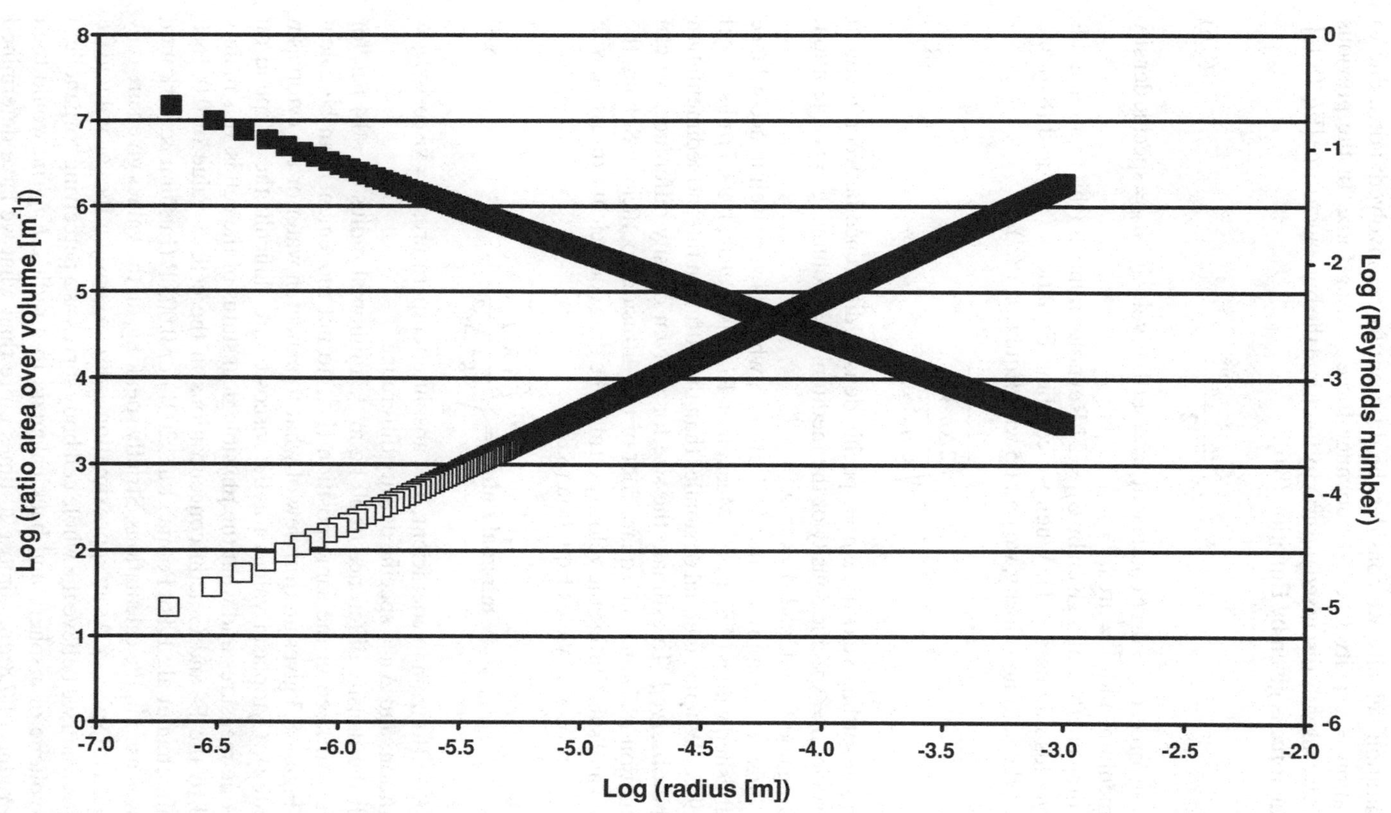

Figure 8.2. Relation between cell radius, the ratio of area over volume (**closed squares**), and the Reynolds number (**open squares**). In the calculations, a velocity of 50 μm s^{-1} was assumed.

sedimentation velocity, we consider the different forces involved: Force due to acceleration = gravity force − buoyancy force − drag force. In the previous sections, we already discussed the gravity force and the buoyancy force [Eq. (8.1)]. The drag force is given by Equation (8.6).

$$F_D = \frac{C_D\, A\, \rho_m\, v_s^2}{2} \tag{8.6}$$

F_D = drag force, C_D = drag coefficient, v_s = settling velocity, ρ_m = specific density of the medium, and A = area.

The drag coefficient is given by $C_D = 24$/Reynolds number (this relation holds for Reynolds numbers ≤ 1). When we combine Equations (8.1) and (8.6) and solve for the settling velocity, we end up with Equation (8.7).

$$v_s = \sqrt{\frac{2\Delta\rho\, Vg}{C_D\, A\, \rho_m}}, \tag{8.7}$$

where v_s = settling velocity, $\Delta\rho$ = specific density difference between medium and particle, ρ = specific density of the medium, V = volume, g = acceleration, C_D = drag coefficient, and A = area.

For *Euglena*, v_s approximates 11.3 μm s^{-1}, which is very well in accordance with published values (9.8 μm s^{-1}; Machemer-Röhnisch et al., 1999, 7 μm/s; Vogel et al., 1993). When v_s is equal or smaller than the diffusion rate, no sedimentation could be observed. To estimate the size for a given density difference, we can use Equation (8.7) and the square root of the diffusion coefficient. Solving the equation and substituting the velocity term in the drag coefficient expression with the square root of D yields Equation (8.8).

$$\textbf{minimal radius} = \sqrt[5]{\frac{6\, K\, T_\eta}{\pi g^2\, \Delta\rho^2}}, \tag{8.8}$$

where K = Boltzmann coefficient, T = absolute temperature, η = viscosity, g = acceleration, and $\Delta\rho$ = specific density difference.

With the density difference of 50 kg m^{-3}, a minimal radius of ~0.4 μm for a particle to keep in the same position (i.e., no net movement) can be calculated. Figure 8.3 gives an overview of what to expect in water or the cytoplasm (simulated). Cytoplasm values for the viscosity are difficult: they depend on where one measures and what the physiological status of the cell is. It is mainly related to the F- and G-actin concentrations in the cell. A value of 0.0016 P (Swaminathan et al., 1996; Potma et al., 2001) or 0.003 P (Fushimi & Verkman, 1991) was used in the calculations. Strictly speaking, all the values given are only calculated for spheres. As an approximation, one can consider the cell volume (in the cases when we believe the whole cell body serves as a perceiving "organelle") as the volume of a sphere and back calculate to a radius. *Euglena* would then have a radius of 7.8 μm, almost 20 times more than what we have determined for the specific density difference of 50 kg m^{-3}. For *Paramecium caudatum*, the virtual radius is 42.7 μm, and we would need 550 nm. When we look at the

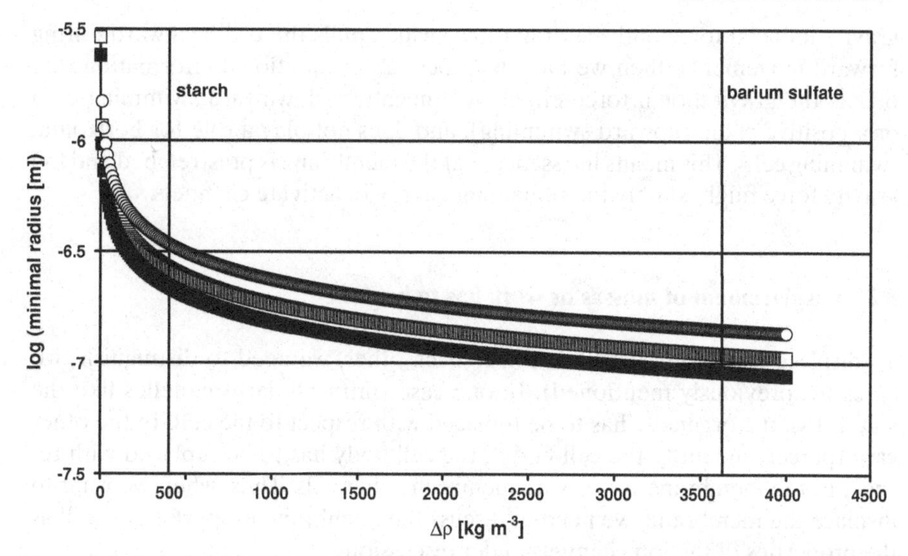

Figure 8.3. Relation between $\Delta\rho$ (specific density difference) and expected minimal radius of particle. Calculations were done for three viscosities: water (**closed squares**), $\eta = 0.001$ P; cytoplasm (**open squares**), $\eta = 0.0016$ P (Swaminathan et al., 1996; Potma et al., 2001); and cytoplasm (**open circles**), $\eta = 0.003$ P (Fushimi & Verkman, 1991).

intracellular organelles like the statoliths, we already have almost spherical bodies. For a typical statolith, we can use a value of 1 μm as the radius ($\Delta\rho = 400$ kg m^{-3}), and the calculation yields 220–270 nm as a threshold, depending on cytoplasmic viscosity. Barium sulfate-filled vesicles in *Chara* have a radius of approximately 500 nm, and the predicted minimal radius is about 100 nm. And, last, but not least, the barium sulfate crystal in the Müller body has a radius of 1 μm and that compares favorably to the predicted minimum of 100 nm ($\Delta\rho = 3370$ kg m^{-3}). We can conclude that all these organelles or cells are well in the range necessary to produce a displacement, which in turn is required to produce the energy to activate receptor proteins. However, one has to distinguish between two cases: firstly, an organelle (like the statoliths or the Müller body) is sedimenting inside a cell; on the other hand, a cell body (i.e., everything inside a cell) is sedimenting and exerts a force on the (lower) membrane. In addition, the drag force (F_D) should be considered. With a very low Reynolds number of ~0.0003, significant forces operate on the membrane, depending on the direction of movement. For a small cell, like *Euglena*, this force amounts to 77 pN [applying Eq. (8.6); gravity force – buoyancy force: 1 pN], while for a larger cell like *Paramecium caudatum* – with a higher Reynolds number – only an additional 3.9 pN (gravity – buoyancy force: 128 pN) will be applicable. These forces are probably drastically underestimated, because the area was estimated for a smooth sphere, and the surface of single cells is not smooth at all (see below). However, the question is whether the drag force will be constitutive background "noise" or includes useful orientational information. When we assume a simple vector addition of the gravity-induced force and the drag force (which pulls the cell backward during forward movement), then we have to expect an orientational information that offsets the gravitational force either by a negative (downward swimming cell) or a positive value (upward swimming), and does not play a role for horizontal swimming cells. This means in essence that the membrane is prestretched and the gravity force might supply the remaining energy to activate channels.

8.2 Displacement of masses or work has to be done

To displace something with relation to each other, we need to distinguish two cases (as previously mentioned). In one case (intracellular organelles like the statoliths), the organelle has to be replaced with respect to the cell. In the other case (perceiving entity the cell body), the cell body has to be replaced with respect to the membrane to activate membrane channels. Thus, when we want to displace the membrane, we need to discuss the membrane properties, as well as the properties of the ion channels under discussion.

Mechanosensitive ion channels can be, besides the ion selectivity, divided into several classes (Morris, 1990; Sachs & Morris, 1998). One class is so-called "stretch-activated" ion channels (SACs) or stretch-inactivated channels, which are less abundant. Essentially, this channel class is activated (which includes an increased or decreased "open" probability) by applying a pressure that, in turn, results in a membrane tension (pressure and membrane tension are connected by the area expansion modulus; see below). Pressure is defined as force per area [Eq. (8.9)]. In a first approximation, the area can be estimated as the horizontal cross-section (i.e., "shadow of the cell"; Machemer-Röhnisch et al., 1999). Taken as a basis, this assumption allows determination of the pressure that can be maximally applied to the cross-section by dividing the force (F) by the area [A; Eq. (8.10)].

$$P = FA^{-1}, \tag{8.9}$$

where P = pressure [Pa; N m^{-2}; F = force (N); A = area (m^2)].

$$A = \pi ab, \tag{8.10}$$

where A = area (m^{-2}), a = long-axis radius, and b = short-axis radius.

For *Euglena*, this would yield a maximum of 0.003 Pa, for *Paramecium caudatum* 0.013 Pa, for *Paramecium tetraurelia* 0.009 Pa, and for *Tetrahymena* 0.006 Pa. All estimates are based on standard size dimensions found in the literature [*Euglena*: 50-μm length and 8-μm width, cf. Table 5.1 (Lebert et al., 1999b); *P. caudatum*: 250 μm and 50 μm (Machemer-Röhnisch et al., 1999); *P. tetraurelia*: 115 μm and 34 μm (Nagel & Machemer, 2000b); and *Tetrahymena*: 70 μm and 25 μm (Machemer-Röhnisch et al., 1999)], a horizontal swimming cell, and the assumption that the cell body projection is in the first approximation an ellipse. In vertically swimming cells, these values increase due to the decreased cross-section to 0.06 Pa for *Euglena*, 0.2 Pa for *Paramecium caudatum*, 0.094 Pa for

P. tetraurelia, and 0.05 Pa for *Tetrahymena* (the cross-section is assumed to be a circle, with area calculation based on cell width).

8.3 The potential role of membranes in graviperception

To activate SACs, these pressures should stretch, expand, shear, or bend the membrane. A typical biological bilayer membrane is a specialized structure in terms of ion conductance and water permeability. Any change in the internal spatial relation of the acyl chains will result in dramatic changes in these previously discribed properties. Thus, one can assume that, during evolution, the structure has been optimized to compensate for many externally or internally applied stresses. One of these adaptations to potentially fast-occurring stresses is that, in many cells, a huge excess of membrane is present on top of the required area to cover the volume of the cell. In red blood cells (the preferred subject in this research area), an excess of 40% is estimated (Evans, 1992), whereas in other cells (lymphocytes, oocytes, etc.), this may be 100% to 1000% in the form of extra membrane structures due to folding (Dulhanty & Franzini-Armstrong, 1975; Sokabe et al., 1991; Evans, 1992; Ross et al., 1994; Zampighi et al., 1995; Zhang & Hamill, 2000). The increase in cell volume, for example, in swelling cells does not require extra membrane insertions (Zhang & Hamill, 2000; Ross et al., 1994; Olson & Li, 1997; Solsona et al., 1998). This membrane excess is also to be expected in the case of *Euglena.* The cytoplasmic membrane covers an underlying structure called the pellicle. The pellicle forms a layer below the cytoplasmic membrane and is composed of helically oriented parallel strips that can slide with respect to each other (Suzaki & Williamson, 1986a–c). The sliding of strips explains the extreme variability of cell forms observed in *Euglena*, as well as the so-called "euglenoid movement" (Suzaki & Williamson, 1986a–c; Mikolajczyk, 1972,1973). The pellicle is somehow connected to the cytoplasmic membrane and by this means forces the membrane in ripple-like structures (Suzaki & Williamson, 1986a–c). These structuring will supply additional membrane material if any mechanical deformation (including euglenoid movements) occurs before any tension develops in the bilayer (see below for tensions to be expected; Hamill & Martinac, 2001). The cytoskeleton network (CSK) supplies the structural organization of the membrane while being very flexible (elasticity and plasticity) at the same time.

In a first approximation, a membrane can be described as an elastic solid. Keeping this simplification in mind, membrane properties can be described by four elasticity constants termed "moduli." These constants, which are dependent on the specifics of the structure as well as the thermal environment, allow one to predict the membrane responses to compression, expansion (as predicted to occur in the models explaining gravitaxis), bending, or extension (Evans & Hochmuth, 1978; Evans & Skalar, 1980). In an elastic system, any force applied will result in an immediate response. Most of the measurements with regard to membrane properties were performed using red blood cells, and, while comparable responses are very likely for unicellular organisms, the relation still needs to be established.

8.3.1 *Membrane compressibility*

Measurements of the membrane compressibility indicate that a biological bilayer behaves like an incompressible fluid (Srinivasan et al., 1974). Evans and Hochmuth (1978) estimate a compressibility modulus of 10^9–10^{10} N m^{-2}. Obviously, membrane compression cannot play a role in graviperception.

8.3.2 *Membrane expansion*

The area expansion modulus describes the resistance of a membrane against an area expansion. The area expansion modulus is a proportionality factor that connects the area expansion with the applied tension [Eq. (8.11); Hooke's law; Evans & Hochmuth, 1978; Petrov & Usherwood, 1994].

$$T_m = K_a(\Delta A/A_0) \tag{8.11}$$

T_m = membrane tension, K_a = area expansion modulus, A_0 = initial area, and ΔA = area change.

The membrane tension can be determined according to the Laplace law, which states that a hydrostatic pressure difference across a membrane creates a lateral membrane tension T_m, depending on the degree of curvature of the bilayer [Eq. (8.12); Petrov & Usherwood, 1994].

$$p = T_m(1/R_1 + 1/R_2) \tag{8.12}$$

p = hydrostatic pressure difference, T_m = membrane tension, and R_1 and R_2 = the two principal radii of curvature of the membrane.

The principal radii of curvature can be estimated on the basis of cell length and width. Imagine the lower half cross-section of a horizontal swimming cell (which is in first approximation an ellipse), then the lowest point is the vertex. The distance from the vertex to the center of curvature is known as the principal radius of curvature. R_1 is perpendicularly oriented with respect to R_2. R_1 and R_2 can be calculated based on the cell width and cell height according to Equation (8.13). Essentially, we are extending the "half-ellipse" to a circle (which is a crude approximation) and determine the expected radius.

$$\boldsymbol{R} = \left(\frac{\rho^2}{\boldsymbol{h}} + \boldsymbol{h}\right) 0.5 \tag{8.13}$$

R = principal radius of curvature, ρ = radius of the assumed circle (R_1 = cell length/2; R_2 = cell width/2), and h = height (R_1 = cell width/2; R_2 = cell height/2).

Calculations for a horizontally swimming *Euglena* reveal $R_1 = 314.5\ \mu$m and $R_2 = 12.82\ \mu$m. With a pressure of 0.003 Pa, the membrane tension can be estimated to a value of 1×10^{-6} N m^{-1}. This value changes to 3.8×10^{-7} N m^{-1} ($R_1 = 12.82\ \mu$m, $R_2 = 12.82\ \mu$m) in a vertically swimming cell. One should keep in mind the previous remarks regarding excess membranes and the CSK that most likely changes the whole picture.

Another example is *Paramecium caudatum*. In this case, the membrane tension can be estimated for a horizontally swimming cell as 6.3×10^{-6} N m^{-1} ($R_1 = 325$ μm, $R_2 = 163$ μm) and for a vertically swimming cell as 1.06×10^{-6} N m^{-1}.

To the best of our knowledge, no area expansion modulus for membranes in unicellular organisms has been determined up to now. The closest examples are red blood cells, with a published value of approximately 500 mN m^{-1}, which is in good correspondence to values from skeletal muscle membranes (Hochmuth & Waugh, 1987; Needham & Hochmuth, 1989; Needham & Nunn, 1990; Nichol & Hutter, 1996). This does not include the area expansion modulus of the CSK network ($K_a \sim 10$ μN m^{-1}). These highly flexible structures will expand, and the excess membrane will allow any possibly evolving tensions to be counteracted, but this adaptation will change the curvature of the membrane (see below). However, following Equation (8.4), the expected area expansion for *Euglena* [$T_{m\,(\text{max. horiz.})} = 0.003$ Pa and $T_{m\,(\text{max. vert.})} = 0.06$ Pa] would be 2×10^{-4}% for a horizontally swimming and 7.6×10^{-5}% for a vertically swimming cell. In any case, these values do not seem likely to activate SACs. For *P. caudatum*, the corresponding values are 1.26×10^{-3}% for a horizontally oriented organism and 2.12×10^{-4}% for a vertically oriented organism. The same conclusion as for *Euglena* can be reached: this is not going to work!

How does that compare with the pressure activation of SACs? In the prokaryotic kingdom, the stretch-sensitive channel MscL (Mechanosensitive channel Large; first found in *Escherichia coli*, but later on homologs were also found in many other bacteria) is well characterized (see Hamill & Martinac, 2001, for an overview). Reported half-maximal activation pressures of MscL vary between 0.6–10 kPa, depending on measuring conditions (low values for applying positive pressure and high values for applying negative pressure). This simply means that the response is not symmetric as expected for the function of a safety valve (see below). The related channels (Mechanosensitive channel Medium and Mechanosensitive channel Small) have lower activation pressures, which are still high, compared with pressure expected in single cells, like *Euglena*. The half-maximal activation pressure will yield an area expansion of more than 3%. These pressures obviously do not have anything to do with gravity perception, but the structures are involved in osmoregulation, where pressures in the range given previously can easily build up. It seems that MscL serves as a safety valve, which is underlined by the differential pressure sensitivity. MscL is activated 10 times more easily by positive pressure (positive pressure would result in a lysis of the cell, while negative pressure can be counterbalanced by a physiological response like the production of osmotically active substances). In addition, in contrast to eukaryotic types of mechanically gated ion channels, no ion selectivity was observed for MscL, which underlines the function as a safety measure. However, genetic dissection experiments build up supporting evidences that the prokaryotic mechanogated ion channels are at least in structure very much related to eukaryotic ones (Hamill & Martinac, 2001). As previously described, eukaryotic mechanogated channels are believed to have lower activation pressures. However, most experiments have been performed in patch-clamp setups, which make it at the current time impossible to determine values due to tension build up in the membrane inside the patch pipette even without the application of any pressure (McBride & Hamill, 1992; Bett & Sachs, 1997; Sachs & Morris, 1998; Hamill & Martinac, 2001). Nevertheless, even if the values are much lower than reported for MscL, they would be orders of magnitude higher than the pressure applied by gravity acting on a cell body. In addition, this assumes that no additional supporting measures are available, like spare membrane and the cytoskeleton that counteracts any tension building up in the membrane. It seems safe to conclude that SACs and the connected membrane tension cannot play a role in graviperception, because under physiological conditions the forces required are not present.

One additional remark is necessary for the K_a. The area extension modulus seems to be dependent on the membrane potential. In red blood cells, K_a varies by ±40% when the membrane potential changes by ±200 mV (Katnik & Waugh, 1990a,b). While the molecular mechanism of this observation is not very well understood, the finding correlates with the behavior of inner ear–outer hair cells that shorten upon depolarization, and hyperpolarization leads to an extension of the cell soma (Brownell et al., 1985; Ashmore, 1987). The proposed mechanism is based on the presence of unknown proteins in the lateral membrane (Gulley & Reese, 1977; Kalinec et al., 1992), which upon changes in polarization expand or contract in the plane of the lateral membrane. This process is independent of ATP hydrolysis and is only powered by the electrical field itself (Hudspeth, 1997, and references therein).

8.3.3 *Membrane thickness elasticity*

If membrane tension does not play a role, which other membrane parameters might be involved in graviperception? The next membrane parameter under investigation is a corresponding value to area expansion: thickness elasticity modulus [Eq. (8.14), K_d; Evans & Hochmuth, 1978].

$$T_d = K_d(\Delta d/d_0) \tag{8.14}$$

T_d = membrane tension, K_d = thickness elasticity modulus, and d = thickness.

K_d and K_a are related according to Equation 8.15 (Evans & Hochmuth, 1978)

$$K_a = K_d\, d_0 \tag{8.15}$$

K_a = area expansion modulus, K_d = thickness elasticity modulus, and d_0 = initial thickness.

This would result in a value of approximately 1.7×10^8 N m^{-2}, when an initial membrane thickness of 3 nm is assumed. Direct measurements yielded a value of 2×10^7 N m^{-2} (Alvarez & Latorre, 1978) which would result in a very low K_a of 70 mN m^{-1}. One possible explanation is the artificial nature of the membranes used in the experiments.

In the following estimates, we will use T_m as T_d, which is a crude approximation. For *Euglena*, we can estimate 5.9×10^{-17}% change in membrane thickness for a horizontally oriented specimen. So, there is no significant change thickness. This holds for the other examples like *Paramecium, Tetrahymena*, and *Loxodes.*

8.3.4 Membrane shearing

Up to now, we could not identify a membrane parameter that is significantly changed by the forces or pressures applicable for single cells. What about membrane shearing? When a bilayer is subjected to a uniaxial lateral tension T_i **and** a uniaxial extension ΔD is applied while the total area is constant (which is a prerequisite to think further on because any area extension seems to be impossible in single cells due to energetic considerations; see above), then the membrane is sheared. In other words, if two counteracting forces are applied to a closely related membrane patch, shearing occurs (the same thing as when one presses the hands together and move them in opposite directions). This modulus is mainly governed by the cytoskeleton and not by the membrane itself, which does not show much shear rigidity above the phase transition temperature. Values for shear modulus for red blood cells vary from ~10 μN m^{-1} (Hamill & Martinac, 2001) to 2.5 μN m^{-1} (Hénon et al., 1999). The latter authors used – in contrast to the common glass-pipette technique (patches of membranes are used with a glass pipette, which may introduce artificial tensions) – an optical tweezer that allows a noninvasive determination of the modulus. The relation between the shear extension and the shear modulus is given in Equation (8.16) (Hénon et al., 1999).

$$\Delta \boldsymbol{D} = \frac{\boldsymbol{F}}{\boldsymbol{2}\,\pi\,\mu} \tag{8.16}$$

D = extension, μ = shear elasticity modulus, and F = force.

For *Euglena*, with a mean applicable force of 1 pN, this will give an extension of 63.7 nm. The same calculation for *Paramecium caudatum* (F = 128 pN) yields a value of 8.15 μm. While both values look promising on first glance, there might be a drawback: First of all, is the shear modulus of red blood cells relevant for *Euglena, Paramecium*, or other protists? In many ciliates, dinoflagellates, and euglenoids (like *Euglena*), the membrane CSK is organized as a discrete cortical structure, the epiplasm (for review, see Bouck & Ngo, 1996). In all described cases, these structures lie either in close proximity to (ciliates and dinoflagellates) or below and in direct contact with the cytoplasmic membrane (*Euglena*, pellicle; see above). As previously described, the rigidity of the CSK mainly governs the shear modulus, as well as the membrane integrity. In red blood cells, a hexagonal organization of spectrins supplies the rigidity of the cell form. The crucial function of the CSK is emphasized by the finding that CSK disruption results in a spontaneous fragmentation in red blood cells (Evans et al., 1994; Hamill et al., 2000). The super protein family of spectrins seems to be universally distributed in nature. Spectrin epitops (by antibody analysis) were detected in many ciliates, including *Paramecium tetraurelia* and *Tetrahymena*; genetic analysis confirmed these results (Williams et al., 1989; Lorenz et al., 1995; Thomas et al., 1997). In contrast, no spectrins could be found in *Euglena*, but analogous protein classes with comparable functions, the articulins and epiplasmins (Dubreuil & Bouck, 1985; Dubreuil et al., 1988; Rosiere et al., 1990; Marrs & Bouck, 1992). The same proteins were also later found in *Paramecium caudatum, P. tetraurelia*, and other ciliates as many constituents of the epiplasm (Huttenlauch et al., 1995, 1998a,b). Interestingly, Bouck and coworkers (Fazio et al., 1995) identified these proteins as targets for threonine and tyrosine kinases. Protists (including flagellates and ciliates) are considered to possess the most sophisticated cell shapes and surface architecture (Grain, 1986; Bouck & Ngo, 1996). It seems that, due to these specifics and the obvious difference between membrane-stabilizing structures, the issue of comparability is not yet settled. However, if the shearing of membranes is in the range of the estimated values, it could be a mechanistical handle for detecting gravity vector information. In addition, the shearing should be highest at an intermediate angle of orientation and minimal at a perfect vertical orientation. The epiplasmic structures will most likely play an important role in graviperception (see below). Shear forces might actually play a role in the hearing process that will be discussed in greater detail (Guharay & Sachs, 1984; Hamill & McBride, 1996; Furness et al., 1997).

8.3.5 Membrane bending and curvature

The last issue in membrane properties to be discussed is bilayer bending and curvature. Bending resistance (i.e., the corresponding bending modulus) is significantly lower than the area expansion resistance (Evans & Hochmuth, 1978) and can be calculated by Equation (8.17).

$$M = K_b(1/R_1 + 1/R_2 - 1/R_0) \tag{8.17}$$

M = torque across the membrane, K_b = curvature/bending modulus, $R_{1/2}$ = principal radii of curvature (see above), and R_0 = initial or "spontaneous" membrane curvature (Helfrich, 1973).

R_0, or the initial curvature, describes the nonstressed curvature of a bilayer. This will be determined by the lipid composition and the area of each monolayer. Furthermore, coupling to the CSK will influence this factor. Measured values for K_b are around 10^{-19} N m (J, 0.7×10^{-19} N m for erythrocyte membranes; Duwe et al., 1987). (*Note*: even when the unit for K_b is N m – which is equivalent to Joule – it is not an energy unit.) Energy is a scalar unit while torque is a vector. This value decreases twofold when the CSK is disrupted (Zilker et al., 1987). The low energy required for bending the membrane makes it an attractive candidate for graviperception. However, the same argument implies a high influence of thermal fluctuations, which would make the signal very noisy (Petrov & Bivas, 1984; Zhelev et al., 1994; Meleard et al., 1998).

What is the torque over the membrane in the model cases of *Euglena* and *Paramecium caudatum* for a horizontally swimming cell? It will depend on the spatial relation between the center of gravity and the center of the cell and the swimming angle with respect to the gravity vector [Eq. (8.18)].

$$M = Fr\cos(\theta) \tag{8.18}$$

M = torque, F = force, r = distance between center of gravity and center of cell, and θ = angle of movement in respect to gravity vector.

Figure 8.4 summarizes expected torque values in relation to the previously described distance. To obtain a torquė of more than 2×10^{-21} N m (Brownian motion), we would need a distance of approximately 44 nm (*Paramecium*) or 300 nm (*Euglena*). Both values seem very reasonable due to the irregular shapes of the cell bodies of both species.

After the journey through the world of membranes, we can conclude that bending and shearing the membrane are both potentially useful mechanisms with respect to graviperception. Both types are in the range of the energy that can be supplied by a unicellular cell body sedimenting inside its membrane and exerting by this means a force. The sedimentation of the cell body should not be mistaken as the sedimentation of the whole cell (which we will discuss later for intracellular organelles, like amyloplasts or barium sulfate crystals). The basic expectation is that the force applied will displace the membrane (i.e., A = constant). So, we are looking at a force over a distance which is per definition work or energy [1 Joule (J) = 1 N m]. We know that the force of a sedimenting cell body is large enough to move the membrane (either in a shear or bending way). A movement of the membrane by 1 nm by a force of 1 pN will yield 1×10^{-21} J. Is this sufficient to activate a mechanosensitive ion channel? That is a difficult question. Firstly, this value is about half the energy supplied permanently (i.e., independent of the position of the cell) by the thermal motion. Integrating over time (e.g., in chemoorientation of bacteria) might enable such a system to supply useful information, and we will see that, at least in some unicellular organisms, it seems likely that this mechanism is implemented. Secondly, in most cases, energy requirements to activate a channel is not known due the restrictions of the measuring technique used. The one example for which some reasonable assumptions are made is MscL. ΔG_0 values range from 14 to 16 kT (5.8×10^{-20} J to 6.6×10^{-20} J at 300 K; Sukharev et al., 1999; Hamill & Martinac, 2001), which would need a displacement of the membrane of 58–66 nm.

8.4 The hearing process as a model for graviperception in single cells

Again, two questions need to be addressed: Firstly, is that a reasonable value with regard to other systems? Secondly, do we have to supply the whole energy (i.e., the displacement at once)? Both questions are connected with regard to the system we are going to discuss next. In the cochlea (inner ear), which is responsible for hearing, bundles of cilia-like structures (stereocilia) were identified as primarily responsible for the physiological response. Figure 8.5 shows a schematic drawing of the structures involved. One stereocilium side is connected to the tip of a second, smaller one by a thin filament. The length, as well as

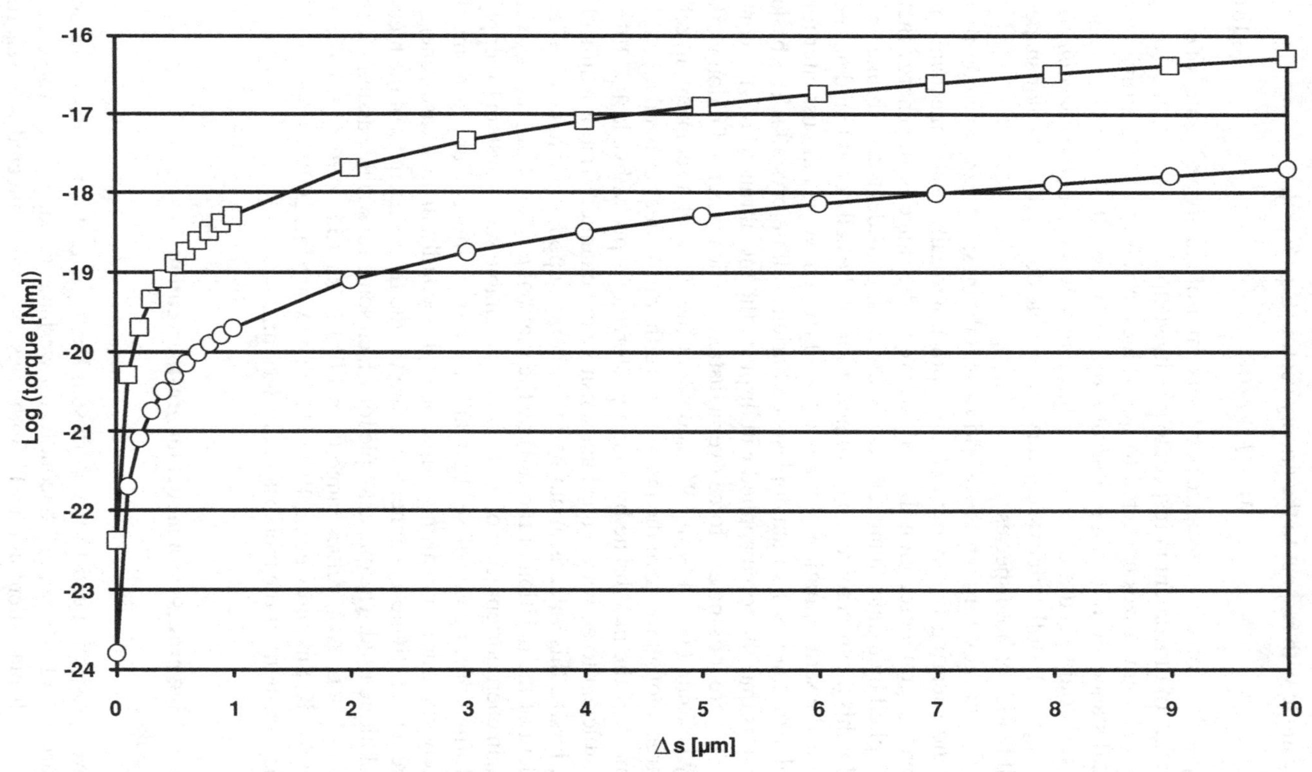

Figure 8.4. Relation of the distance of center of gravity and center of cell (Δs) with rotational torque. **Circles**: *Euglena gracilis*; **squares**: *Paramecium caudatum.*

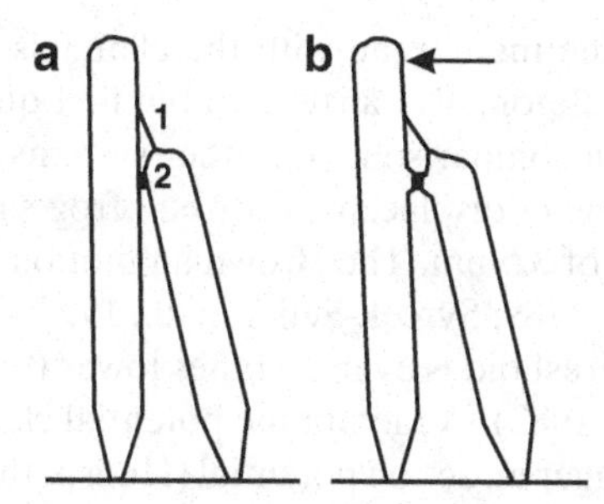

Figure 8.5. Schematic drawing of a pair of stereocilia. Relaxed **(a)** and in a forced **(arrow)** position **(b)**. 1, tip-link; 2, horizontal connections. See text for details. (Redrawn after Hamill & Martinac, 2001.)

the rigidity of the filament ("tip-link"), leads to a pull on the membrane (Fig. 8.5a), which supplies part of the energy (displacement) required for the activation of the channels. A localized force application (Fig. 8.5b) pulls the membrane of the second cilium via the tip-link further out and by this means leads to an activation of mechanosensitive ion channels. The channels have a high potassium conductance and a smaller conductance for calcium. During activation, the potassium influx depolarizes the membrane. Adaptation can influence the gating at least at three points: firstly, gating of channels causes a decrease in ciliary bundle stiffness, secondly, calcium ions might bind to a secondary activity control site of the mechanosensitive channel and by this means decrease the open probability; and thirdly, calcium might influence myosin, which is present at the anchoring point of the tip-link and change the set-point of the filament (Hudspeth, 1997; Hudspeth et al., 2000, and references therein). Essentially, the whole system works as a mechanical resonator, which is tuned to a specific resonance frequency. Only this frequency will force vibrations of the bundle. This is essentially the model which is – except some modifications – still valid for explaining the hearing process (see Markin & Hudspeth, 1995, for a review). It turned out that only a few such channels are required, and only a few of those with a high conductance are responsible for almost all the membrane potential changes (Holton & Hudspeth, 1986). The hearing process and the entities involved are optimized for fast responses and fast adaptations. In contrast, gravity is a static force and adaptation of gravitational responses still needs to be shown. These facts were taken as arguments that the hearing process cannot be a model for gravity perception (Björkman, 1992). This is a valid point for intracellular perception organelles in more static organisms, like higher plants, but not for motile organisms where we discuss the whole cell body as a perceiving entity. In this case, the cells are not only moving in time in three dimensions, but at the same time often rotate with a constant frequency around their long axis. Thus, in principle, it might be possible that part of the energy required for activation of channels by extracellular or intracellular structures might be contributed by the rotational energy. This line of thinking supplies us with the answer to the second question: yes, there is an alternative mechanism for changing the membrane conductance besides directly manipulating the bilayer – tethering the membrane with the channels to intracellular structures, such as the cytoskeleton. The answer to the first question is no; 60 nm is not a reasonable value in comparison with other systems. Denk and coworkers (1989) determined the mean displacement of bull frog sacculus hair cell stereocilia to be in the range of 3.5 nm. The Brownian motion of stereocilia is in the same range (Denk et al., 1989; Svrcek-Seiler et al., 1998; Gebeshuber & Rattay, 2001), but the hearing threshold is even 10 times lower (~500 pm; Svrcek-Seiler et al., 1998; Sellick et al., 1982). A membrane potential change of approximately 0.1 mV is sufficient to trigger an action potential (Hudspeth, 1989). Working close to the thermal noise threshold, these features imply that many "false" signals are triggered. Around 100 action potentials can be measured per second, even if no stimulation is present (Relkin & Doucet, 1991). The incredible sensitivity of outer hair cells is further emphasized by the finding that the cells themselves vibrate at a frequency close to the maximal spectral sensitivity of the specific cell with an amplitude of almost 40 nm, especially when the ciliary bundle is deflected (Hudspeth, 1997). How can such a system work reliably? Hair cells take advantage of the sinusoidal, repetitive nature of the stimulus, sound (or in an other context, motile cells take advantage of the periodic variation of the gravity forces exerting on the membrane during rotation). Even before reaching the threshold, a phase lock of the random thermal variations can be observed (Gebeshuber & Rattay, 2001). At the threshold, a coincidence between a maximum of the sound wave and the Brownian motion-dependent self-stimulation of the ciliary bundle results in an action potential. The sound stimulus leads to an accumulation of energy over time (a couple of milliseconds) via a mechanic resonance mechanisms of the hair cells (Hudspeth, 1997). At the same time, thermal noise is averaged out. But there is a price to pay: the high sensitivity is accompanied by a lowered temporal resolution.

One can calculate the minimum gating force of the mechanosensitive ion channels involved in the hearing process on the basis of literature values (Hudspeth, 1997). It is approximately 0.67 pN, which corresponds to an activation energy to 2.35 zJ (zepto Joule = 10^{-21} J) at a gating distance of 3.5 nm. That is the same as the thermal noise! In the case of *Euglena gracilis*, the applicable force of 1 pN would yield 3.5 zJ and *Paramecium caudatum* 448 zJ at a gating distance of 3.5 nm. The comparison with the values determined for outer hair cells shows clearly that, in both cases, a displacement mechanism would allow graviperception, but only if the cell is rotating, which allows a sinusoidal modulation of the input stimulus. A conclusion at this point is that, for small cells which use a cell body sedimentation mechanism, a rotation around one cell body axis is an indispensable prerequisite for gravireaction. Consequently, gliding cells like *Loxodes* (which shows a free-swimming behavior, but in contrast to other cells does not show a smooth swimming) or immotile cells (up to the size where applicable forces are big enough (e.g., *Chara* stem cells) have to use one or more intracellular organelle(s) for graviperception.

What else do we know about the mechanosensitive channel properties in the hearing process? One interesting feature is the steep increase in open probability of the channel. The dependence of the displacement can be estimated by

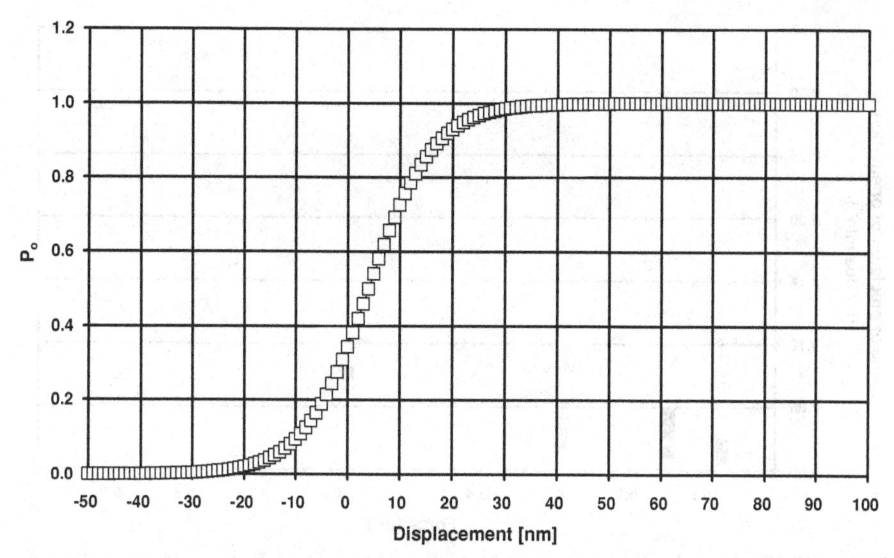

Figure 8.6. Idealized Boltzmann distribution of the open probability (P_0) of a mechanosensitive ion channel, which mimics very well the behavior of a displacement-sensitive channel in the hearing process. $z = 0.67$ pN, $X_0 = 4$ nm (depends very much on the adaptational status of the outer hair cell). For details, see text (Hudspeth et al., 2000).

Equation (8.19).

$$p_0 = \frac{1}{1 + e^{-z(x-x_0)/(KT)}} \tag{8.19}$$

p_0 = open probability, z = single-channel gating force, X = displacement, X_0 = displacement at which open probability is one-half, k = Boltzmann constant, and T = absolute temperature.

Figure 8.6 shows the distribution that mimics very well the open probability of the hearing process (Hudspeth et al., 2000). One should note that slight changes in the setpoint of the thin filaments might result in drastic changes of the overall amplification. In addition, the open probability of a channel is significant, even if no force is applied. In the calculations, a change from −5 nm displacement to +5 nm results in a increase of the open probability from 18% to 40%. Such displacements are reasonable to assume for single cells like *Euglena gracilis* or *Paramecium caudatum*. Thus, the rotation of the cells around the long axis (when the cells swim horizontally) would result in a significant change in channel activities. This hypothesis only holds when we assume an asymmetric distribution of the mechanogated ion channels, which is one of the most essential elements of the models for graviorientation in single cells. Electrophyiological studies revealed the predicted distribution in the case of several ciliates; additionally, the graviceptor potential could be measured in dependence of the spatial orientation of the cell (Gebauer et al., 1999; Krause, 2003). How does the assumption of an analogous type of channel compare with reality? In Figure 8.7, the applied esti-

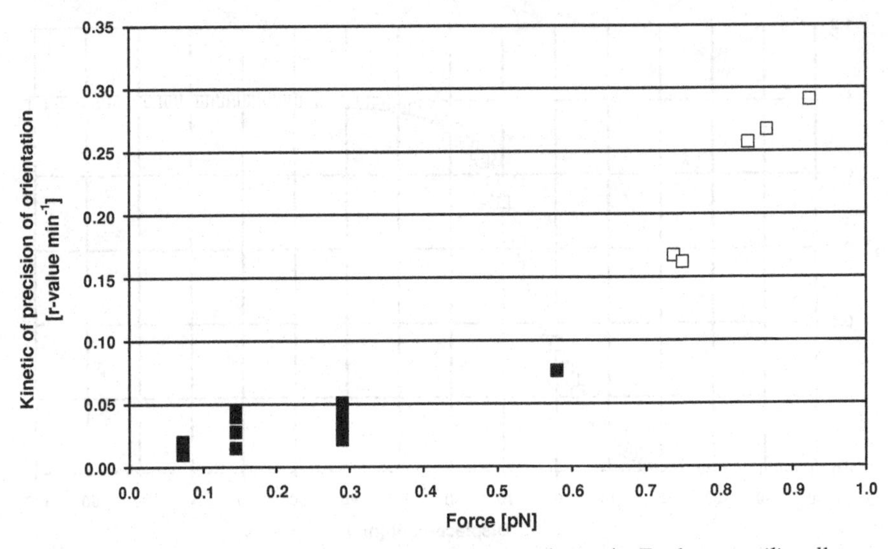

Figure 8.7. Estimated force exerted on the lower membrane in *Euglena gracilis* cells versus orientational kinetics. **Closed symbols** reevaluated experimental results of the IML-2 mission. **Open symbols** determined for cells of different sizes and specific densities (Lebert et al., 1999b).

mated force on the lower membrane is plotted against the orientational kinetics, which should – in a first approximation – be related to the channel activities. As in the theoretical estimate, we can observe a slow increase in reactivity that above a certain threshold (~0.6 pN) turns to a much steeper slope. While this can and is not proven, both relations (Figs. 8.6 and 8.7) seem to be correlated.

In the hearing process, the integration was previously described as one factor to enhance sensitivity. It was also mentioned that the payoff for this is a lower temporal resolution (Hudspeth, 1997). Thus, there is most likely an optimum integration time in relation to the velocity and rotation frequency in single cells. Modern technologies like motion analysis systems (cf. Chapter 3) allow analysis of the reorientational behavior of cells with a high temporal resolution. Careful analysis revealed at least in the case of *Euglena* a short but significant lag time before the first reaction of the cells can be observed (3–6 s, ~5–10 rotations; Lebert & Häder, 1999c; Lebert et al., 1999b). This result can be interpreted as an integration time. After that first reaction during reorientation, another period of a few seconds passed before the next reaction was detected. This is in the line of argumentation that the hearing process can be a model for graviorientation.

The situation in the case of the outer cells seems to be kind of straightforward: a tip-link tethers one cilium to another. What can tether a channel to what in a single cell without any obvious extracellular structures? One possibility is a link of a cytoplasmatic domain of a channel to the cytoskeleton CSK (see above) either directly or via additional proteins like in *Acheta domesticus* (house cricket) or *Chaenorhabditis elegans* (Thurm et al., 1983; Thurm, 1983). A shearing or bending of the membrane would then, when the CSK keeps its position, be activated. An influx of potassium accompanied by a calcium influx could then be the first step of the signal transduction chain. Calcium as a ubiquitous signal for all cases could then in turn deactivate the channels and influence the CSK. During reorientation, a change in cell form was observed in *Euglena gracilis* (cf. Chapter 9; Lebert & Häder, 1999c). A positioning of the channels close at the flagellum root would also allow a direct influence on the flagellum beating pattern.

It can be stated that we could identify potential physical and physiological mechanisms that allow even in the case of a small organism use of the weak force gravity to control signal transduction chains. In addition, it seems that the hearing process might be a model for the understanding of graviorientation in single cells.

Up to now, we concentrated on single cells. In the following, a few issues on intracellular gravity-perceiving organelles will be discussed. In these cases, a sinusoidal modulation of the stimulus input gravity is not applicable. In the case of *Chara* rhizoids (Sievers & Volkmann, 1979), barium sulfate vesicles are considered as the gravity-perceiving organelles. The vesicle sediment in 2–5 min to the lower membrane if the rhizoids are exposed horizontally. Following sedimentation, it was thought that the area of the bottom membrane covered by the vesicles was protected against further incorporation of membrane vesicles responsible for tip growth. In light of new results, this hypothesis does not hold anymore (M. Braun, 2003, personal communication). It seems that at least a few of the barium sulfate vesicles have to have direct contact with the lower membrane. This causes a calcium influx (sic!), which in turn somehow inhibits the incorporation of further vesicles. At least for the moment, it does not seem that, for the activation, a direct interaction with the CSK is necessary. For estimation of the forces and pressures involved, we can use the values from Table 8.1. A single statolith can apply a force of 0.14 pN. About 50–60 are present per cell, which would yield ~7.7 pN. Applied to an area of 225 μm^2 (Sievers & Volkmann, 1979), this would result in a maximal pressure of 30 mPa. This is in the same range as for a vertically swimming *Euglena* (60 mPa, see above). Most likely, the pressure will not be sufficient to put significant tension on the membrane. Alternatively, binding of the barium sulfate statoliths to membrane entities could directly induce a conformational change of channels (i.e., activation).

Loxodes striatus shows a pronounced positive or bimodal gravitaxis. In a bimodal gravitaxis, approximately one-half of the population moves upward and the other half downward; thus, no net change in the position of the cell population can be observed. The species can be found either in the sediment where cells are gliding on the substrate or free-swimming. In contrast to other ciliates or flagellates, the free-swimming involves a substantial amount of tumbling or spontaneous reorientations of the cells. In addition, with respect to physiology, the situation is more complicated due to a strong influence of the oxygen tension (Fenchel & Finlay, 1984). A *Loxodes striatus* cell contains one to four Müller bodies, which are considered to act as a gravity-perceiving organelle. Nothing is known about the necessary cross-talk between the bodies to control the motor output, though the lesion of only one of them resulted in a loss of the graviresponse (Hemmersbach et al., 1996a; Donath, 1999). In each body, a barium sulfate crystal is fixed by a rudimentary cilium ("stalk"; Fenchel & Finlay, 1986a) to one side of the vesicle. The crystal has a diameter of 3–4 μm and the statocystoid vesicle a width of 7–8 μm (Fenchel & Finlay, 1984). Excursions of the Müller body by about 1.5 μm were observed in cells moving in a horizontal plane, corresponding to a deviation of 10° at the base of the stalk. No larger excursions have been seen in gliding cells moving in the vertical plane (Neugebauer & Machemer, 1997). This finding devaluates the proposed gravity-perceiving mechanism that was based on the assumption that the crystal deviates from the unstimulated position until some contact with the surrounding membrane is achieved (Fenchel & Finlay, 1984). The forces involved can be calculated according to Equation (8.1) and range from 0.5 pN to 1.1 pN (Table 8.1). The torque applied to the end of the stalk can be estimated by Equation (8.18). When the crystal is located in the middle of the Müller body, the length from crystal midpoint and the stalk basis is approximately 3–4 μm. At that distance, an energy of 1.5×10^{-18} J to 4.4×10^{-18} J would be supplied. These energies are even sufficient to activate MscL (ΔG_0 values = 14–16 kT (5.8×10^{-20} J to 6.6×10^{-20} J at 300 K); Sukharev et al., 1999; Hamill & Martinac, 2001), which is thought to be most unlikely to act as a gravity-triggered, mechanogated ion channel due to insensitivity. The settling velocity of the crystals will be mainly determined by the forces involved with the complication of the (unknown) stiffness of the stalk. When we use Equation (8.7) and ignore the stalk stiffness, a sedimentation velocity of 2–3 μm/s can be determined. To achieve the full observed excursion of the barium sulfate crystal, less than 1 s would be necessary. Thus, from the energetic point of view, the properties of the Müller body make them a very good candidate for graviperception – but only for gliding cells. Free-swimming cells, while rotating around the long axis, show a disturbed forward movement with frequent tumbles which (as discussed) will not allow for sinusoidal, periodic variation of the gravity stimulus identified as a basis for cells using the whole cell body as perceiving organ. One possible explanation would be that, in contrast to, for example, *Euglena* or *Paramecium tetraurelia*, the cells cannot detect where is up and down, but instead are able to decide whether they are swimming in accordance with the gravity vector or at an angle. In every case, in organisms moving with some deviation to the gravity vector, a tumbling occurs and the forward-swimming resumes. This mechanism would be in the same line of argumentation as the chemoorientation of bacteria.

Physics and biophysics are helpful disciplines to figure out what kind of mechanism for graviorientation is possible and what kind of mechanism is not. During the calculations based on the most basic assumptions and measurements (e.g., cell width, length, and specific density), we could rule out stretch-sensitive ion channels as possible candidates for gravity perception. Instead, it seems that displacement-sensitive ion channels – like in the hearing process – are the most promising candidates. Energy supplied by physical effects on the membrane (either by bending or shearing) should be sufficient to activate these kinds of channels. Finally, the lower size value for detection of gravity by a mechanism based on cell body sedimentation is possibly much lower than currently thought.

9

Models for Graviperception

Whether a pure physical mechanism is sufficient to describe gravi-related motility phenomena is a question discussed for more than 100 years. The scope of this chapter is not the energetic considerations (cf. Chapter 8 for an in-depth discussion of energetics), but the description of historical and recent models explaining gravitaxis and gravikinesis in ciliates and flagellates. In general, it can be stated that, in most systems, where enough information for a detailed model is available, most likely gravi-related behavior is a combination of both: a physical component and a physiological component.

Since the first discovery of gravitational effects on motile, free-swimming, unicellular organisms, scientists discussed the underlying mechanisms and principles (Schwarz, 1884; Verworn, 1889b; Jennings, 1906; cf. Section 1.1). In general, since the early days, two schools claimed to understand gravity-related phenomena. The "physics" group tried to explain gravitaxis – to the best of our knowledge no physical model for gravikinesis exists – first detected by Dembowski (1929b) as a pure physical phenomenon. The "physiology" group thought of gravitaxis as a typical signal transduction-based cellular response. As usual, the truth will be somewhere in the middle, as we will see later.

Schwarz (1884) was the first who expressed these as a first-glance contradictory hypotheses based on his results with *Euglena viridis* (a close relative of *Euglena gracilis*). For the cylindrical to elongated, ellipsoidal *E. gracilis*, the author assumed an active, physiological mechanism for gravitaxis. In contrast, for *E. viridis* (which has about same size as *E. gracilis*, but a pointed posterior end) and *Chlamydomonas pulviculus*, he considered that either gravity acts as a stimulus like light (which implies that a coupled signal transduction chain is involved) or that the center of gravity is eccentrically located in the cell, which as a result passively pulls the cell in an upward orientation. He favored the first idea due to his findings that, for example, dead cells were not sedimenting in a preferred orientation – a phenomenon that was also stated by Kuznicki in the case of immobilized *Paramecium* cells (Kuznicki, 1968).

Cells from older cultures of *E. viridis* tend to form resting stages. Before these cells became immotile, he observed that they did no longer accumulate at the culture surface. Finally, he found that, at temperatures below 5°–6°C, cells were still motile, but again did not accumulate at the surface.

Four years later, Aderhold (1888) published his results that essentially confirmed Schwarz's findings of gravitaxis in protists. He explained the earlier observation by Stahl (1884), who could not find gravitaxis in *Euglena* by an experimental flaw: the cells were fixed in gelatin to prevent thermal convection of the medium in the observation chamber. Aderhold, in contrast to Schwarz, could not find a temperature dependence of gravitaxis, but found a strong attraction of the organisms by air.

One year later, Verworn (1889b) was the first to connect the flagellar/ciliary beating pattern with reorientation. Jensen (1893) criticized all earlier authors (some things never change) because the fixation methods used might have a subtle influence on the cell form. When he used iodine for fixation – which did not seem to have any effect on the cell form of *Paramecium* – he observed a sedimentation with the front end down. Consequently, he excluded a pure passive alignment of the organisms, but instead thought that *Paramecium* could sense the hydrostatic pressure difference between the cell poles and aims at a position of lower pressure difference.

Almost 20 years later, Wager (1911) found that *Euglena viridis* cells – when fixed with hot water, osmium, or iodine – sediment preferentially with the rear end down. He explains this observation with the location of heavier particles in the rear end (nucleus, paramylon). Thus, this short overview of the first findings leaves us a little bit at a loss. Many authors, many contradictory results. Many of these contradictory findings can be explained by culture conditions, as well as the strong circadian rhythm of gravitaxis (Lebert et al., 1999a). However, the early authors showed a strong tendency for a true physiological explanation of the phenomenon "gravitaxis," even if they considered physically based mechanisms. This picture changed in the younger literature. The gravity buoyancy theory originally discussed by Schwarz and Verworn was revived (see above; Wager, 1911; Dembowski, 1929a,b; Fukui & Asai, 1985; Brinkmann, 1968). Later, Roberts (1970, 1981) introduced another model: the drag-gravity theory. Subsequently, Jahn and coworkers published the propulsion-gravity hypothesis (Jahn & Votta, 1972; Winet & Jahn, 1974). Both models consider only a physical basis of gravitaxis. The discovery of specialized gravity-sensing organelles in some ciliates (e.g., *Loxodes*; Müller organelles; cf. Section 4.2 for details) were the first cases for which a physiological mechanism could be clearly shown. The underlying theory is known as "statolith hypothesis." However, in most unicellular species, no such organelles or statoliths could be found so far; but, recently, some authors consider the whole protoplast as a "statolith," which might deform the lower membrane by sedimentation (due to a higher specific density as compared with the medium) and by this means acts as a gravity sensor (Machemer et al., 1991; Schatz et al., 1996; Lebert & Häder, 1996; Lebert et al., 1997). In contrast to gravitaxis, a physical basis of gravikinesis was never under discussion due to its intrinsic nature (see below).

Some of the previously described models were ruled out by simple experiments. The drag-gravity hypothesis (Roberts, 1970) – also known as shape-hydrodynamic hypothesis – considers hydrodynamic effects of the cell shape as important for gravitaxis (i.e., the wider rear end sediments faster than the narrower front end). The model could not explain that no obvious changes in cell shape are detectable for cells showing positive, negative, or no gravitaxis at all. In addition, the rate of upward reorientations is not independent from velocity (Winet & Jahn, 1974; Taneda et al., 1987; Nowakowska & Grebecki, 1977).

According to the lifting-force hypothesis (Nowakowska & Grebecki, 1977), translational components are the key forces in gravitaxis, and drag and lift should contribute to torque. However, at low Reynolds numbers, shape-dependent generation of lifting forces should not apply to an actively moving body (Bean, 1984).

One of the first proposed physiological mechanisms was the hydrostatic-pressure hypothesis, where cells should sense a pressure difference between the upper and lower parts of the cell (Jensen, 1893). The facts that negative gravitaxis persisted in inverted hydrostatic pressure gradients and that the swimming velocities did not differ at different hydrostatic pressures do not support this hypothesis (Kanda, 1914; Taneda, 1987; Schaefer, 1922).

According to the resistance hypothesis (Davenport, 1897), the cells should swim in the direction of increased resistance to achieve their highest energy consumption. However, negative gravitaxis persisted even if the density gradient was inversed by the application of gum arabic or heavy water to 1.1 g cm^{-3} (Platt, 1899; Lyon, 1905; Taneda, 1987).

In the following, we will discuss in detail some of the models outlined above and compare them with the current views of the field. In addition, due to the fast developments in computer-based motion analysis, it was in some cases possible to compare "real-world," high-resolution measurements with model predictions and by this means allow us to judge the validity of these theories.

9.1 Gravity-buoyancy model

An object that is partially (ship) or totally submersed (submarine) has more than one center. One is the center of gravity, the other the center of buoyancy. The center of gravity is the imaginary point where gravity works (on sedimentation). Because the medium has an homogeneous density distribution, the center of buoyancy is identical with the geometrical center of the object. The stability of a position of an object in a given orientation in space depends on the relation of both centers with respect to each other. In a stable position, the center of buoyancy and the center of gravity are aligned with the gravity vector. In this case, the center of gravity is located below the center of buoyancy. In any other orientation, the center of gravity will be pulled down with respect to the center of buoyancy. In a homogenous spherical body – where both centers coincide – no movement will be observed. In any other case, the heavier end will be pulled down, irrespective of the positioning at the front or rear end. Such uneven mass distribution could be caused by heavier organelles (like mitochondria, chloroplasts, or the nucleus) predominately located in the rear end. The speed of turning directly depends on the distance between both centers and the drag forces applicable. Due to the small size and very low Reynolds numbers applicable for small protists like *Euglena* or *Paramecium* (cf. Chapter 8 for details), inertia can be neglected and therefore the speed of reorientation is directly (and not by the square) proportional to the forces applied. In other words: the rotational force is proportional to the angle between swimming direction and the gravity vector. A special case is a directly downward swimming cell. While according to the vector addition, the resulting forces are again zero, the orientation is unstable (comparable with a pendulum that points upward) because the helical path of swimming protists will result in small displacements from the "ideal" direction of movement. In any case, an orientational mechanism like the buoyancy principle could not be defined as true "taxis," which is defined as an active orientational steering with respect to a stimulus (Brinkmann, 1968). One example of this kind of gravi-"taxis" would be *Chlamydomonas* (Hill & Häder, 1997). In every "real-world" experiment, some cells are directly swimming downward. This is due to the small forces applicable that only slowly pull the cells in an upward swimming direction. While in fact the rotational speed of *Paramecium caudatum* is slow, it is much faster in the case of *Euglena gracilis*.

Fukui and Asai (1985) presented a model based on the principles outlined previously that sufficiently describe gravitaxis in *Paramecium caudatum*, as well as in *P. multimicronucleatum, P. tetraurelia*, and *Tetrahymena pyriformis*. Equations (9.1–9.3) show the mathematical description of the model.

$$x - x_0 = nP\,T^{-1}\,R\cos\Theta \tag{9.1}$$

$$y - y_0 = -nP\,T^{-1}\,R\ln\cos\Theta - nmg\,T^{-1}\,R\ln\,\tan(0.5\Theta + 0.25\pi) \tag{9.2}$$

$$\tan(0.5\Theta + 0.25\pi)\tan(0.5\Theta_0 + 0.25\pi)^{-1} = e^{Ttn^{-1}} \tag{9.3}$$

g = gravity constant (9.81 m s^{-2}), m = effective mass of the organism in water, n = coefficient of friction for rotational movement around the center of gravity of the cell, P = propulsive force of swimming, R = coefficient of friction for translational motion of the center of gravity of the cell, T = the torque produced by the posterior density bias in the cell body, Θ = current angle (rad), Θ_0 = initial angle of movement with respect to the gravity vector (rad), x = horizontal component of the motion, x_0 = initial horizontal component of motion, y = vertical component of motion, and y_0 = initial vertical component of motion.

For a detailed comparison between the model predictions and the "real-world" movements of *Euglena*, cells were allowed to orient with respect to gravity. After this initial period, the cuvette was turned by −90°. As an example, two tracks of *Euglena gracilis* analyzed in detail are shown in Fig. 9.1.

To apply the model to the gravitactic behavior of *Euglena gracilis*, some organism-specific parameters must be determined. The effective mass of the organism in water (m) is given by the volume of a cell (2 × 10^{-12} liters) multiplied by the specific density difference between the cell body (1.05 g ml^{-1}) and the

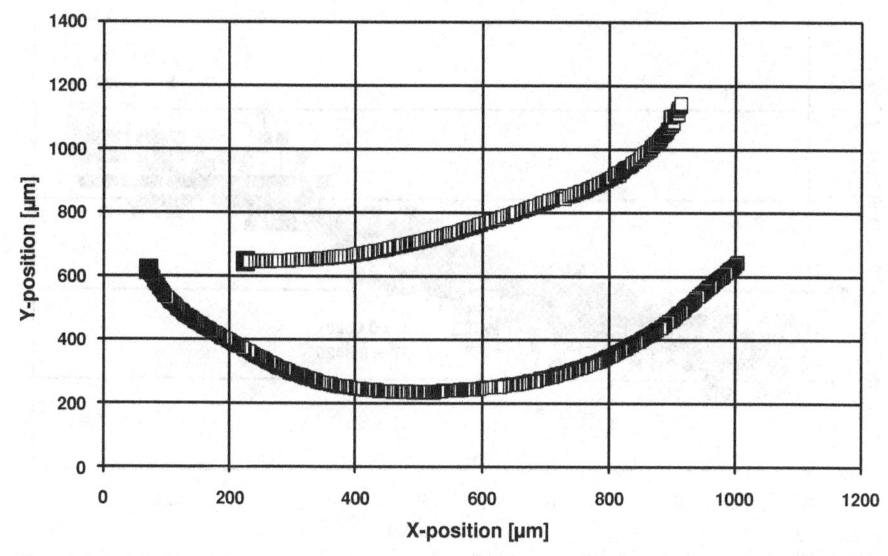

Figure 9.1. High-resolution path reconstruction of reorienting *Euglena gracilis* cells. Cells were allowed to orient for 3 min in a vertical cuvette. Then, the cuvette was turned by −90° and cell tracking started. **Leftmost squares** indicate starting points (**closed squares**). (Redrawn after Lebert & Häder, 1999c.)

medium (water; 1.000 g ml^{-1}). This results in a value of 1.10^{-13} kg for the effective mass *m*. The coefficient of friction for translational motion of the center of gravity of the cell (*R*) was determined from Stokes approximation by assuming the cell to have a diameter of 8 μm. The propulsive force of swimming (*P*) is given as the product of *R* and the mean swimming speed of the cells in the horizontal plane (55 $\mu m\, s^{-1}$). A value of 4.15×10^{-6} dyn was used in the calculations. *T* (the torque produced by the posterior density bias in the cell body), as well as *n* (coefficient of friction for rotational movement around the center of gravity of the cell), cannot be determined directly, but the linearized change of the angle of orientation over time gives the ratio $T\,n^{-1}$ as the slope of the best-fitting lines. Figure 9.2 shows the result of the analysis. Values of 0.0328 and 0.0474 s^{-1} were determined for both analyzed tracks. The variability depends on the initial angle of movement; values between 0.022 s^{-1} (60°) and 0.058 s^{-1} (−90°) were found. Calculations were performed with a value of 0.04 s^{-1}.

The prediction on the angular change can be simplified. The translational torque, a result of an uneven mass distribution, depends on the angle of movement. In principle, it should be possible to simplify the model by using $T\,n^{-1}$ as a calibration. This allows us to predict the *y*-displacement and by this means calculate the direction of movement [Equation (9.4)].

$$\Theta = \Theta_0 + T\,n^{-1} \cos \Theta_{t-1}(t_n - t_{n-1}), \qquad (9.4)$$

where Θ_0 = initial angle of movement (rad), Θ = angle of movement (rad), and t = time.

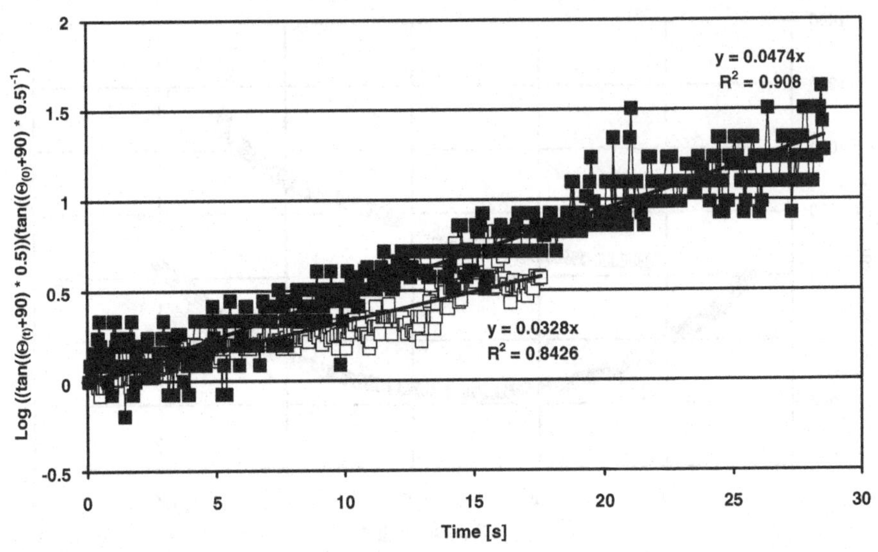

Figure 9.2. Linearized paths of the tracks shown in Fig. 9.1. See text for details. (Redrawn after Lebert & Häder, 1999c.)

Using Equation (9.4) and a value of 0.13 s^{-1} Tn^{-1} (for *Paramecium caudatum*; Fukui & Asai, 1985), an angular change was determined for different initial angles and compared with actual measurements (Fig. 9.3, heavy lines). An almost perfect match can be observed, indicating that in fact the simplified model [Equation (9.4); Fig. 9.3, open squares] can be used to fit data obtained for gravitactic orientation of *Paramecium*.

To test this model in the case of *Euglena*, the angular change during reorientation must be determined. Figure 9.4 shows the results. Initially, no change of direction of movement can be observed. This period lasts for about 3 s. While a smooth orientation can be observed in the case of *Paramecium caudatum*, during reorientation of *Euglena* step functions were found (Fig. 9.4, boxed regions). During these periods, no significant net angular change is detectable.

These data are compared with the model predictions based on Equation (9.4) (Fig. 9.5). $Tn^{-1} = 0.04$ s^{-1} is as calculated in Fig. 9.3, and initial angles +10° and −65° are used for calculating the time-dependent angular change. In both cases, the model predicts a final angle of 90° after ca. 150 s (−65°) and 112 s (10°). As can be seen in Fig. 9.5, this time course is not comparable with the measured values. This finding rules out essentially that a gravity-buoyancy mechanism alone is sufficient to explain gravitaxis in *Euglena* (Lebert & Häder, 1999c). However, *Euglena* does not have a static form, but might change its mass distribution during reorientation. If this is the case, Tn^{-1} would not be constant as in *Paramecium*, and the angular change might vary during reactions. In fact, a change of form can be observed during reorientation (Lebert & Häder, 1999c). On the other hand, this mechanism would require an active recognition of gravity. In addition, the whole reorientation also depends on (or is at least modified by) an active movement

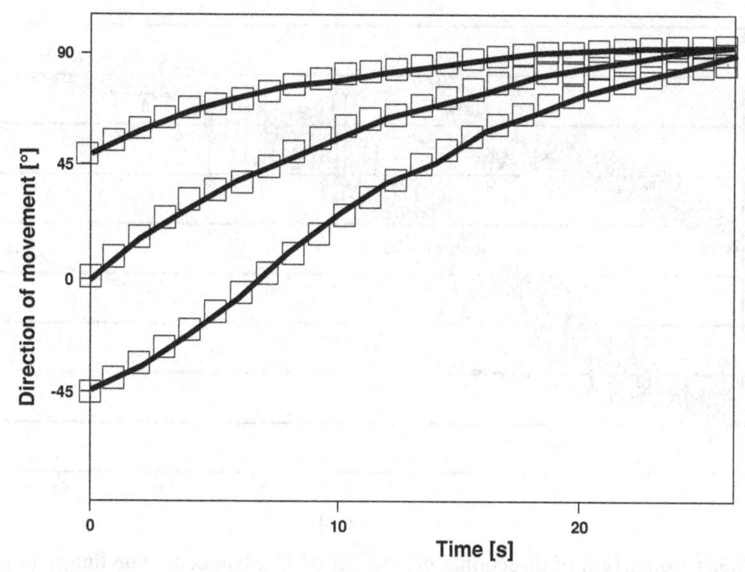

Figure 9.3. Comparison of *Paramecium* cell tracks (**solid lines**; redrawn from Fukui & Asai, 1985) and a path prediction based on a simplified model (**open squares**). See text for details.

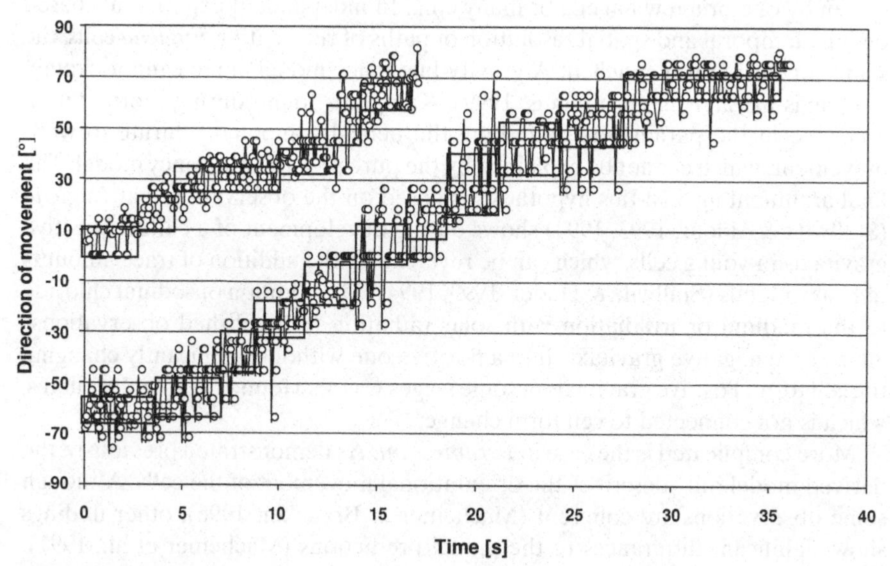

Figure 9.4. Time-resolved movement analysis of tracks shown in Fig. 9.1. **Boxed regions** indicate periods with no significant change in direction.

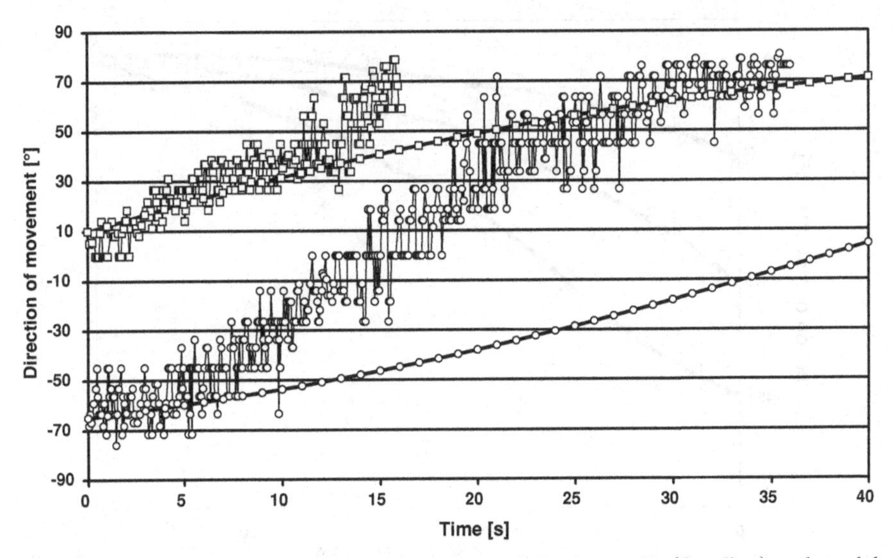

Figure 9.5. Comparison of directional movement of *Euglena* cells (**fine line**) and model predictions (**heavy lines**) calculated on the adapted version of the model presented by Fukui and Asai (1985) for the orientation of *Paramecium caudatum*. Corresponding symbols refer to the same organism.

driven by one or a few flagella or many cilia. In independent experiments based on high temporal and spatial resolution of paths of reorienting *Euglena* cells, the same conclusion was reached: A gravity-buoyancy model alone cannot explain gravitaxis in *Euglena* (Kamphuis, 1999). Kamphuis found during reorientation a change in the parameters describing the helical component during forward movement, which cannot be explained by the pure physical buoyancy model. The final argument against this hypothesis is based on the observation that *Euglena* (Stallwitz & Häder, 1993, 1994) shows during development of a culture positive gravitaxis in young cells, which can be reversed by the addition of trace amounts of heavy metals (Stallwitz & Häder, 1993, 1994). The addition of sodium chloride to the medium or irradiation with solar radiation (unpublished observations) altered the negative gravitaxis into a positive one without significantly changing the cell form. Positive gravitaxis in some stages was also found in other flagellates, which is not connected to cell form change.

More complicated is the case in *Paramecium*. As demonstrated previously, the derived model fully describes the orientational movement of the cells. Although some observations are coherent (Machemer & Bräucker, 1996), other findings show significant differences to the model predictions (Machemer et al., 1991). Finally, Taneda and Miyata (1995) took the effort to analyze in detail reorientational movements of *Paramecium*. A frame by frame analysis of swimming paths revealed a dependence of the reorientational rate on the swimming velocity. Although this finding does not rule out the involvement of a buoyancy mechanism in gravitaxis, it certainly makes it very unlikely as the sole cause.

9.2 Drag-gravity model

In 1970, Roberts proposed an alternative physical model, which could explain gravity-dependent orientation in microorganisms (Roberts, 1970; Roberts & Deacon, 2002). In the discussion of the gravity-buoyancy model, it was assumed that, for the sedimentation only, one Reynolds number sufficiently describes the conditions encountered by a cell in the medium. That is essentially where the drag-gravity model comes in. It assumes that for nonspherical, asymmetrical bodies like *Paramecium* or *Euglena*, at least two Reynolds numbers have to be applied and that these bodies have to be viewed as an assembly of two or more coupled spheres. In this case, Stokes's law [Equation (9.5)] predicts that a bigger rear end (which in the assembly represents a bigger sphere), sediments faster than the smaller front end (the smaller sphere), even when no uneven mass distribution exists:

$$v = \frac{2(\rho_b - \rho_m)gr^2}{9\eta} \tag{9.5}$$

v = velocity, ρ_b = specific density of the body, ρ_m = specific density of the medium, g = acceleration (9.81 m s^{-2}), r = radius, and η = viscosity of the medium [cP].

Roberts performed experiments with model bodies that sedimented in glycerol to compensate size differences and the related Reynolds numbers between his models and cells. He compared his findings with artificial models of immobilized *Paramecium* cells and found a very good correlation. (*Note*: immobilized cells especially in the case of ciliates might not reflect accurately the conditions in freely moving cells due to space requirements of beating cilia and the change of apparent form related to these movements.) The immobilized *Paramecium* cells sedimented with 10–20 μm s^{-1} in Robert's experiments (more recently measured values are closer to 80 μm s^{-1}) and turned into a front end-up orientation with a speed of 0.5°–3° s^{-1}. He found a linear relation between the square root of the sedimentation speed and the size of the models (as expected by application of Stokes's law). In addition, the sedimentation speed varied by about 30%, depending on the orientation of the bodies. Following his line of argumentation, cells would accumulate in the upper parts of a water column due to the rotation and the active movement (as long as the sedimentation speed is overcompensated), if not too many spontaneous reorientations occur. The rotational speed (in other words, the reorientational speed) directly depends on the deviation of the long cell axis on the vertical orientation [Equation (9.6)].

$$\frac{d\Theta}{dt} = -\beta \sin\Theta \tag{9.6}$$

Θ = deviation of the long axis from a vertical orientation, and β = rotational rate.

The velocity of the cells should not have any influence on the rotational rate (as in the gravity-buoyancy model). In contrast, empirical measurements with *Paramecium* suggest such a dependence (Taneda & Miyata, 1995). For this reason, it seems unlikely that both models (drag-gravity or buoyancy-gravity) could sufficiently explain the findings related to gravitaxis in *Paramecium*. The same holds for *Euglena*. The previously described high-resolution analysis of *Euglena* paths (Kamphuis, 1999) showed a change in helical parameters during forward movement, which is not in accordance with the drag-gravity model.

9.3 Propulsion-gravity model

Both models discussed up to now do not consider the helical path followed by moving cells. The propulsion-gravity model takes this specific feature as an important factor for the upward swimming of ciliates (Winet & Jahn, 1974). Specifically, the authors mentioned that, in flagellates – which show positive gravitaxis at some developmental stages without a change in cell form – a greatly reduced velocity can be observed. The previously described experiments by Taneda and Miyata (1995), including the detailed analysis of the findings, point in the same direction. The basis of the model is the observation that, in most ciliates, the front end rotates on a bigger radius around the helical axis, compared with the rear end. This can be explained by a distance between the center of effort (which is the vectorial sum of all vertical forces) applied by the cilia of the organism and the geometric center of the cell. The center of effort is closer to the front end than the geometric center. By this means, a torque is applied that moves the front pole of a horizontal moving cell up and down in a symmetrical way. But this is not fully symmetrical, because gravity applies an additional force that causes different effects during the up-pointing half period of the helical movement, compared with the down-pointing half. The sedimentation (i.e., gravity) is counteracted by the sedimentation resistance of the liquid (essentially the effect of the low Reynolds number), which acts on the whole cell and is directed upward. The vertical force applied by the cilia is proportional to the sine of the angle deviation between the cell's long axis and the horizontal horizon, and is counteracted by the friction resistance of the medium. In contrast, during the upward swing, it supports the sedimentation resistance. In the other direction, the forward forces counteract the sedimentation resistance. Thus, essentially, less force is available for the rotational movement (cilia energy output must be constant during rotation) during the downward pointing half period of the helix. This will lead to a slightly upward turned helical axis and eventually to an upward movement of the cell.

The critical point in this model is the positioning of the center of effort in front of the geometric center. This can be achieved by an asymmetrically beating frequency of cilia comparing the front to the rear end or an asymmetric shape of the cell. In fact, the asymmetric shape seems to be an important factor. Consequently, the center of effort was found to be placed closer to the front end than the geometric center in a typical ciliate, *Tetrahymena pyriformis* (Winet & Jahn, 1974). A placement of the center of effort like this can also be assumed for flagellates, such as *Euglena*, with one flagellum originating at the front end.

In contrast to the other models discussed, the propulsion-gravity model is based on an active movement of the cells. The model seems to explain very well all observations made with *Paramecium*. At the moment, it seems rather unlikely that this model could fully describe gravitaxis in *Euglena*. Even high-resolution analysis of paths did not reveal a strong correlation between upward swing during helical movement and reorientation (Kamphuis, 1999). Thus, it seems, that again a purely physical approach is not sufficient to explain the observations in *Euglena*.

9.4 Physiological models – statocyst model

9.4.1 Gravitaxis

As described in the introductory part of the chapter, earlier authors prefer a physiological basis (signal perception and amplification). The early hypotheses – detection of the maximal energy consumption (Davenport, 1908) or detection of a hydrostatic pressure difference (Jensen, 1893) could be ruled out. However, which mechanism could be applicable and to be more precise, which mechanism can allow detection of changes introduced by gravity in an aqueous environment? As noted in Chapter 8, gravity as a force has to interact with a mass to be detectable. There is no reason whatsoever to believe that gravitaxis in flagellates or ciliates is not supported by the pure physical models outlined previously. This means that a controversial discussion – which tries to decide between either "physics" or "physiology" – is missing the point. Finally, a physical mechanism will guide the cell upward, but much slower than a combined system of physical and physiological mechanisms. Depending on the direction of movement with respect to the gravity vector, one or the other will be predominant. In any case, an interaction with a mass is required. This can be either intracellular organelles acting as statoliths like amyloplasts (higher plants), the nucleus, or barium-sulfate-containing vacuoles [e.g., in *Chara* or within the Müller organelles (*Loxodes*)]. It might also be speculated that the position of the centriole pairs, occurring aligned at right angles to each, may function as a gravisensor (Cogoli, personal communication). Or, what becomes more and more likely: the whole cell body interacts with gravity and at this point introduces changes detectable by the cell. This hypothesis is somehow supported by the finding of Kiss and coworkers (1989) who analyzed starchless or starch-reduced mutants in *Arabidopsis thaliana*.

This group showed that starchless plants still show a gravitropism of the roots, but on a much longer time scale than wild-type plants. A straightforward interpretation would be that the whole cell body of the statenchyma cells acts as a receptor. The use of high-density particles like amyloplasts makes this process faster and more reliable. One interpretation would then be that the mechanism used by flagellates and ciliates is more ancient than the other ones previously described. Hydrostatic pressure of the entire protoplast (gravitational pressure model) has been suggested to trigger stimulation transformation at the plasma membrane in the case of the gravity-caused polarity of cytoplasmic streaming in Characean internodal cells (Wayne et al., 1990; Staves et al., 1992). In contrast to statenchyma cells, these are much bigger. In this model, the cell perceives the

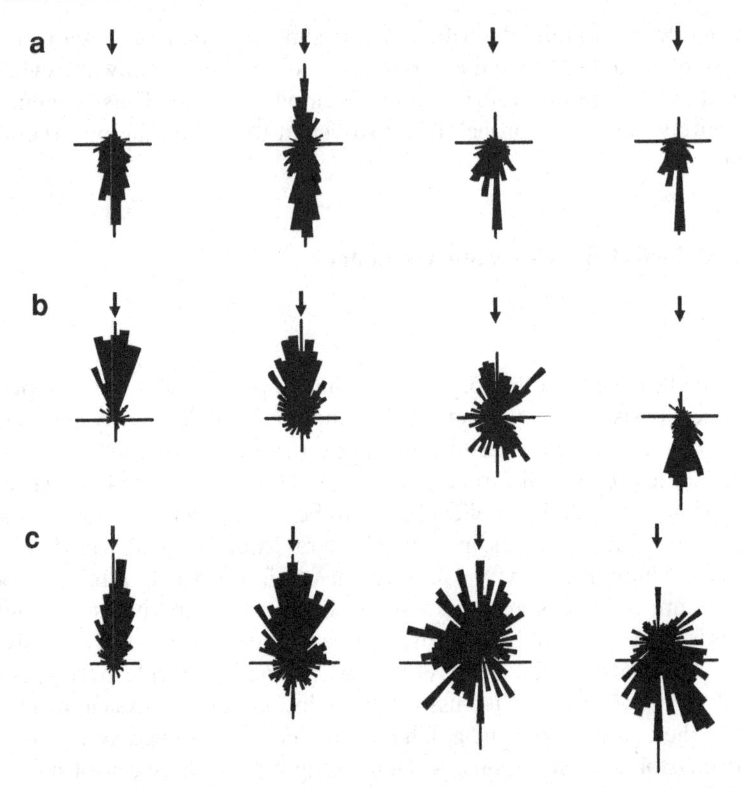

Figure 9.6. Effect of manipulation of the specific density of the medium on gravitaxis in *Loxodes* (**a**; specific density 1.03 g ml^{-1}), *Paramecium biaurelia* (**b**; specific density 1.04 g ml^{-1}), and *Euglena gracilis* (**c**; specific density 1.05 g ml^{-1}). **Arrows** indicate gravity vector. **a** [from left to right; specific density of the medium (g ml^{-1})]: 1.00, 1.02, 1.05, 1.08; **b**: 1.00, 1.02, 1.05, 1.08 (Hemmersbach et al., 1998); and **c**: 1.00, 1.02, 1.04, 1.08. (Redrawn after Lebert & Häder, 1996.)

gravity vector by sensing the differential tension and compression between the plasma membrane and the extracellular matrix at the top and bottom of the cell, respectively (Staves, 1997).

In this section, we will concentrate on *Paramecium*, *Loxodes*, and *Euglena* – which are by far the best-known systems. Two of the systems, *Paramecium* and *Euglena* seem to use the whole cell body for gravity detection, whereas *Loxodes* uses intracellular organelles. This can be demonstrated by the manipulation of the specific density of the medium (Fig. 9.6). Although gravitactic orientation is independent of the specific density of the medium in *Loxodes* (Fig. 9.6a), gravitaxis disappears when the specific density of the medium approaches the specific density of the cell body in the cases of *Paramecium* (Fig. 9.6b) and *Euglena* (Fig. 9.6c). In ecological respect, this finding makes sense. *Paramecium* and *Euglena* are free-swimming organisms with no or negligible contact to solid surfaces. In contrast, *Loxodes* lives most of the time in the sediments of lakes, where mechanical disturbances during the gliding movement is typical and where the

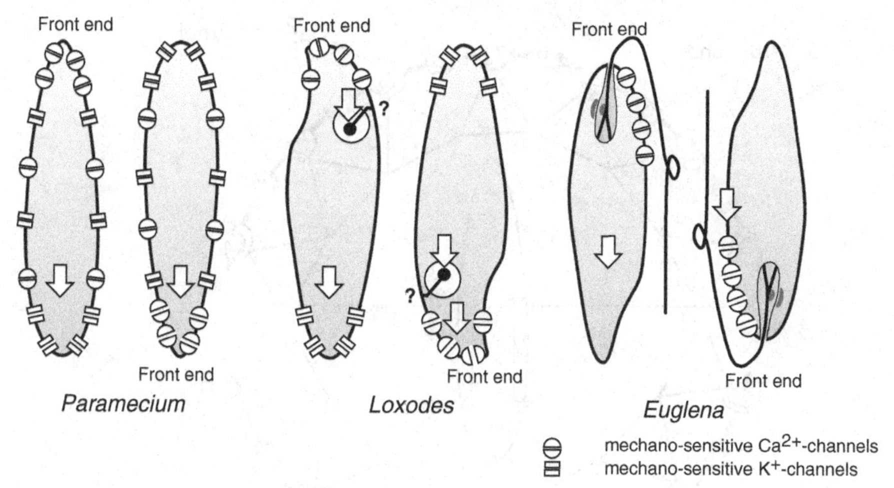

Figure 9.7. Models of graviperception in three microorganisms. (See text for details; using data from Hemmersbach & Bräucker, 2002.)

density of the medium might often exceed the cellular density. Thus, mechanoreceptors in the cell membrane should be often stimulated by external contacts and might have favored the development of an intracellular gravisensor.

In *Paramecium*, the existence of two species of mechanoreceptor channels is proposed, which are incorporated in the cell membrane in a gradient-like distribution (Baba et al., 1991; Machemer et al., 1991), thus being ideally suited for gravisensation. According to this hypothesis, calcium mechanoreceptors prevail at the anterior hemisphere and potassium mechanoreceptors dominate in the posterior hemisphere (Fig. 9.7). Depending on the swimming direction and the position during a helical turn, different parts of the cell membrane will encounter a force caused by the sedimenting cell mass. During the upward swing of the helical path, potassium channels will open more frequently than calcium channels, resulting in a small hyperpolarization and a corresponding increase in swimming rate. In contrast, during the downward pointing helical period, calcium channels will open – which by a depolarization decrease the swimming rate. In integrating over several helical turns during the forward movement, the helical path will finally point upward (Fukui & Asai, 1985; Taneda, 1998; Taneda & Miyata, 1995). This model is supported by the measurement of gravireceptor potentials in *Paramecium* (Gebauer et al., 1999) and *Stylonychia* (Krause, 2003; cf. Chapter 4). Essentially, this mechanism seems to be reasonable as an additional, physiologically based supporting system for the proposed propulsion-gravity model discussed previously. Unfortunately, a clear distinction between the impact of the propulsion-gravity model and the effect of the uneven distribution of mechanosensitive potassium and calcium channels cannot be made due to lack of specific inhibitors or knock-out mutants. In addition, it is still unclear (while very likely) whether the channels proposed as responsible for gravikinesis (see below) are identical with the channels involved in gravitaxis. Further

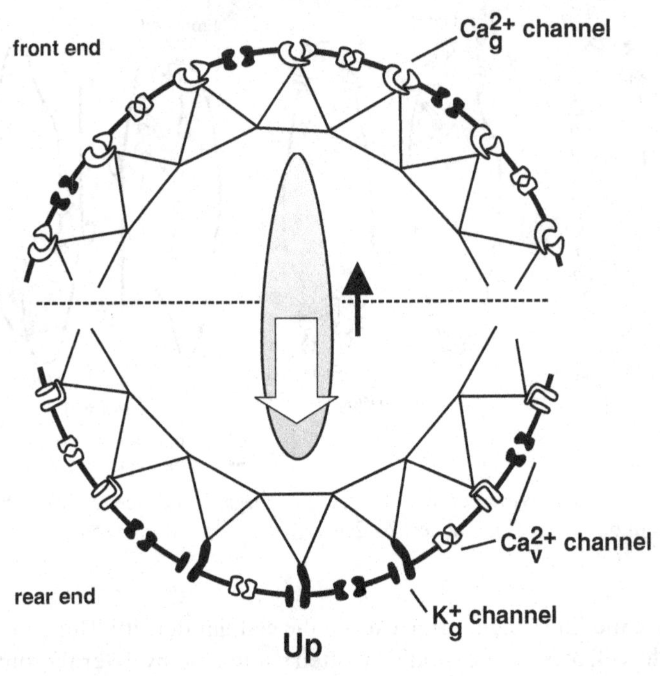

Figure 9.8. Schematic model visualizing a peripheral cytoskeletal network connecting to the proposed gravisensory channels, regulating the opening of the channels. In an upward swimming cell, hyperpolarizing potassium channels are opened (open channels, **black**; closed channels, **white**). Ca_g^{2+} channel = gravity-dependent calcium channel; Ca_v^{2+} = voltage-gated calcium channel; K_g^+ = gravity-dependent potassium channel. (Redrawn and modified after Machemer-Röhnisch et al., 1996, with permission.)

elements are discussed concerning their role in graviperception. For signal amplification and channel-gating, cytoskeletal elements are discussed (Fig. 9.8). For fine-tuning of the ciliary beat and channel properties, an involvement of second messengers can be considered (cf. Section 4.1.4; Pech, 1995). In an upward swimming *Paramecium*, the pressure on the lower cell membrane should increase under hypergravity, thus inducing hyperpolarization and faster swimming and slower swimming in microgravity. Because cAMP is thought to be the second messenger coupled to hyperpolarization, the level of cAMP might increase in hypergravity and decrease in microgravity if this second messenger is involved in graviperception. Consequently, the level of cGMP coupled to depolarizing stimuli should remain unaffected. Recent results support this hypothesis though the temporal link between membrane potential and second messenger production has yet to be clarified (Hemmersbach et al., 2002). It was shown earlier that the electrophysiological properties (Nagel, 1993) in *Loxodes* are similar to those in *Paramecium*, with respect to the receptor channel distribution. Consequently, *Loxodes* shows a similar gravikinesis. However, *Loxodes* additionally bears the Müller organelles, which are obviously involved in gravidetection (Fig. 9.7; cf. Section 4.2). An artificial disruption of the stalk holding the barium sulfate crystal

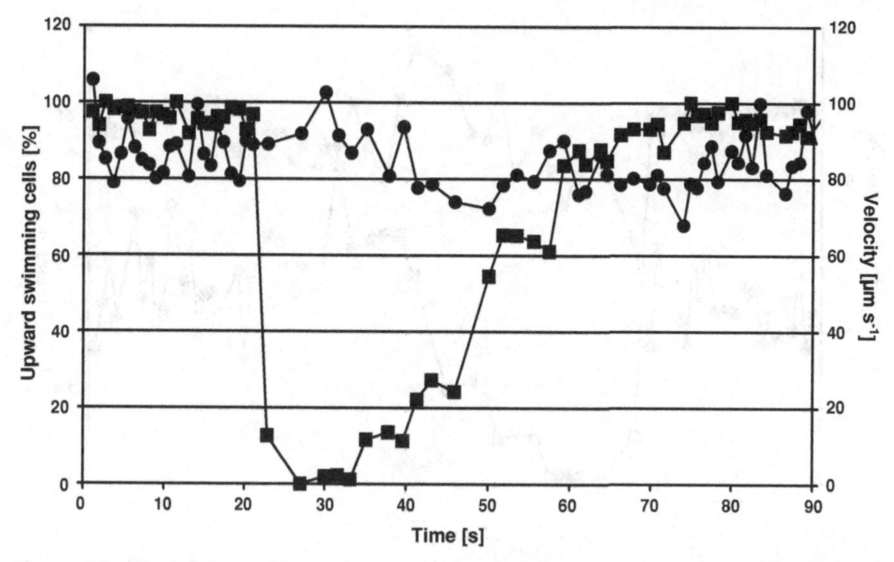

Figure 9.9. Comparison of rate of reorientation (**squares**; percentage of upward moving cells) with the mean velocity of the cells during reorientation (**circles**).

by a laser pulse results in the disappearance of gravitaxis (Hemmersbach et al., 1998). However, involvement of the Müller body is not resolved up to now. One possibility would be that the mechanical load on the stalk might easily be amplified by the lever function of the stalk to directly activate ion channels. To trigger reorientational beating pattern changes of the cilia, a direct connection between the Müller body and the cytoplasm membrane would be required. Alternatively, second messengers like cAMP or cGMP might be involved.

Although in the case of *Euglena gracilis* electrophysiological data are not available, many data sets exists that helped to physiologically define the gravitactic response. As previously described in Chapter 8, an integrating mechanism seems to be involved. In Fig. 9.4, the boxed regions indicate periods of no significant orientational change. The same can be observed in population measurements (Fig. 9.9). After *Euglena* cells are allowed to orient with respect to gravity the observation chamber was turned by 180° (indicated by a sharp drop in the percentage of upward moving cells). During the first 8 s, no obvious change in the percentage of upward moving cells can be observed. In the following 35 s, the cells reorient. But, it is quite noticeable that the reorientation is done in a stepwise function. When a step is reached, some time passes before the next significant increase can be detected. This correlates with individual cell observations. After 35 s, steps are not that clearly identifiable anymore. It seems that turning the culture somehow synchronizes the behavior. This synchronization is lost after some time. Figure 9.9 includes information on the corresponding mean velocity of the population. The slight increase in the mean velocity of the population is most likely due to the addition of the forward velocity of the cells with the sedimentation rate.

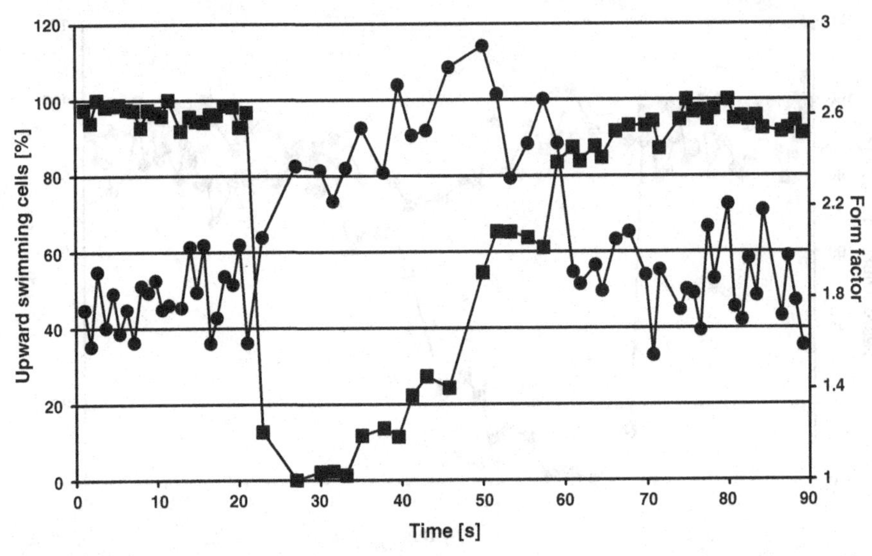

Figure 9.10. Comparison of reorientation in a *Euglena* culture after rotation of a cuvette by 180° (percentage of cells moving upward, **squares**) with the form factor changes measured in parallel (**circles**). Every data point summarizes at least 400 analyzed tracks. The form factor is defined as the perimeter of cell squared divided by $4 * \pi *$ area. A perfect circle has a form factor of 1.0. Any distortion will result in a higher value.

To analyze the gravitactic orientation in more detail, the form factor (perimeter of the cell divided by the area normalized to the circle) was plotted against time in comparison with a reorientational parameter (Fig. 9.10). Clearly, the form factor increases during reorientation of the cells and comes back to its initial values after the reorientation is completed. What can be seen in the population method can also be detected in single cell measurements (Lebert & Häder, 1999c).

In a simplified form, the model for gravitactic orientation of *Euglena gracilis* is based on the finding that the cell body has a higher specific density than the surrounding medium. Energetic considerations make it very likely that the forces that act on the membrane due to sedimentation during the rotation around the long axis of the cell are not sufficient to directly *stretch* the membrane to open mechanosensitive channels. Instead, it seems possible that displacement-sensitive ion channels are involved. These kinds of channels were first identified in the hearing process. The activation energy for these channels is very close to the Brownian motion. This means that numerous false signals will be produced even if no stimulus is present. To overcome this problem, an integrating mechanomembrane potential system is implemented. The observation that cells do not immediately react to an encountered disorientation might suggest a similar kind of mechanism. The cells rotate several times before a reaction occurs. Then, in the next round, again some rotations are required. The current model predicts that, during a rotation in a distinct position of the cell body, mechanosensitive channels are activated (see Fig. 9.7 for proposed positioning of channels). During each activation, an intracellular signal builds up (calcium or cAMP?) until a threshold

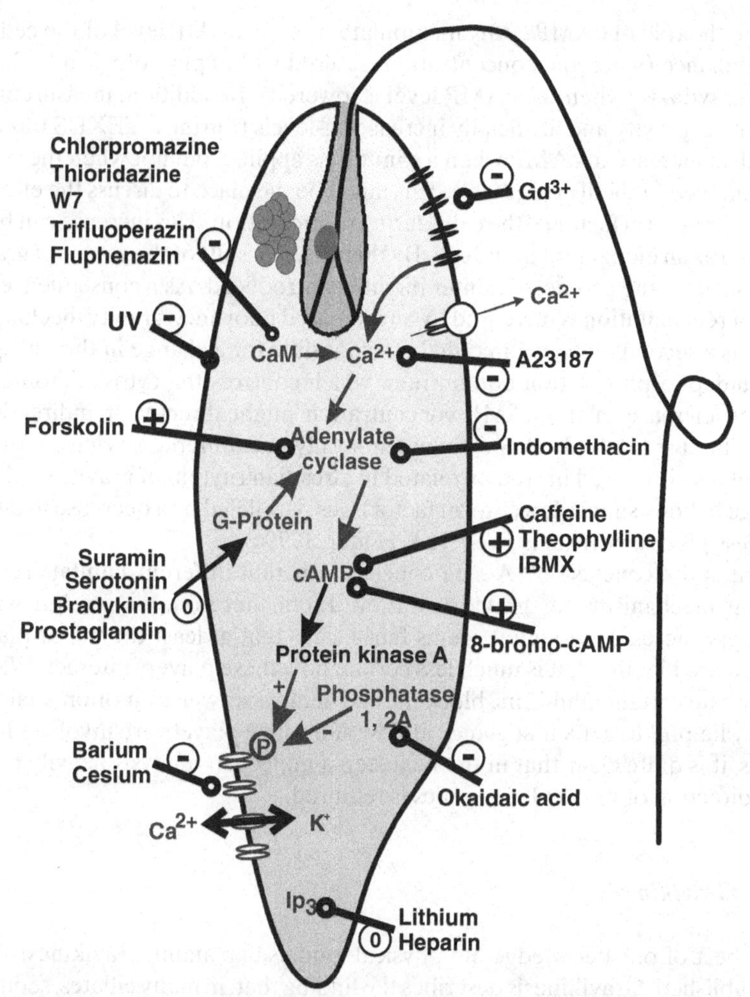

Figure 9.11. Summary of inhibitor, blocker, and analog studies done in *Euglena gracilis*. **Lines with circles** indicate inhibition or augmentation of gravitactic response indicated by – and +, respectively, 0 equals no response.

value is reached. At this point, a massive calcium influx can be detected (Richter et al., 2001b). This is paralleled by a depolarization of the cells (Richter et al., 2001a). Unfortunately, no single cell measurements for these events are available. It seems that after every reaction, the membrane potential is reset, but nothing yet is known in detail. What is known in addition? Figure 9.11 summarizes most inhibitor and blocker studies done up to now. A block of the mechanosensitive ion channels (gadolinium) inhibits all other events. Any impairment of the membrane potential also impairs gravitaxis. So, it seems pretty safe to conclude that the membrane potential is directly involved in gravitactic orientation of *Euglena gracilis* and relatives. The prominent role of calcium is reemphasized by the findings related to calcium blockers, inhibitors, and analogs (Fig. 9.11). What could be the role of cAMP? Any manipulation of the cAMP level of the cell will either enhance (when the concentration is raised within physiological limits) or impair gravitaxis (when the cAMP level is lowered). In addition, measurements under microgravity and artificially increased g-levels (during a TEXUS mission) showed an increase of cAMP when a stimulus is applied, but not when the stimulus is removed (Tahedl et al., 1998). This might be the place to discuss the effect of the change in form factor of the cells during reorientation. This increase can be interpreted as an elongation of the cell. By this means, a shift of the center of gravity with respect to the geometric center might be introduced. As a consequence, the speed of reorientation is increased by an increased buoyancy-gravity mechanism. cAMP as a second messenger could be responsible for a change in the phosphorylation/dephosphorylation equilibrium, which controls the cytoskeleton form. Thus, an increase in the cAMP concentration might directly or indirectly be responsible for the body shape change not only encountered during reorientation, but also observed in studies related to circadian rhythm of gravitaxis. There a change in body shape (lower form factor) was paralleled by a decrease in cAMP and a less precise gravitaxis (Lebert & Häder, 1999c).

What is the conclusion? A safe conclusion is that different habitats require different mechanisms for graviorientation. From there on, we are left with a lot of hypotheses. Although it seems fairly clear that at least the major players involved are identified, it is much less certain how these players interact. What is next? While certain inhibitors, blockers, and analogs as well as motion analysis – are very helpful to get a first glance at how and which players are involved in the process, it is quite clear that in the next step a molecular analysis heavily biased in the direction of molecular genetics is required.

9.4.2 Gravikinesis

To the best of our knowledge, no physical models explaining gravikinesis have been published. Gravikinesis describes the finding that, in many ciliates, sedimentation is partially or fully compensated by a change in the forward velocity (cf. Chapter 1.2). Figure 9.12 summarizes findings in some ciliates. It is obvious that the degree of compensation varies considerably between species. In the few cases of the detailed analysis of this phenomenon, the hypothesis is that an uneven distribution of ion channels might be responsible (Fig. 9.7; Machemer et al., 1991). Depolarizing calcium channels are located in the front end and hyperpolarizing potassium channels are positioned in the rear half of the cell. An upward swimming cell will activate the potassium channels by the mechanical load due to sedimentation and by this means increase the forward swimming velocity. This increase will partially, fully, or even overcompensate the sedimentation (Fig. 9.12). In the reverse direction, cells will be slowed down by the activation of calcium channels, which depolarize the cell (cf. Chapter 4).

Two reports are known for gravikinesis in *Euglena*. One states that no gravikinesis can be detected (Vogel et al., 1993). In a more recent report, a clear gravikinesis was found (Machemer-Röhnisch et al., 1999). However, in this paper, the

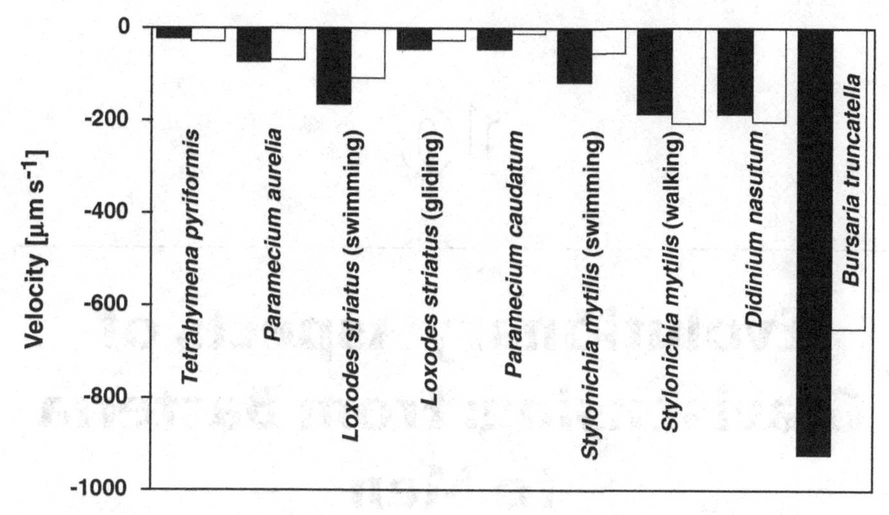

Figure 9.12. Comparison of sedimentation rates (**filled bars**) and gravikinetic values (**open bars**) in seven ciliate species. All species tend to counterbalance sedimentation by an orientation-dependent active change in velocity (gravikinesis). (Data from Hemmersbach & Bräucker, 2002.)

methods for the observation of the cells were seriously flawed. The observation light was photosynthetically active, which in the case of *Euglena* should lead to the production of oxygen during the experiment. As a consequence, oxygen and not gravity might trigger the behavioral response as *Euglena* cells and speed up its swimming velocity in the direction of an oxygen source. In addition, light is known to induce photokinesis, an increase in swimming velocity at a given light intensity, which is known in this organism (Wolken & Shin, 1958).

10

Evolutionary Aspects of Gravisensing: From Bacteria to Men

Gravity has been a pervasive factor throughout evolutionary history, and biological systems are assumed to have used this external clue for orientation rather early to find and stay in optimal living conditions, which offer ecological advantages. Consequently, already early during evolution, unicellular organisms developed organelles for active movement and sensors for diverse environmental stimuli. This chapter summarizes some aspects and ideas concerning evolution and refinement in gravisensing, and focuses on the fundamental question concerning common origins in the underlying signal transduction chains.

10.1 Development of gravisensing during evolution

As discussed in several chapters: To introduce a detectable change in the cell, gravity has to interact with a mass. Consequently, size matters! To be more precise: size and specific density difference matters. Published values for the specific density of prokaryotes are scarce. Guerrero and coworkers (1984) published a value of 1.16 g ml^{-1} for *Allochromatium vinosum*, a purple sulfur bacterium. This value might vary drastically due to living conditions, feeding and metabolic status, etc. The sedimentation rate is only 0.12 $\mu m\ s^{-1}$. In contrast, sedimentation rates in eukaryotes are generally one to two orders of magnitude higher. Size makes the difference. Prokaryotes are typically at least one order of magnitude smaller than eukaryotes. Brownian motion, as well as low Reynolds numbers, prohibit any active steering mechanism in free-swimming species. But, there is a way out: limiting diffusion dimensions. The only true tactic responses that obey the strict definition are observed in gliding prokaryotes (phototaxis in cyanobacteria). Effectively, gliding limits the influence of Brownian motion and allows an effective, active steering toward or away from a light source. Although, to the best of our knowledge no report is published up to now about a gravitactic orientation of a prokaryotic organism, there is in theory no reason to believe that this is not possible.

Again, we discussed in several chapters the potential or proven involvement of mechanosensitive ion channels in gravitaxis-related signal transduction. When we believe in the evolutionary pathway from prokaryotes to simple eukaryotes to multicellular plants, fungi, and animals, there are probably at least one or more ancestors in the prokaryotic world, including the Archaea to possess channels similar to those in eukaryotes. Kloda and Martinac (2002) discuss certain possibilities. It seems that, whereas the mechanosensitive and conductivity properties of the mechanochannels differ significantly between the Archaea and the Eubacteria, they share a common ancestor – at least for mechanosensitive channels of large and small conductivity. They share a similar gating mechanism. In addition, sequence homologies to eukaryotic proteins, thought to be functioning as mechano-gated channels, suggest that these ancestors also bridge the gap between mechano-gated channels in eukaryotes and prokaryotes (Kloda & Martinac, 2002, and references therein). The general function of these channels is related to osmo- and pressure control. Sudden changes in the osmolarity in the medium poses a severe challenge, which under some circumstances, might only be matched by a quick release of osmotic-active substances from the cell. However, the issue is not resolved how these osmo-channels are related to gravity-related conductivity changes in the cytoplasmic membrane. Energies and the related forces required to activate these channels are orders of magnitude higher than anything gravity might induce in membranes by interacting with the cell mass.

What are the differences between unicellular and multicellular eukaryotic organisms? This is not directly related to sessile versus free swimming as one might think. It is in contrast related to size. As size exceeded a certain limit during evolution, sedimenting, heavy particles are favored instead of the cell body mass. Thus, there are exceptions that are most likely related to the ecological niche (e.g., *Loxodes*; cf. Section 4.2). But, the "older" mechanism seems to be still present as the findings with starch-free *Arabidopsis* mutants suggest (Kiss et al., 1998). Also, in freely motile animals, sedimenting heavy particles are responsible for gravity perception. One additional notion is that, at the time when size became an issue, specialization of cell types organized in tissues developed in parallel. Not the whole plant (or every cell) triggers and controls gravitropism, but it is only a handful of cells in the root cap that perform this function. In fact, it is arguable whether size or the related specialization are the driving force. In any case, are there other differences between unicellular and multicellular organisms besides size and specialization? One obvious, additional difference is that, in multicellular organisms, not one but a few or many masses interact with gravity. When one amyloplast interacts with gravity, the signal-to-noise ratio might be bad, but the picture changes when more bodies are involved and the signal outputs of each are integrated – that will make the overall performance very solid (i.e., the signal-to-noise ratio is very good). In contrast, unicellular organisms will need to integrate over some time to distinguish between signal and noise. However, the following steps share many similarities, allowing use of single cells as model systems.

One of the interesting findings in gravitaxis research of *Euglena gracilis* is the interaction between phototaxis and gravitaxis. Although it is not clear yet where exactly both signal transduction chains merge, it seems clear that it is either on the calcium or the cAMP level (i.e., very early in the chain). A similar interaction between light and gravity as stimulus can be found in plants. In an astrobiology workshop (NASA Ames Research Center, 1996), Lewis Feldman followed up the idea of an early merging and extended it to gravitropism of higher plants. He argues that even very early prokaryotes have most likely evolved light avoidance reactions. This was necessary because high-energy radiation was reaching the Earth's surface due to the lack of an ozone shield. After photosynthesis was developed by later prokaryotes, light searching mechanisms were evolved on the basis of the light avoidance systems to supply the organisms with light energy by reaching spots optimal for growth and reproduction. One way or the other, half of the extant prokaryotes are capable of directed movements. After motility was coupled with orientation, it seemed feasible to develop a gravity-dependent system by simply plugging in a "module" that could somehow manipulate the phototactic pathway. This is only necessary when it is dark or illumination is diffuse. At other times, light-guiding as a stimulus essential for photosynthetic organism takes over. What arguments support this hypothesis? In motile, unicellular plants, gravitaxis is highly impaired by ultraviolet (UV) radiation, which might suggest that gravitaxis evolved after high-energy UV was screened by the ozone shield. The argument is questionable because even oxygenic photosynthesis must have developed before the ozone shield was built. Nevertheless, the photosynthetic apparatus as assembled by proteins is sensitive to UV radiation. Another argument is made around the flagellar responses triggered by light in some flagellates (*Chlamydomonas reinhardtii*). Calcium and calcium-mediated membrane events are central to this phenomenon. On the other hand, it is unclear if, and to what extent, calcium-dependent membrane potential changes are involved in the phototaxis of *Euglena*. Recently, the photoreceptor proteins were linked to adenylyl cyclase activity (Iseki et al., 2002). This enzyme can control the cAMP concentration, a second messenger also involved in gravitaxis.

In higher plants, its seems that calcium plays a central role in gravity-related signal transduction (at least in root gravitropism). In many plants, roots show a "complete" gravitropism and grow vertically downward. This happens both in light and in darkness. However, lateral plant roots grow almost parallel to the soil surface in darkness. When the plants are illuminated, these roots start growing downward. Thus, the idea is that these plants use an ancient reaction where light and gravity act on the same pathway. During evolution, both pathways became more and more independent until a phenomenon known as the "complete" gravitropism evolved. Although this hypothesis might be worth considering, a lot of work has to be done to make it more plausible.

10.2 Primary receptor for gravity

It is clear that the first interaction of gravity with a biological system is of a physical nature. This might result in either a pure passive torque and thus passive alignment in the gravity field or it might be coupled to a signal transduction chain, leading to a physiologically guided response. Current hypotheses (cf. Chapter 9) and energetic considerations (cf. Chapter 8) favor physiologically guided mechanisms, which might be supported by physical forces (e.g., buoy mechanism). Thus, different questions arise: At what level of organization does gravity becomes effective? What is the nature of a cellular gravireceptor? Can gravity be perceived only by cells that have developed specialized gravisensors or is graviperception a common capacity of cells? Do chemical and biochemical reactions already depend on gravity? It has been suggested that biological systems might "depend on gravity by way of bifurcation properties of certain types of non-linear chemical reactions that are far-from-equilibrium" (Tabony et al., 2002). In this context, bifurcation is a decision at a specific time point in a continuous process where even subtle changes in chemical or physical parameters will shift the reaction equilibrium one way or another. Consequently, the presence or absence of gravity at this critical step of self-organization might influence the ensuing morphology (Kondepudi & Prigogine, 1981). To test this hypothesis, the in vitro formation of microtubules, a major component of the cytoskeleton, has been studied in the absence of gravity. Under appropriate conditions, in vitro preparations of microtubules spontaneously self-organize themselves by reaction-diffusion processes, and their pattern formation depends on the gravity vector. Experiments in microgravity revealed that gravity triggers self-organization processes of microtubules when bifurcation occurs (Papaseit et al., 2000). Therefore, it has been concluded that gravity might affect fundamental cellular processes, which depend on microtubule self-organization and might provide an explanation for changes in which cytoskeletal elements are involved. The cytoskeleton of cells – whose major components are actin filaments, microtubules, and intermediate filaments – plays an essential role in different cellular functions, such as maintenance of cell shape, cell motility, signal transduction, cell division, and maintenance of a proper spatial position of cell organelles with respect to each other. Even "heavy" cell organelles, such as the nucleus, do not sediment to the lower cell end, but are found in the middle or upper part of a cell – embedded and held by the cytoskeleton. The complex organization and the functioning of the cytoskeletal network depend on various associated proteins (for review, see Lewis, 2002). Disruption of this machinery impacts, for example, cell growth, metabolism, and will finally lead to apoptosis (programmed cell death). Data from experiments with different cell cultures revealed that many, but not all, space-flown cells exhibit cytoskeletal anomalies (for review, see Lewis, 2002). Examples are the findings of changes in cell shape and an increase in F-actin in A431 epidermoid carcinoma cells (Boonstra, 1999), bundling of vimentin and tubulin filaments in T-lymphocytes (Sciola et al., 1999), a decreased actin filament linearity in *Xenopus* embryonic muscle cells (Gruener et al., 1993), a reduction in the number of stress fibers in mouse MC3T3-E1 osteoblasts (Hughes-Fulford & Lewis, 1996; Lewis et al., 1998), coalesced microtubules that do not extend to the cell membrane in a T-lymphocyte leukemic cell line (Lewis et al., 1998), and altered microtubules in human breast cancer cells [Vassy et al., 2001; for further details, see review by Lewis (2002)]. The cytoskeleton is also involved in signal transduction processes by regulation of ion channel activation of key enzymes (such as protein kinase C; PKC) and by regulation of gene expression (for review, see Lewis, 2002). In this context, two findings are of importance: changes of the expression of different genes in microgravity and hypergravity e.g., microgravity decreased epidermal growth factor (EGF)-induced proto-oncogene *c-fos* expression by 20% in A431 epidermoid carcinoma cells (De Groot et al., 1991) and sensitivity of PKC to gravitational changes as revealed by the kinetics of translocation and the cellular quantity of PKC in human leukocytes (Hatton et al., 1999). The previously described examples reveal the cytoskeleton as an appropriate candidate in mediating cellular gravitational effects. It may be involved in signal enhancement in small cellular systems or even serve as gravisensor. The latter function was proposed by Sievers and coworkers (1991), who calculated that actin bundles with a diameter of >25 nm even have enough mass for graviperception.

It is astonishing and still an open question why gravity obviously does not seem to be required for normal onto- and embryogenesis in vertebrates and invertebrates, as revealed by developmental studies with different model systems (amphibians, fish, insects, and sea urchins) performed in microgravity so far (for review, see Marthy, 2002). It has to be proven on the Space Station whether regulation capacities of developing organisms will diminish under continued microgravity conditions.

In the following, we refer to examples where gravity interactions are directly visible, as they are used for orientation.

10.3 Graviorientation in microorganisms

From a historical point of view, much of our current understanding of sensory transduction pathways comes from studies with microorganisms. Examples are the *chemotaxis* in the bacterium *Escherichia coli*, mechanisms of aggregation and morphogenesis in the cellular slime mold *Dictyostelium*, and the swimming behavior of the ciliate *Paramecium*. Due to the information presently available, microorganisms provide clues to the stimulus-response systems in mammalian cells – especially at the level of genetic manipulations, microorganisms have clear advantages. According to Kincaid (1991), "It would not be surprising if the keys to understanding many of the complex events of mammalian physiology are provided by some of our earliest ancestors."

Although prokaryotes are commonly small enough to withstand sedimentation, larger and more complex organisms had to overcome sedimentation. Thus, graviorientation was acquired early in phylogeny for survival reasons (e.g., optimal food and oxygen supplies); thus, it is already found in ancestral unicellular eukaryotes. Recent studies favor the hypothesis that, in free-living cellular systems, gravity is perceived either by intracellular receptors (statocyst-like organelles) and/or by sensing the cell mass by means of gravisensitive channels located in the cell membrane (cf. Chapter 9). Protists show that these different graviperception mechanisms have been evolved in parallel. In the case of *Loxodes*, an intracellular gravisensing mechanism is supported by a subsidiary membrane-located one, while – as far as we know – *Paramecium* and *Euglena* only have a membrane-located gravisensing mechanism that might be supported by a physical one. Ciliates are comparatively large unicellular organisms (cell lengths are about 300 μm), which arose on Earth more than 1.5 billion years ago. There are numerous similarities between ciliated protozoa and vertebrate and invertebrate neurons. The action potentials of ciliates are similar to those of nerve and muscle cells; therefore, ciliates have been apostrophized as "swimming neurones," and ion channels must have evolved early. Furthermore, they clearly show that the general principle of mechanoreceptors in metazoa is already represented in unicellular organisms. Perception of gravity is obviously linked to the known principles of mechanosensation. No case of absence of mechanosensitivity has been stated. The bipolar distribution of two types of mechanosensitive ion channels seem to be ideal and a monopolar arrangement (e.g., found in *Didinium*; cf. Chapter 4) might be an archetype of mechanosensitivity. Interestingly, no single existence of hyperpolarisation mechanosensitivity has been stated, but solely depolarization mechanosensitivity. An evolutional question is whether the distribution and development of mechano-(gravi)receptors in single cells depend on the presence of gravity – an answer can be found by multigeneration experiments under the condition of long-term weightlessness (e.g., on the International Space Station).

Although graviresponses of plants (gravitropism) and protists (gravitaxis and gravikinesis) differ in their final steps, there is growing evidence that they share common mechanisms in the transformation of the physical stimulus gravity into a biological signal (i.e., in the steps of perception and transduction). Examples for gravitropism in lower and higher plants are given in the following section.

10.4 Gravitropism in lower and higher plants

The orientation of organs of sessile organisms in the gravitational field is defined as gravitropism. As detailed in Chapter 1 (cf. Section 1.2), main shoots usually bend away from the center of gravity (negative gravitropism), primary roots orient toward the center of the Earth, and lateral branches and roots – as well as leaves – orient perpendicular or at another angle to the gravitational field. In the following, the orientation of several model systems is summarized, with emphasis on the evolutionary aspects and common features with gravitaxis in microorganisms.

10.4.1 Gravitropism in fungi

The mold *Phycomyces blakesleeanus* has been studied mainly to elucidate the mechanism of phototropism. In addition, the unicellular sporangiophores also show a pronounced negative orthogravitropic (upward growth away from the center of the Earth) orientation. When placed horizontally, the 100-μm-diameter aerial hyphae start bending upward after about 20 min and reach a vertical position within about 10–12 h (Ootaki et al., 1995). Octahedral protein crystals and an aggregate of lipid globules in the apical cytoplasm seem to be involved in graviperception. The paracrystalline protein crystals are 3–5 μm in diameter (Wolken, 1969) and are found both inside and outside the growing zone (uppermost 2 mm of the sporangiophores below the sporangium), but most are found in an apical aggregation of small vacuoles (20–40 μm in diameter) in the top 1 mm of the sporangiophore. Because the crystals can form large clusters of up to forty inside vacuoles, they have been suggested to play a role as statoliths (Horie et al., 1998; Schimek et al., 1999). In contrast, the spherical aggregate of lipid globules seems to be less dense than the cytoplasm and floats upward to settle at the apex of stage 1 (without sporangium) sporangiophores (Thornton, 1968). Both structures (octahedral crystals and lipid globules) redistribute upon reorientation of the sporangiophore (Schimek et al., 1999). After a 90° turn, the crystals sediment at a speed of 0.5–2 $\mu m\ s^{-1}$ traversing the horizontal sporangiophores in 50–200 s. Another indication of their role in gravitaxis is that crystal-devoid mutants bend only 40°–50° upward. Also, the lipid globules, which are normally located about 50 μm below the apex of stage 1 sporangiophores, floated upward to a new position after placing the organ horizontally. These results were interpreted to confirm the roles of both octahedral crystals and lipid globules in graviperception (Schimek et al., 1999). Measuring phototropism on Earth always involves an interaction with gravitropism. When a vertical *Phycomyces* sporangiophore is irradiated with a lateral beam of light, it will bend toward the light source (positive phototropism). As soon as the tip bends out of the vertical, it is stimulated by the gravitational field of the Earth, and the gravitropic response counteracts the phototropic bending (Gressel & Horwitz, 1981; Grolig et al., 2000). This phenomenon has been defined as photogravitropism. The photoreceptor is a blue light-absorbing pigment, and action spectra for photogravitropism have been determined (Campuzano et al., 1996; Ensminger et al., 1990). Whereas the gravitropism obeys a sine law, photogravitropism follows an exponential law (Galland et al., 2002). Recently, a new method has been developed to measure gravity-induced absorption changes in *Phycomyces* sporangiophores to determine the primary events of graviperception (Schmidt & Galland, 2000), but no definite mechanism has been proposed yet.

Even though gravitropism mutants have been isolated (genotype *mad*, named after Max Delbrück), the investigation of this phenomenon has not been as intensive as that on phototropism. Because these mutations are also retarded in their phototropism and show a subdued light-growth response (Bergman, 1972), the defect in these mutants seems to be located in the early transduction chain of phototropism and gravitropism rather than in the graviperception mechanism itself. So far, no specific gravitropism mutants have been identified.

Also, basidiomycete fungi (mushrooms) show a pronounced gravitropism that has been noted more than 100 years ago (Czapek, 1895; Pfeffer, 1881). The stalk of the fruiting body orients negative orthogravitropically and the cap perpendicularly. In this case, the ecological significance is not an improvement of the photosynthetic activity, which fungi do not have, but it facilitates release of the spores from the lamellae underneath the cap. A deviation by as little as 5° from the vertical, 50% of the spores are not liberated. After placing the fruiting body of *Flammulina velutipes* into a horizontal position, reorientation is finished after 12 h. Under microgravity conditions (during the D2 mission on the Space Shuttle Columbia), the fruiting bodies first grew radially (away from the substratum) and then in random directions, whereas they showed a clear-cut negative gravitropism on the $1 \times g$ reference centrifuge (Kern & Hock, 1993).

The curvature is limited to the top 3 mm under the cap (Kern et al., 1997). The fruiting body is composed of parallel hyphae. If the stalk is separated into longitudinal segments, each shows gravitropic bending – indicating that each hypha is capable of gravity perception. There are no obvious statoliths in the hyphae; the only heavy elements are the nuclei that are suspended in a network of actin filaments. The current model assumes that the nuclei exert a force on the actin cables and that there is an information exchange between adjacent hyphae to warrant a concerted growth response of the fruiting body in which the lower flank grows faster and the upper one is retarded. Whether or not a chemical transmitter is involved is currently under investigation.

The mechanics of differential growth relies on the formation of microvesicles in the cytoplasm, which subsequently merge with the tonoplast. There is a pronounced gradient in the density of microvesicles with almost none in the top hyphae (6% of the area in electron microscopic images) to many in the bottom hyphae (37%).

10.4.2 Gravitropism in Chara *rhizoids*

The multicellular green alga *Chara* produces cylindrical unicellular rhizoids that originate from oospores or nodal cells and that can reach the length of several centimeters at a diameter of about 30 μm. At room temperature, the rhizoids grow strictly at their tips at about 180 μm/h in a positive orthogravitropic direction (Sievers & Volkmann, 1979). The basal portion of the rhizoid contains a large vacuole and holds the nucleus. Gravitropic perception and response are restricted to the ellipsoid-shaped apex, which contain 50–60 vesicles filled with electron dense crystals of $BaSO_4$ (Fig. 10.1). These statoliths do not sediment all the way into the tip of the rhizoid, but are suspended 10–30 μm above the apex and held in place by an array of thin actin filaments that interact with the myosin-coated statoliths (Braun, 2001). When statoliths are forced closer to the apex by centrifugation, complete sedimentation is prevented by the actomyosin system. In microgravity (Space Shuttle flight STS 65 and parabolic flights on TEXUS and MAXUS rockets), they retreated in a basipetal direction (Braun, 2002).

When the rhizoids are placed horizontally, the $BaSO_4$ vacuoles sediment within a few minutes onto the lower membrane. An older hypothesis assumed that the settled statoliths mechanically prevent the insertion of Golgi vesicles, carrying cell wall material, into the lower cytoplasmic membrane, so that the upper cell wall grows faster and thus the rhizoid bends downward (Sievers & Volkmann, 1979). In vertical rhizoids, the statoliths can also be forced into a lateral position by an optical tweezer using a noninvasive infrared laser micromanipulation technique (Braun, 2002). A change in the growth direction occurred only when at least two

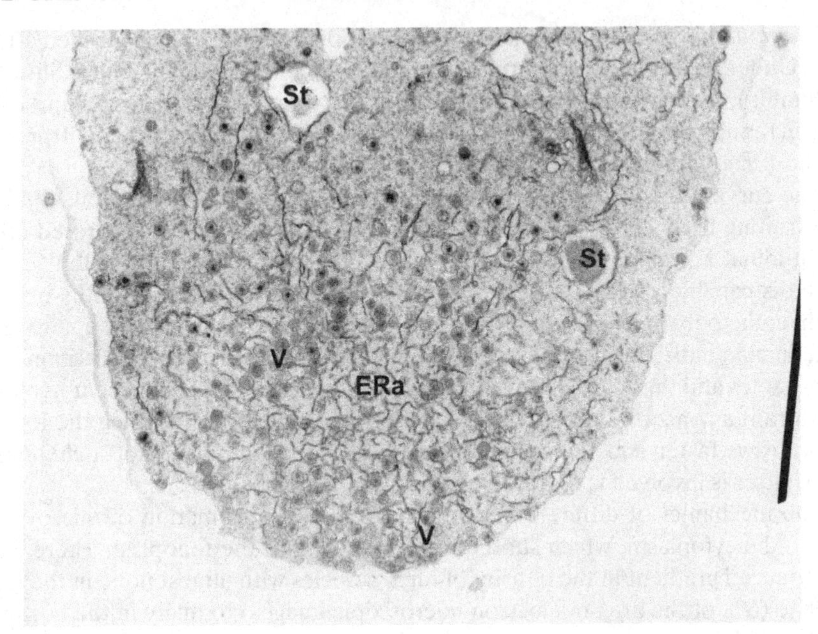

Figure 10.1. Electron microscopic view of the tip of a *Chara* rhizoid with an array of electron dense $BaSO_4$ crystals. St, statolith; V, secretory vesicle; ERa, ER aggregate. The cell wall is barely visible after 3% glutaraldehyde fixation. (Courtesy of M. Braun, Bonn, Germany.)

or three statoliths settled on a specific area of the plasma membrane, and the pure physical process of sedimentation was not sufficient to initiate asymmetric cell growth. Therefore, the author concluded that specific plasma membrane areas are sensitive to statolith interaction; they may be the primary receptors for graviperception and play a pivotal role in initiating the signal transduction pathway of gravitropism in *Chara* rhizoids. Recently, in downward-growing rhizoids, a steep calcium gradient was documented, using the calcium indicator Calcium Crimson, as well as a symmetrical localization of dihydropyridine (DHP) receptors, putative calcium channels, by using fluorescent labeling (Braun & Richter, 1999). After horizontal positioning, the rhizoids showed continuous bowing downward due to differential flank growth, but both the calcium gradient and the DHP receptors remained symmetrically localized in the tip at the center of growth.

In contrast, in the likewise unicellular *Chara* protonemata, which also arise from nodal cells – but show negative gravitropism – a pronounced displacement of the calcium gradient and the fluorescently labeled DHP receptors to the upper flank of the apical dome were observed. The authors conclude from these results that the positive gravitropic curvature in rhizoids is caused by differential growth limited to the opposite subapical flanks of the tip and does not require displacement of the growth center, the calcium gradient, or calcium channels. In contrast, in protonemata, a statolith-induced asymmetric relocalization of the calcium gradient and calcium channels precedes – and might mediate – the rearrangement of the center of growth that may be due to the displacement of the "Spitzenkörper" (located between the statoliths and the apical tip) to the upper flank. The recent results indicate that the sedimentation of the statoliths is not sufficient for graviperception, but that the cytoskeleton plays a pivotal role in the process.

10.4.3 Gravitropism in higher plants

We do not intend to provide a comprehensive review of gravitropism in higher plants in this section (for recent reviews in this field, we refer to, for example, Ranjeva et al., 1999; Sievers, 1999; Kiss, 2000; Perbal et al., 2002; Chen et al., 1999; Wagner, 1998; Hemmersbach et al., 1999b; Takahashi, 1997). Rather, we will summarize the recent knowledge on the early events in graviperception and signal transduction to compare these to those identified in motile microorganisms.

While in *Chara* graviperception and response are located within the same cell, in higher plants perception and bending are separated by many cells. In shoots of higher plants, graviperception is assumed to be the site of graviperception and in roots the columella cells in the root cap (calyptra) (Weise et al., 2000; Kiss, 2000). Surgical removal of the root cap (Juniper et al., 1966) or laser ablation of specific cells within the columella affect the gravitropic responsiveness (Blancaflor et al., 1998). During regeneration of the root cap gravitropism is restored.

The graviperception mechanism seems to involve the action of statoliths in all higher plants analyzed so far. In the roots of *Lepidium, Arabidopsis, Zea*, and others, amyloplasts function as heavy particles, and starch-deficient mutants show a much impaired graviresponse (Kiss et al., 1996, 1997). Indeed, electron microscopy, and in some cases even light microscopy, reveal that the amyloplasts sediment to the lower cell wall and rest on a cushion of endoplasmatic reticulum vesicles (Fig. 10.2a). In microgravity the statoliths are distributed throughout the cell (Fig. 10.2b). Upon reorientation of the plant organ (root or shoot), the amyloplasts sediment to the respective lower cell wall (Volkmann & Sievers, 1979). These findings and several other pieces of evidence led to the **statolith sedimentation hypothesis**. The degree of graviresponsiveness is proportional to the total mass of the amyloplasts per cell (Kiss et al., 1997).

Regarding the mass of cell organelles and their density difference to cytoplasm, it was calculated that the function of mitochondria as gravisensor seems unlikely (Audus, 1962; Björkman, 1988), whereas amyloplasts in the statocytes of higher plants (Volkmann & Sievers, 1979), $BaSO_4$ vacuoles in the *Chara* rhizoid (Sievers & Volkmann, 1979), Müller organelles in representatives of Loxodidae (Fenchel & Finlay, 1984), and also the nucleus [e.g., postulated for *Flammulina velutipes* (Monzer, 1995)] are suitable candidates.

The energy of a sedimenting statolith exceeds the thermal noise within a cell, assumed to be on the order of 4×10^{-21} J, by a factor of more than 10 (Hasenstein, 1999). But, Wayne and Staves (1996) calculated the kinetic energy of a sedimenting statolith to be seven orders of magnitude lower – based on different assumptions, – and therefore concluded that this energy is too low to be recognized by the cell as a signal above the thermal noise.

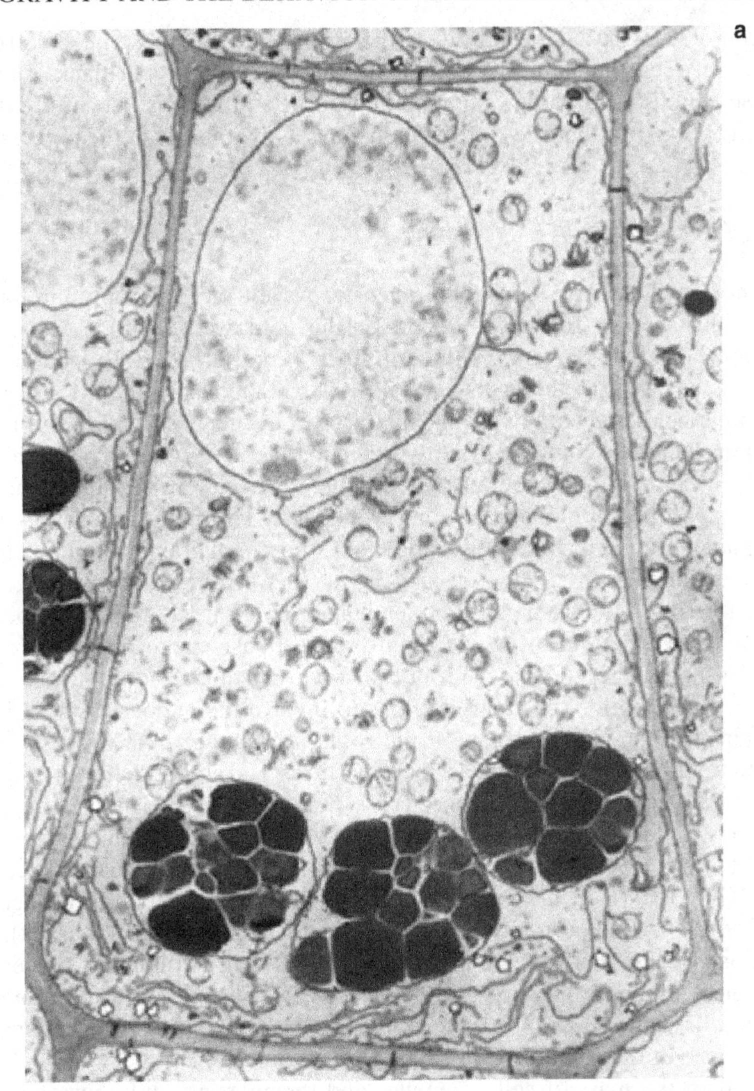

Figure 10.2. Electron microscopic views of a longitudinal section through the root cap of *Lepidium* showing the statocytes with the amyloplasts sedimented onto a cushion of endoplasmatic reticulum at $1 \times g$ **(a)** and amyloplasts scattered throughout the cell in microgravity **(b)**. N, nucleus; ER, endoplasmatic reticulum; bar, 1 μm. (Courtesy of D. Volkmann, Bonn, Germany.)

However, even mutants of higher plants (such as corn, lacking amyloplasts altogether) show a weak or retarded, but obvious, gravitropism. This has led to a second hypothesis to explain graviperception: not the sedimentation of statoliths, but the entire protoplast within the cell might be responsible for the response (Sack, 1997) [**protoplast pressure hypothesis**; (Pickard & Thimann, 1966)]. The heavy plastids may either increase the specific density of the cytoplasmic cell contents or alternatively exert a force on the cytoskeleton elements they are

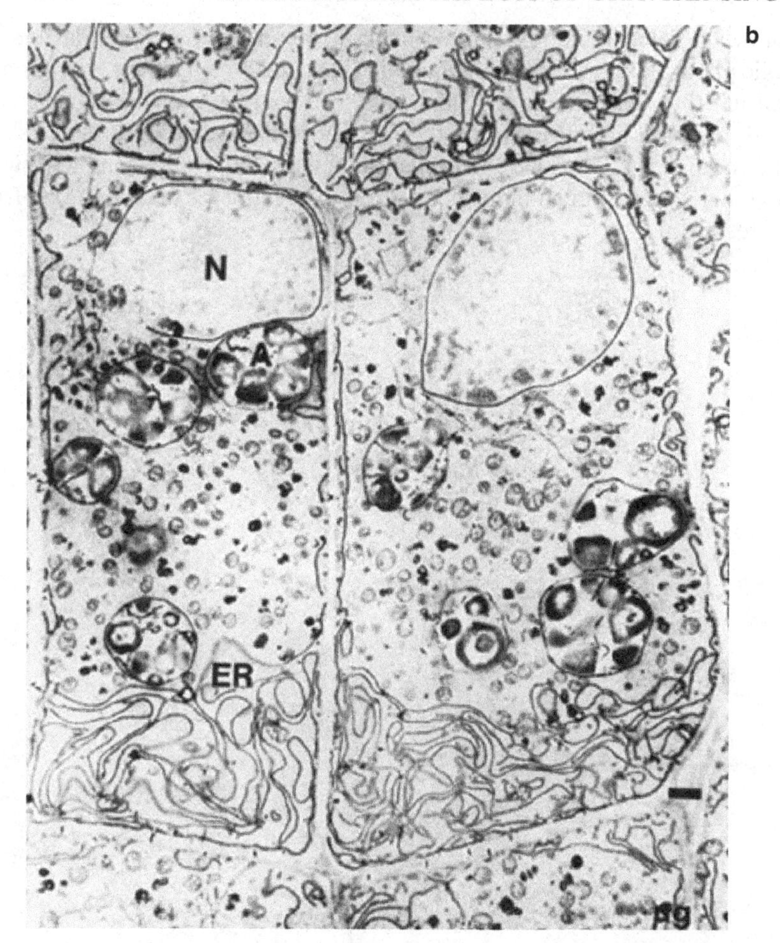

Figure 10.2. (Continue)

attached to. It has been speculated that the latter could, in turn, transduce the force to the membrane to which they are anchored [e.g., via integrins (Lynch et al., 1998)]. Changes in the density of the outer medium alter the gravity-dependent cytoplasmic streaming in internodal cells of *Chara*, which – in contrast to the rhizoids and protonema cells – do not contain statoliths. The density of the external medium also seems to affect gravitropism in rice, but not in the tip-growing cells of the moss *Ceratodon purpureus* (Sack, 2002). Although the energy of the entire sedimenting protoplast is higher than that of the individual amyloplasts, Björkman (1988) doubts that this is energetically sufficient to stimulate the cell.

Although it is clear that the discussion on the possible mechanism of graviperception is far from settled, there is growing evidence that multiple mechanisms exist in different organisms (Barlow, 1995). There may be even parallel mechanisms in the same organism that would warrant a response to the stimulus if one of the mechanisms were disabled by a mutation.

Following gravistimulation, rapid changes in the membrane potential and the cytosolic pH are observed in the columella cells of the root cap in *Arabidopsis*, but not in other cells (Scott & Allen, 1999). Transmission of the signal to the site of differential growth involves the phytohormone auxin both in the shoot and the root (Firn et al., 2000). However, understanding of the complete signal transduction chain from stimulus perception to gravitropic bending is far from complete.

11

Perspectives

What is the future of gravitational biology? In this chapter, we consider some fundamental questions that may be answered in the near future due to the development of new techniques and hardware with respect to long-term studies on the Space Station. One outcome will be the technology for future regenerative life support systems, which is the basis for long-term manned space missions (e.g., a lunar base or a mission to Mars).

Space biology has evolved dramatically during the last few decades and claims a competitive position in the field of life sciences. Because different methods have been developed to enhance or reduce the influence of gravity [e.g., by means of clinostats, centrifuges, and space experiments (cf. Chapter 2)], a large amount of experimental data obtained by studies with different biological systems is now available. After a period of phenomenological studies, recent cell biological studies concentrate on signal transduction processes, especially the primary interaction of gravity with a sensory (perception) system in single cells. Because these responses occur rather quickly within seconds or minutes, they are ideally suited for short-term experiments in microgravity. However, knowledge is lacking; how will long-term microgravity under controlled experimental conditions affect biological systems. Interesting and fundamental questions in gravitational cell biology are, e.g.,

- Does prolonged microgravity affect the behavior, motility, and vitality of cells – free-living ones and those building up tissues and organs?
- How do gravitational changes alter signal transduction mechanisms?
- Does prolonged microgravity alter threshold values and thus the responsiveness of biological systems?
- Does microgravity affect proliferation, aging, differentiation, and apoptosis?

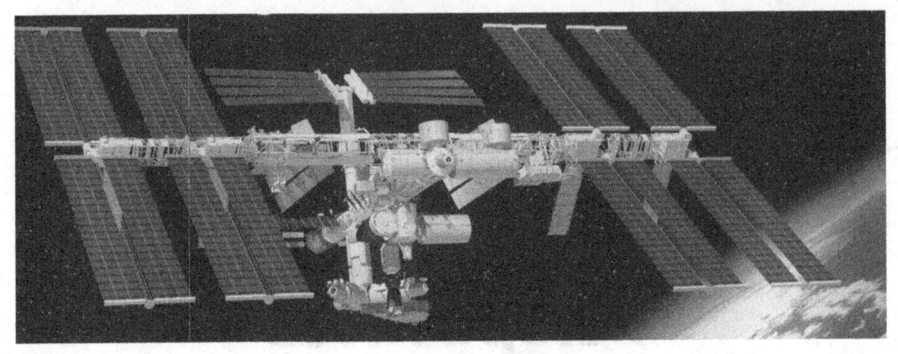

Figure 11.1. The ISS: an international endeavor. (Courtesy of D. Ducrois, European Space Agency, Paris, France.)

- Does prolonged microgravity affect the structure and function of gravisensors and the cytoskeleton that counterbalance sedimentation at $>1\times$ g conditions?
- What happens if animal and plant systems are exposed to microgravity over many generations?

More regular and routine access to space enables the setup of comprehensive research programs by using the International Space Station (ISS; Fig. 11.1), sounding rockets, shuttles, and unmanned carriers – depending on the scientific question (cf. Chapter 2). Dedicated hardware – which takes the specific experimental conditions in space into account – and the possibility to repeat experiments are prerequisites for suitable and significant results. To study gravitational effects on small organisms – ranging from bacteria, yeasts, cell cultures, frog and sea urchin larvae to fruit-flies, nematodes, and plant seedlings – the multiuser laboratory facility Biorack was developed by the European Space Agency (Brillouet & Brinckmann, 1999). Biorack was flown on six missions, thus being in space for 1,402 hours and by this enabling the performance of eighty-two experiments. As the philosophy of Biorack was successfully proven during its lifetime from 1985 to 1997, the successor Biolab has been constructed for ISS and will be integrated into the Columbus Laboratory (Fig. 11.2) (http://www.spaceflight.esa.int/users/biolab). With respect to experimentation on ISS, problems of long-term storage and sample preparation in microgravity have to be taken into account. Standard experiment containers and a controlled environment, by means of a life support system with selectable air composition and humidity, allow a wide variety of experiments. The biological samples will be transported into space either already within the experiment containers or in vials, if they require freezing (–80°C). The experiment containers include the experiment-specific hardware and provide the interface to, for example, the life support system, power, data, and video. Once in orbit, those experiment containers with samples can be directly inserted into the incubator. Further preparation of the samples can be performed in the Bioglovebox, providing a clean, controlled, and closed environment for manual operations of the samples. By means of an Experiment Preparation Unit, it will be possible to thaw samples under controlled gravitational conditions ($1 \times g$ to

Figure 11.2. Biolab – a multiuser laboratory facility for ISS – for studies on small biological systems. (Courtesy of D. Ducrois, European Space Agency, Paris, France.)

$100 \times g$; including removal of cryprotectant and washing procedures) to avoid microgravity influences before the start of the experiment. After manual insertion of the experiment containers into the two Biolab centrifuge rotors, one of them usually static, the automatic experiment run can start. In addition, control experiments are performed on ground in a "Science Reference Model," a model providing operational possibilities as in the flight version of Biolab. Biolab on ISS offers the possibility of a simultaneous processing of samples under identical environmental conditions in microgravity and on a $1 \times g$ reference centrifuge. The centrifuges can be also used for threshold studies, because they offer acceleration profiles of forty selectable steps between $0.001 \times g$ up to $2 \times g$. At the end of the experiment, a "Handling Mechanism" will automatically transport the samples to the analysis instruments (microscopes, spectrophotometer) within the Biolab and/or to the automatic stowage in temperature-controlled or ambient units. The handling mechanism manipulates fluids and can activate mechanisms within the experiment containers by pulling, pushing, or rotating. By means of telecommands, the scientist participates in the experiment control and can change the experimental protocol, if required. In addition to the previously described example of Biolab, a variety of experimental payloads for dedicated gravitational biological studies (e.g., the European Modular Cultivation System with the focus on multigeneration experiments in plants) will be installed on ISS (see the homepages of the national agencies). Besides the opportunity to use ISS and the unique experimental conditions for basic research, there is also strong interest with respect to commercialization and spin-on activities. One of these approaches is biotechnology (Grigoriev et al., 2002; Binot, 2002). In this context, space bioreactors (Walter, 2002; Freed & Vunajak-Novakovic, 2002) and artificial ecosystems are needed for the cultivation of cells, unicellular organisms, and higher systems. One activity in biotechnology is the development of three-dimensional mammalian cell cultures (three-dimensional bioreactor) to understand the fundamental principles guiding tissue differentiation and cell–cell interactions (Binot, 2002). Different hardware has been developed for studies on the swimming and mating behavior and the development of small aquatic systems. Examples are the STATEX centrifuge (derived from STATolithen-EXperiment), a pressurized container containing a $1 \times g$ reference centrifuge and place for microgravity samples – originally used for studies on the development of the frog (tadpoles) of *Xenopus laevis* (Neubert et al., 1991) and also later on cichlid fish larvae or small aquaria (the Japanese Aquatic Animal Experimental Unit) for the freshwater killifish "Medaka" (*Oryzias latipes*) (Ijiri, 1995). Because fish (as some other organisms) loop in microgravity, a particular nonlooping strain of Medaka was taken for the space experiment to enable mating. These devices had life support systems restricted to respiratory gas exchange and thus being suited for short-term missions. According to the theme of our book, we will refer to examples with respect to the long-term cultivation of unicellular organisms. In the following, examples are given based on the demands of the scientific community for multigeneration experiments. The minimized ecosystem MICROPOND has been developed for long-term cultivation and observation of ciliates. The inhabitants are the heterotrophic ciliate *Stylonychia mytilus* and the mixotrophic alga *Chlorogonium elongatum.* The ecosystem is automatically computer-controlled. The longest lasting experiment was stable for 48 d, with a pH value of about 6.8, an oxygen concentration between 70% and 100% air saturation, and temperature of about 22°C (Bräucker et al., 1999). Another minimodule, the CEBAS (Closed-Equilibrated Biological Aquatic System) fulfills the requirements of a freshwater habitat that allows culturing of different aquatic species in a self-sustaining artificial ecosystem (Blüm et al., 1994; Slenzka, 2002). Scientific aim of this ecosystem is to study the effect of microgravity on physiology, morphology, biochemistry, and the interaction of different species – such as higher plants (water plant: hornweed, *Ceratophyllum demersum*), the pulmonate water snail (*Biomphalaria glabrata*), and the ovoviviparous teleost fish (swordtails: *Xiphophorus helleri*; cichlids: *Oreochromis mossambicus*). Further aim is to test how far such a system can provide astronauts with oxygen, water, and food, and can deposit

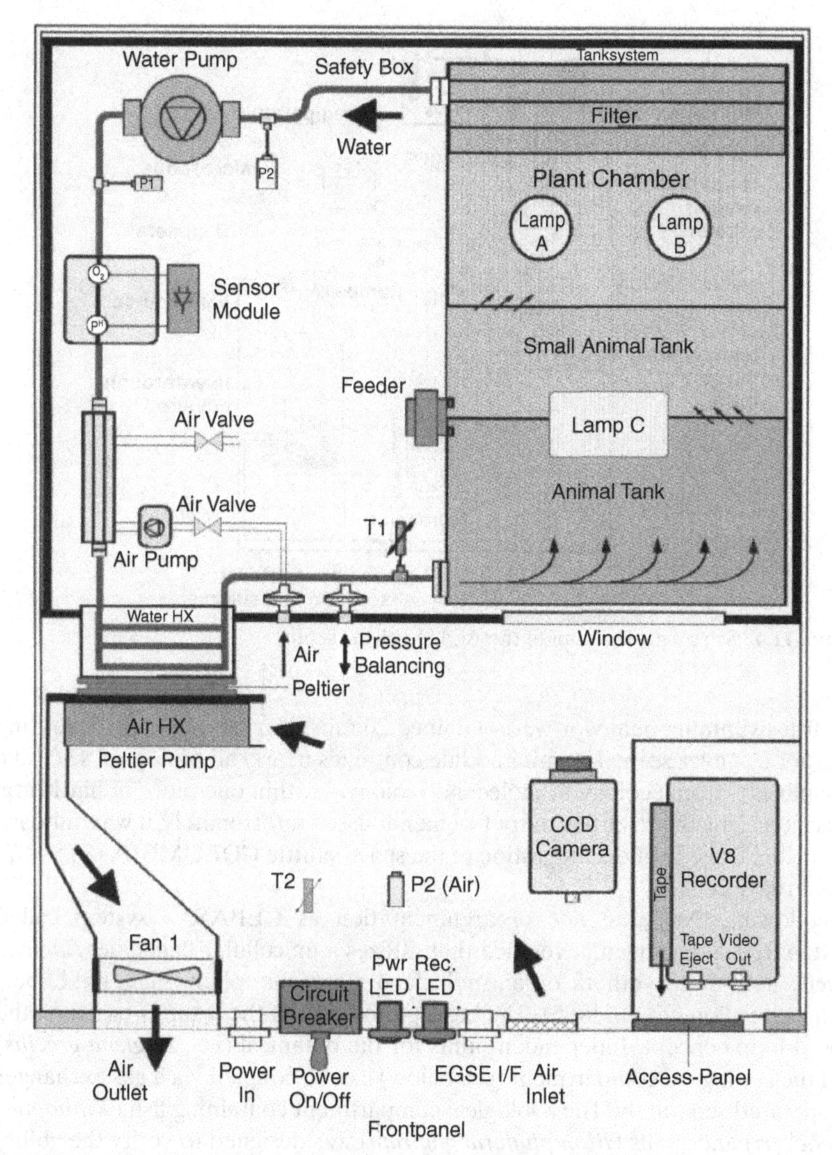

Figure 11.3. The CEBAS Minimodule – functional diagram demonstrating the major components. (Courtesy of K. Slenzka, OHB-System GmbH, Bremen, Germany.)

waste. It consists of a modular 8.6-liter aquarium system, with up to four separate compartments (experiment module) and a support subsystem that controls water flow, thermal situation, and illumination cycles, and monitors housekeeping data like pH, oxygen, and temperature (Fig. 11.3) (Slenzka, 2002). A video system enables the scientist to observe two compartments in the actual setup. The animals are automatically fed with dry food. The CEBAS minimodule is the only German payload flown successfully 3 times in space. For the first time, video sequences of

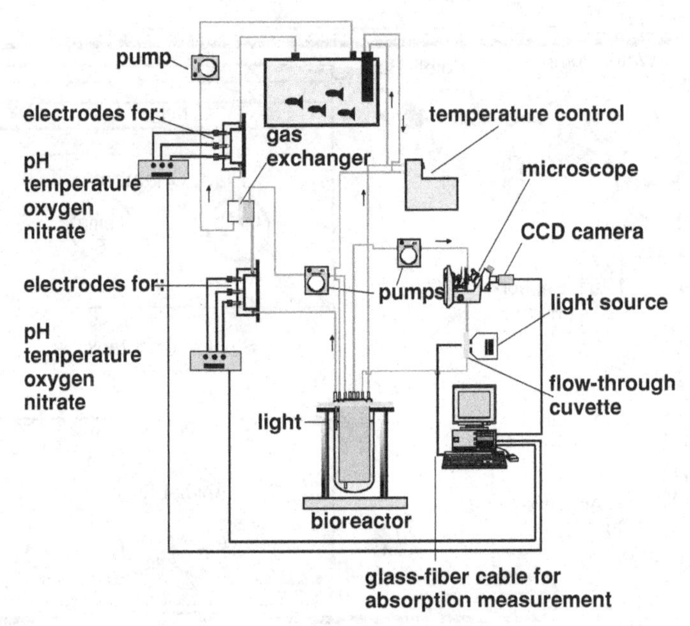

Figure 11.4. Schematic drawing of the AQUARACK setup.

the fish swimming behavior were obtained 20 min after landing, demonstrating alterations never seen. The minimodule combines nearly all biological scientific disciplines – from ecology to molecular biology – within one piece of hardware, thus increasing the scientific output tremendously. Unfortunately, it was onboard during the tragedy of disintegration of the space shuttle COLUMBIA (STS-107) in February 2003.

Following the same line of argumentation as CEBAS, a system called AQUARACK has been developed that utilizes a unicellular flagellate, *Euglena gracilis* as a photosynthetic organism instead of a higher plant (Fig. 11.4; Lebert et al., 1995; Porst et al., 1996). The basic philosophy of the setup is based on the closed-loop concept. Independent units for the botanical (i.e., *Euglena gracilis*) and the zoological compartment (see below) can be coupled via a gas exchanger or operated separately. The zoological compartment containing fish (*Xiphophorus helleri*) and snails (*Biomphalaria glabrata*) was designed to verify the ability of the algae to utilize carbon dioxide and nitrate produced in the zoological compartment and, in turn, supply the animals with oxygen. To conduct experiments under microgravity on a space station, conditions for long-term cultivation were established. Besides a stand-alone culture of *Euglena gracilis*, which has been stable for more than 700 d, tests confirmed that the coupled units are operational for more than 3 weeks, which is longer than the typical CEBAS experimental period. In addition to monitoring housekeeping data (oxygen, temperature, pH, etc.), cellular motility parameters – like cell number, motility, velocity, cell form, and precision of orientation – can be automatically determined in the botanical compartment.

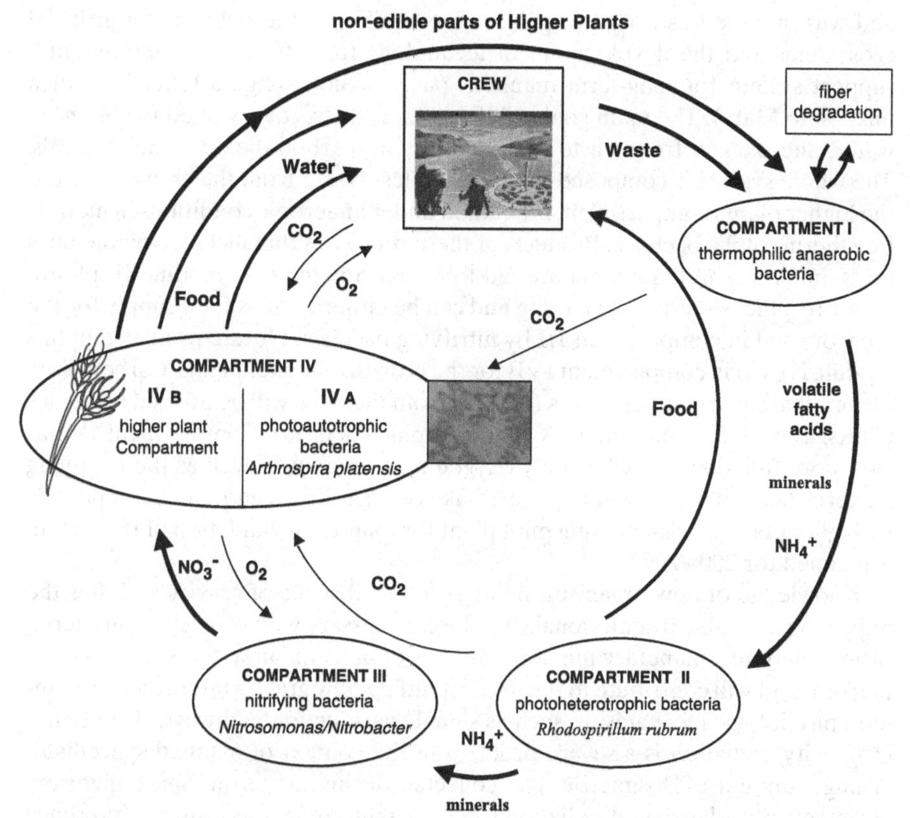

Figure 11.5. Schematic overview of the compartment interaction in MELiSSA. (Courtesy of Christophe Lasseur, ESTEC, AG Noordwijk, The Netherlands.)

A miniaturized AQUARACK system without a zoological compartment (AQUACELLS) was scheduled for the FOTON-M1 mission (a recoverable and unmanned capsule) in autumn 2002. In this system, behavioral responses of the cells are recorded on a military-rated videocassette recorder (VCR). Housekeeping data, such as temperature, are monitored and data storage, as well as the pump for the cell suspension and the VCR, are controlled by an embedded microprocessor. Data obtained with that system during a 14-d space travel were supposed to be compared with a ground control reference system. Major scientific goals, besides the technical challenge of an autonomous, microprocessor-controlled experiment, were questions related to culture development, aging, and the long-term effects of microgravity on the unicellular organism. Unfortunately, the rocket exploded seconds after lift-off.

All three systems (CEBAS, AQUARACK, and AQUACELLS) were designed primarily for scientific reasons. In the following, a setup primarily developed as a life-support system will be discussed. MELiSSA (*M*icro-*E*cological *Li*fe *S*upport *S*ystem *A*lternative; Fig. 11.5; http://www.estec.esa.nl/ecls/melissa/melissa.html) has been conceived as an ecosystem based on microorganisms and higher plants, and was intended as a tool to gain understanding of the behavior of artificial ecosystems and the development of technology for a future regenerative life-support system for long-term manned space missions (e.g., a lunar base or a mission to Mars). The main goal of MELiSSA is the recovery of edible biomass, water, and oxygen from waste (feces and urea), carbon dioxide, and minerals. The whole system is composed of five modules. Waste from the crew, as well as the higher plants compartment, is liquified under anaerobic conditions in module I by thermophilic bacteria. Products of these processes that include volatile fatty acids, minerals, and ammonia are fed to compartment II. In module II, photo-heterotrophic bacteria are growing and can be either used as food supply for the crew or used in compartment III by nitrifying bacteria. Nitrate produced in this module is used in compartment IVB for the growth of higher plants. Carbon dioxide originating in compartments I, II, and from the crew will be utilized by higher plants, as well as in module IVA (phototrophic bacteria). Compartment IV has two more functions: It will supply oxygen for the crew, as well as the nitrifying bacteria, and will clean water by the uptake of organic and anorganic compounds by higher plants. A functioning pilot plant for long-term validation of the system is planned for 2006/2007.

Knowledge of how organisms manage in the absence of gravity and thus the impact on a reliable orientational stimulus is necessary with respect to long-term, orbital, and interplanetary missions. Studies in unicellular systems are easier to perform and will contribute to the understanding how gravity interacts with common physiologic mechanisms, such as signal transduction pathways. Besides microgravity, radiation is a severe problem in the context of manned spaceflight. A huge amount of dosimetric data collected during numerous space flight experiments reveal elevated radiation levels and different compositions (presence of high-energy heavy ions), compared with the Earth's surface. For data and hypotheses of the impact of these parameters on biological systems, as well as a proposed synergistic action of microgravity and radiation, see, for example, Horneck et al. (2002a). Detection of meteorites, which originated from Mars, raises questions concerning life on Mars, the interplanetary transfer of life (Panspermia hypothesis; Horneck et al., 2002b), and, finally, the limits of life.

Interdisciplinary and interactive research between scientists all over the world will help to understand how long-term exposure to microgravity, cosmic radiation, and psychological stress – which are characteristics of spaceflight – influence and can be tolerated by biological systems (ranging from molecules to man), thus providing the basics for further exploration of the universe, such as travel to other planets.

References

Abel, J. H., Haack, D. W. & Price, R. W. (1971). Effects of weightlessness on the nutrition and growth of *Pelomyxa carolinesis*, in *The Experiments of Biosatellite II NASA SP-204*. Ed., J. F. Saunder. NASA, Washington, pp. 291–308.

Able, K. P. & Able, M. A. (1990). Calibration of the magnetic compass of a migratory bird by celestial rotation. *Nature*, **347**, 378–9.

Able, K. P. & Able, M. A. (1993). Daytime calibration of magnetic orientation in a migratory bird requires a view of skylight polarization. *Nature*, **364**, 523–5.

Aderhold, R. (1888). Beiträge zur Kenntnis richtender Kräfte bei der Bewegung niederer Organismen. *Jenaische Zeitschr. f. Med. u. Naturwisschft.*, **22**, 311–42.

Adler, J. (1996). The use and abuse of look up tables. *Europ. Microsc. Anal.*, **39**, 7–9.

Aizawa, S.-I., Harwood, C. S. & Kadner, R. J. (2000). Signaling components in bacterial locomotion and sensory reception. *J. Bacteriol.*, **182**, 1459–71.

Akerman, A. (1910). Über die Chemotaxis der *Marchantia*-Spermatozoiden. *Z. Bot.*, **2**, 94–103.

Akitaya, T., Hirose, T., Ueda, T. & Kobatake, Y. (1984). Variation of intracellular cyclic AMP and cyclic GMP following chemical stimulation in relation to contractility in *Physarum polycephalum*. *J. Gen. Microbiol.*, **130**, 549–56.

Alani, R. & Pan, M. (2001). *In situ* transmission electron microscopy studies and real-time digital imaging. *J. Microsc.*, **203**, 128–33.

Albrecht-Bühler, G. (1991). Possible mechanisms of indirect gravity sensing by cells. *ASGSB Bull.*, **4**, 25–34.

Allen, R. D. (1985). New directions and refinements in video-enhanced microscopy applied to problems in cell motility. *Adv. Microsc.*, **196**, 3–11.

Allen, R. D. (1988). Cytology, in *Paramecium*. Ed., H.-D. Görtz. Springer-Verlag, Berlin, pp. 4–40.

Allen, R. N. & Newhook, F. J. (1973). Chemotaxis of zoospores of *Phytophthora cinnamomi* to ethanol in capillaries of soil pore dimensions. *Trans. Br. Mycol. Soc.*, **61**, 287–302.

Almagor, M., Ron, A. & Bar-Tana, J. (1981). Chemotaxis in *Tetrahymena thermophila*. *Cell Motil.*, **1**, 261–8.

Alvarez, O. & Latorre, R. (1978). Voltage-dependent capacitance in lipid bilayers made from monolayers. *Biophys. J.*, **21**, 1–17.

Amos, L. (1987). Movements made visible by microchip technology. *Nature*, **330**, 211–2.

Andersen, R. A., Barr, D. J. S., Lynn, D. H., Melkonian, M., Moestrup, O. & Sleigh, M. A. (1991). Terminology and nomenclature of the cytoskeletal elements associated with the flagellar/ciliary apparatus in protists. *Protoplasma*, **164**, 1–8.

Anderson, J. B. (1993). Magnetic orientation in the fire ant, *Solenopsis invicta*. *Naturwissenschaften*, **80**, 568–70.

Anderson, J. D. (1951). Galvanotaxis of slime mold. *J. Gen. Physiol.*, **35**, 1–16.

Andre, E., Brink, M., Gerisch, G., Isenberg, G., Noegel, A., Schleicher, M., Segall, J. E. & Wallraff, E. (1989). A *Dictyostelium* mutant deficient in severin, an F-actin fragmenting protein, shows normal motility and chemotaxis. *J. Cell Biol.*, **108**, 985–95.

Antipa, G. A., Martin, K. & Rintz, M. T. (1983). A note on the possible ecological significance of chemotaxis in certain ciliated protozoa. *J. Protozool.*, **30**, 55–7.

Armitage, J. P. & Evans, C. W. (1981). Comparison of the carotenoid bandshift and oxonol dyes to measure membrane potential changes during chemotactic stimulation of *Rhodopseudomonas sphaeroides* and *Escherichia coli*. *FEBS Lett.*, **126**, 98–102.

Ascoli, C. (1975). New techniques in photomotion methodology, in *Biophysics of Photoreceptors and Photobehaviour of Microorganisms*. Ed., G. Colombetti. Lito Felici, Pisa, Italy, pp. 109–20.

Ashmore, J. F. (1987). A fast motile response in guinea-pig outer hair cells: the cellular basis of the cochlear amplifier. *J. Physiol.*, **388**, 323–47.

Ayed, M., Pironneau, O., Planel, H., Gasset, G. & Richoilley, G. (1992). Theoretical and experimental investigations on the fast rotating clinostat. *Micrograv. Sci. Technol.*, **5**, 98–102.

Baba, S. A. & Mogami, Y. (1985). An approach to digital image analysis of bending shapes of eukaryotic flagella and cilia. *Cell Motil.*, **5**, 475–89.

Baba, S. A., Tatematsu, R. & Mogami, Y. (1991). A new hypothesis concerning graviperception of single cells, and supporting simulated experiments. *Biol. Sci. Space*, **5**, 290–1.

Baeuerlein, E. & Schueler, D. (1995). Biomineralisation: iron transport and magnetite crystal formation of *Magnetospirillum gryphiswaldense*. *J. Inorgan. Biochem.*, **59**, 107.

Baker, R. R., Mather, J. G. & Kennaugh, J. H. (1983). Magnetic bones in human sinuses. *Nature*, **301**, 78–80.

Balkwill, D. L., Maratea, D. & Blakemore, R. P. (1980). Ultrastructure of a magnetotactic *Spirillum*. *J. Bacteriol.*, **141**, 1399–408.

Baluska, F. & Hasenstein, K. H. (1997). Root cytoskeleton: its role in perception of and response to gravity. *Planta*, **203**, 69–78.

Bancroft, F. W. (1913). Heliotropism, differential sensibility, and galvanotropism in *Euglena*. *J. Exp. Zool.*, **15**, 383–428.

Bareis, D. L., Hirata, F., Schiffmann, E. & Axelrod, J. (1982). Phospholipid metabolism, calcium flux, and the receptor-mediated induction of chemotaxis in rabbit neutrophils. *J. Cell Biol.*, **93**, 690–7.

Barghigiani, C., Colombetti, G., Lenci, F., Banchetti, R. & Bizzaro, M. P. (1979a). Photosensory transduction in *Euglena gracilis*: effect of some metabolic drugs on the photophobic response. *Arch. Microbiol.*, **120**, 239–45.

Barghigiani, C., Colombetti, G., Tranchini, B. & Lenci, F. (1979b). Photobehavior of *Euglena gracilis*: action spectrum for the stepdown photophobic response of individual cells. *Photochem. Photobiol.*, **29**, 1015–9.

Barlow, P. W. (1995). Gravity perception in plants: a multiplicity of systems derived by evolution? *Plant Cell Environ.*, **18**, 951–62.

Bashford, C. L., Chance, B. & Prince, R. C. (1979). Oxonol dyes as monitors of membrane potential: their behavior in photosynthetic bacteria. *Biochim. Biophys. Acta*, **545**, 46–57.

Batschelet, E. (1981). *Circular Statistics in Biology.* Academic Press, New York.

Bazylinski, D. A., Frankel, R. B. & Jannasch, H. W. (1988). Anaerobic magnetite production by a marine, magnetotactic bacterium. *Nature*, **334**, 518–9.

Bean, B. (1975). Geotaxis in *Chlamydomonas*. *J. Cell. Biol.*, **67**, 24a.

Bean, B. (1977). Geotactic behavior of *Chlamydomonas*. *J. Protozool.*, **24**, 394–401.

Bean, B. (1984). Microbial geotaxis, in *Membranes and Sensory Transduction*. Eds., G. Colombetti & F. Lenci. Plenum Press, New York, pp. 163–98.

Bean, B., Brandt, J. & Prevelige, P. (1978). Selective effects of cupric ion on the motility and photophobic responses of *Chlamydomonas*. *J. Cell. Biol.*, **79**, 292a.

Bean, B. & Harris, A. (1978). Selective inhibition of flagellar activity in *Chlamydomonas* by nickel. *J. Protozool.*, **26**, 235–40.

Bean, B. & Yussen, P. (1979). Photoresponses of *Chlamydomonas*: differential inhibitions of phototactic and photophobic responses by low concentrations of Cu^{2+}. *J. Cell Biol.*, **83**, 351a.

Beck, C. & Uhl, R. (1994). On the localization of voltage-sensitive calcium channels in the flagella of *Chlamydomonas reinhardtii*. *J. Cell Biol.*, **125**, 1119–25.

Beckmann, M. & Hegemann, P. (1991). In vitro identification of rhodopsin in the green alga *Chlamydomonas*. *Biochemistry*, **30**, 3692–7.

Bedini, C., Lanfranchi, A. & Nobili, R. (1973). The ultrastructure of the Müller body in *Remanella*. *J. Protozool.*, **24**, 394–401.

Bees, M. A. & Hill, N. A. (1999). Non-linear bioconvection in a deep suspension of gyrotactic swimming micro-organisms. *J. Math. Biol.*, **38**, 135–68.

Behrenfeld, M., Hardy, J., Gucinski, H., Hanneman, A. & Lee, H., II. (1993). Effects of ultraviolet-B radiation on primary production along longitudinal transects in the South Pacific Ocean. *Mar. Environ. Res.*, **35**, 349–63.

Bengtsson, E., Nordin, B. & Pedersen, F. (1994). MUSE – a new tool for interactive image analysis and segmentation based on multivariate statistics. *Comput. Meth. Prog. Biomed.*, **42**, 181–200.

Berg, H. C. (2000). Motile behavior of bacteria. *Phys. Today*, 24–9.

Bergman, K. (1972). Blue-light control of sporangiophore initiation in *Phycomyces*. *Planta*, **107**, 53–67.

Berman, T. & Rodhe, W. (1971). Distribution and migration of *Peridinium* in Lake Kinneret. *Mitt. Internat. Verein. Limnol.*, **19**, 266–76.

Bett, G. C. L. & Sachs, F. (1997). Cardiac mechanosensitivity and stretch-activated ion channels. *Trends Cardiovasc. Med.*, **7**, 4–8.

Binot, R. A. (2002). The coming space science programme in animal cell/tissue research, in *Cell Biology and Biotechnology in Space*. Ed., A. Cogoli. Elsevier, Amsterdam, pp. 237–48.

Björkman, T. (1988). Gravity sensing by plants. *Adv. Bot. Res.*, **15**, 1–41.

Björkman, T. (1992). Perception of gravity by plants. *Adv. Space Res.*, **12**, 195–201.

Blakemore, R. P. (1975). Magnetotactic bacteria. *Science*, **190**, 377–9.

Blakemore, R. P. (1982). Magnetotactic bacteria. *Ann. Rev. Microbiol.*, **36**, 217–38.

Blakemore, R. P., Frankel, R. B. & Kalmijn, A. J. (1980). South-seeking magnetotactic bacteria in the southern hemisphere. *Nature*, **286**, 384–5.

Blancaflor, E. B., Fasano, J. M. & Gilroy, S. (1998). Mapping the role of cap cells in root gravitropism. *Plant Phys.*, **116**, 213–22.

Blasiak, J., Mulcahy, D. L. & Musgrave, M. E. (2001). Oxytropism: a new twist in pollen tube orientation. *Planta*, **213**, 318–22.

Block, J., Briegleb, W., Sobick, V. & Wohlfarth-Bottermann, K. E. (1986a). Confirmation of gravisensitivity in the slime mold *Physarum polycephalum* under near weightlessness. *Adv. Space Res.*, **6**, 143.

Block, I., Briegleb, W. & Wohlfarth-Bottermann, K.-E. (1986b). Gravisensitivity of the acellular slime mold *Physarum polycephalum* demonstrated on the fast-rotating clinostat. *Europ. J. Cell Biol.*, **41**, 44–50.

Block, I., Briegleb, W. & Wohlfarth-Bottermann, E. (1996). Acceleration-sensitivity threshold of *Physarum*. *J. Biotechnol.*, **47**, 239–44.

Block, I., Freiberger, N., Gavrilova, O. & Hemmersbach, R. (1999). Putative graviperception mechanisms of protists. *Adv. Space Res.*, **24**, 877–82.

Block, I., Rabien, H. & Ivanova, K. (1998). Involvement of the second messenger cAMP in the gravity-signal transduction in *Physarum*. *Adv. Space Res.*, **21**, 1311–4.

Blum, J. J., Hayes, A., Jamieson, G. A., Jr. & Vanaman, T. C. (1980). Calmodulin confers calcium sensitivity on ciliary dynein ATPase. *J. Cell Biol.*, **87**, 386–97.

Blüm, V. E., Stretzke, E. & Kreuzberg, K. (1994). C.E.B.A.S.-Aquarack Project: the mini-module as tool in artificial ecosystem research. *Acta Astronautica*, **33**, 167–77.

Bogatina, N. I., Litvin, V. M. & Travkin, M. P. (1986). Wheat roots orientation under the effect of geomagnetic field. *Biofizika*, **31**, 886–91.

Boland, W., Hoever, F.-P. & Krüger, B.-W. (1989). Application of molecular modelling techniques to pheromones of the marine brown algae *Cutleria multifida* and *Ectocarpus siliculosus* (phaeophyceae). Metalloproteins as chemoreceptors? *Z. Naturf.*, **44c**, 829–37.

Boland, W., Marner, F.-J., Jaenicke, L., Müller, D. G. & Fälster, E. (1983). Comparative receptor study in gamete chemotaxis of the seaweeds *Ectocarpus siliculosus* and *Cutleria multifida*. An approach to interspecific communication of algal gametes. *Europ. J. Biochem.*, **134**, 97–103.

Boland, W., Terlinden, R., Jaenicke, L. & Müller, D. G. (1982). Binding-mechanism and sensitivity in gamete chemotaxis of the phaeophyte *Cutleria multifida*. *Europ. J. Biochem.*, **126**, 173–9.

Bolte, E. (1920). Über die Wirkung von Licht und Kohlensäure auf die Beweglichkeit grüner und farbloser Schwärmzellen. *Jahrb. Wiss. Bot.*, **59**, 287–324.

Bonini, N. M., Gustin, M. C. & Nelson, D. L. (1986). Regulation of ciliary motility by membrane potential in *Paramecium*: a role for cyclic AMP. *Cell Motil. Cytoskel.*, **6**, 256–72.

Bonner, J. T. (1959). *The Cellular Slime Molds*. Princeton University Press, Princeton.

Bonner, J. T., Barkley, D. S., Hall, E. M., Konijn, T. M., Mason, J. W., O'Keefe, G. I. & Wolfe, P. B. (1969). Acrasin, acrasinase, and the sensitivity to acrasin in *Dictyostelium discoideum*. *Devel. Biol.*, **20**, 72–87.

Boonstra, J. (1999). Growth factor-induced signal transduction in adherent mammalian cells is sensitive to gravity. *FASEB J.*, **13**, S35–S42.

Bootman, M. D., Collins, T. J., Peppiatt, C. M., Prothero, L. S., MacKenzie, L., De Smet, P., Travers, M., Tovey, S. C., Seo, J. T., Berridge, M. J., Ciccolini, F. & Lipp, P. (2001). Calcium signalling – an overview. *Semin. Cell Devel. Biol.*, **12**, 3–10.

Borgers, J. A. & Kitching, J. A. (1956). Reaction of the flagellate *Astasia longa* in gradients of dissolved carbon dioxide. *Proc. Roy. Soc. Lond. B*, **144**, 506–19.

Borle, A. B. (1994). Ca^{2+}-bioluminescent indicators. *Meth. Toxicol.*, **19**, 315.

Boscov, J. S. (1974). Responses of *Chlamydomonas* to single flashes of light. Thesis, Tufts University.

Boscov, J. S. & Feinleib, M. E. (1979). Phototactic response of *Chlamydomonas* to flashes of light. II. Response of individual cells. *Photochem. Photobiol.*, **30**, 499–505.

Bouck, G. B. & Ngo, H. (1996). Cortical structure and function in euglenoids with reference to trypanosomes, ciliates, and dinoflagellates. *Int. Rev. Cytol.*, **169**, 267–318.

Bound, K. E. & Tollin, G. (1967). Phototactic response of *Euglena gracilis* to polarized light. *Nature*, **216**, 1042–4.

Bowne, S. W. Jr. & Bowne, G. D. (1967). Taxis in *Euglena*. *Exp. Cell Res.*, **47**, 545–53.

Bowne, S. W. & Bowne, G. D. (1968). Phototaxis. *Naturwiss. Rdschau*, **21**, 394–5.

Bozzaro, S. & Ponte, E. (1995). Cell adhesion in the life cycle of *Dictyostelium*. *Experientia*, **51**, 1175–88.

Bräucker, R., Cogoli, A. & Hemmersbach, R. (2001). Graviperception and graviresponse at the cellular level, in *Astrobiology: The Quest for the Conditions of Life*. Eds., G. G. Horneck & C. Baumstark-Khan. Springer-Verlag, Berlin, pp. 284–97.

Bräucker, R. & Machemer, H. (2002). CECILIA, a versatile research tool for cellular responses to gravity. *Micrograv. Sci. Technol.*, **XIII**, 3–13.

Bräucker, R., Machemer, H. & Fahn, C. (1999). A closed artificial ecosystem for ciliates. *Proceedings of the 2nd Symposium on the Utilisation of the International Space Station*, ESTEC, Noordwijk, pp. 511–5.

Bräucker, R., Machemer-Röhnisch, S. & Machemer, H. (1994). Graviresponses in *Paramecium caudatum* and *Didinium nasutum* examined under varied hypergravity conditions. *J. Exp. Biol.*, **197**, 271–94.

Bräucker, R., Machemer-Röhnisch, S., Machemer, H. & Murakami, A. (1992). Gravity-controlled gliding velocity in *Loxodes*. *Europ. J. Protistol.*, **28**, 238–45.

Bräucker, R., Murakami, A., Ikegaya, K., Yoshimura, K., Takahashi, K., Machemer-Röhnisch, S. & Machemer, H. (1998). Relaxation and activation of graviresponses in *Paramecium caudatum*. *J. Exp. Biol.*, **201**, 2103–13.

Braun, F.-J. & Hegemann, P. (1999). Direct measurement of cytosolic calcium and pH in living *Chlamydomonas reinhardtii* cells. *Europ. J. Cell Biol.*, **78**, 199–208.

Braun, M. (2001). Association of spectrin-like proteins with the actin-organized aggregate of endoplasmic reticulum in the Spitzenkörper of gravitropically tip-growing plant cells. *Plant Physiol.*, **125**, 1611–9.

Braun, M. (2002). Gravity perception requires statoliths settled on specific plasma membrane areas in characean rhizoids and protonemata. *Protoplasma*, **219**, 150–9.

Braun, M., Buchen, B. & Sievers, A. (2002). Actomyosin-mediated statolith positioning in gravisensing plant cells studied in microgravity. *J. Plant Growth Regul.*, **21**, 137–45.

Braun, M. & Richter, P. (1999). Relocalization of the calcium gradient and a dihydropyridine receptor is involved in upward bending by bulging of *Chara* protonemata, but not in downward bending by bowing of *Chara* rhizoids. *Planta*, **209**, 414–23.

Braun, M. & Sievers, A. (1993). Centrifugation causes adaptation of microfilaments: studies on the transport of statoliths in gravity sensing *Chara* rhizoids. *Protoplasma*, **174**, 50–61.

Braun, M. & Wasteneys, G. O. (1998). Distribution and dynamics of the cytoskeleton in graviresponding protonemata and rhizoids of characean algae: exclusion of microtubles and a convergence of actin filaments in the apex suggest an actin mediated gravitropism. *Planta*, **205**, 39.

Brehm, P. & Eckert, R. (1978). Calcium entry leads to inactivation of calcium channel in *Paramecium*. *Science*, **202**, 1203–6.

Briegleb, W. (1988). Ground-borne methods and results in gravitational cell biology. *The Physiologist*, **31**, S44–S47.

Briegleb, W. (1992). Some quantitative aspects of the fast-rotating clinostat as a research tool. *ASGSB Bull.*, **5**, 23–30.

Briggs, W. R. & Christie, J. M. (2002). Phototropins 1 and 2: versatile plant blue-light receptors. *Trends Plant Sci.*, **7**, 204–10.

Brillouet, C. & Brinckmann, E. (1999). Biorack facility performance and experiment operations on three spacehab Shuttle-to-Mir missions, in *Biorack on Spacehab*. ESA SP-1222, pp. 3–24.

Brinkmann, K. (1968). Keine Geotaxis bei *Euglena*. *Z. Pflanzenphys.*, **59**, 12–6.

Brodhun, B. & Häder, D.-P. (1990). Photoreceptor proteins and pigments in the paraflagellar body of the flagellate *Euglena gracilis*. *Photochem. Photobiol.*, **52**, 865–71.

Brodhun, B. & Häder, D.-P. (1993). UV-induced damage of photoreceptor proteins in the paraflagellar body of *Euglena gracilis*. *Photochem. Photobiol.*, **58**, 270–4.

Brown, A. H. (1992). Centrifuges: evolution of their uses in plant gravitational biology and new directions for research on the ground and in spaceflight. *ASGSB Bull.*, **5**, 43–57.

Brown, F. A., Jr. (1962). Responses of the *Planarium, Dugesia* and the Protozoan, *Paramecium*, to very weak horizontal magnetic fields. *Biol. Bull.*, **123**, 264–81.

Brownell, W. E., Bader, C. R., Bertrand, D. & de Ribaupierre, Y. (1985). Evoked mechanical responses of isolated cochlear outer hair cells. *Science*, **227**, 194–6.

Bruick, R. K. & McKnight, S. L. (2002). Oxygen sensing gets a second wind. *Science*, **295**, 807–8.

Buchanan, C. & Goldberg, B. (1981). The action spectrum of *Daphnia magna* (Crustacea) phototaxis in a simulated natural environment. *Photochem. Photobiol.*, **34**, 711–7.

Bucher, G., Scholten, J. & Klingler, M. (2002). Parental RNAi in *Tribolium* (Coleoptera). *Curr. Biol.*, **12**, R85–R86.

Buder, J. (1919). Zur Kenntnis der phototaktischen Richtungsbewegungen. *Jahrb. Wiss. Bot.*, **58**, 105–220.

Buechner, M., Delcour, A. H., Martinac, B., Adler, J. & Kung, C. (1990). Ion channel activities in the *Escherichia coli* outer membrane. *Biochim. Biophys. Acta*, **1024**, 111–21.

Bünning, E. & Schneiderhöhn, G. (1956). Über das Aktionsspektrum der phototaktischen Reaktionen von *Euglena*. *Arch. Microbiol.*, **24**, 80–90.

Burda, H., Marhold, S., Westenberger, T., Wiltschko, R. & Wiltschko, W. (1990). Magnetic compass orientation in the subterranean rodent *Cryptomys hottentotus* (Bathyergidae). *Experientia*, **46**, 528–30.

Burgess, J. G., Kawaguchi, R., Sakaguchi, T., Thornhill, R. H. & Matsunaga, T. (1993). Evolutionary relationships among *Magnetospirillum* strains inferred from phylogenetic analysis of 16S rDNA sequences. *J. Bacteriol.*, **175**, 6689–94.

Burton, J. L., Law, P. & Bank, H. L. (1986). Video analysis of chemotactic locomotion of stored human polymorphonuclear leukocytes. *Cell Motil. Cytoskel.*, **6**, 485–91.

Byrne, T. E., Wells, M. R. & Johnson, C. H. (1992). Circadian rhythms of chemotaxis to ammonium and of methylammonium uptake in *Chlamydomonas*. *Plant Physiol.*, **98**, 879–86.

Cachon, J., Cachon, M., Cosson, M.-P. & Cosson, J. (1988). The paraflagellar rod: a structure in search of a function. *Biol. Cell*, **63**, 169–81.

Cai, W. M., Braun, M. & Sievers, A. (1997). Displacement of statoliths in *Chara* rhizoids during horizontal rotation on clinostats. *Acta Biol. Exp. Sin.*, **30**, 147–55.

Calenberg, M., Brohsonn, U., Zedlacher, M. & Kreimer, G. (1998). Light- and Ca^{2+}-modulated heterotrimeric GTPases in the eyespot apparatus of a flagellate green alga. *Plant Cell*, **10**, 91–103.

Cameron, J. N. & Carlile, M. J. (1977). Negative geotaxis of zoospores of the fungus *Phytophthora*. *J. Gen. Microbiol.*, **98**, 599–602.

Cameron, J. N. & Carlile, M. J. (1980). Negative chemotaxis of zoospores of the fungus *Phytophthora palmivora*. *J. Gen. Microbiol.*, **120**, 347–53.

Camlitepe, Y. & Stradling, D. J. (1995). Wood ants orient to magnetic fields. *Proc. R. Soc. Lond. B*, **261**, 37–41.

Campuzano, V., Galland, P., Alvarez, M. I. & Eslava, A. P. (1996). Blue-light receptor requirement for gravitropism, autochemotropism and ethylene response in *Phycomyces*. *Photochem. Photobiol.*, **63**, 686–94.

Cantatore, G., Ascoli, C., Colombetti, G. & Frediani, C. (1989). Doppler velocimetry measurements of phototactic response in flagellated algae. *Biosci. Rep.*, **9**, 475–80.

Carre, I. A., Laval-Martin, D. L. & Edmunds, L. N., Jr. (1989). Circadian changes in cyclic AMP levels in synchronously dividing and stationary-phase cultures of the achlorophyllous ZC mutant of *Euglena gracilis*. *J. Cell Sci.*, **94**, 267–72.

Castillo, X., Yorkgitis, D. & Preston, K., Jr. (1982). A study of multidimensional multicolor images. *IEEE Trans. Biomed. Eng.*, **29**, 111–20.

Checcucci, A. (1976). Molecular sensory physiology of *Euglena*. *Naturwissenschaften*, **63**, 412–7.

Checcucci, A., Colombetti, G., del Carratore, G., Ferrara, R. & Lenci, F. (1974). Red light induced accumulation of *Euglena gracilis*. *Photochem. Photobiol.*, **19**, 223–6.

Checcucci, A., Colombetti, G., Ferrara, R. & Lenci, F. (1976). Action spectra for photoaccumulation of green and colorless *Euglena*: evidence for identification of receptor pigments. *Photochem. Photobiol.*, **23**, 51–4.

Chen, R., Rosen, E. & Masson, P. H. (1999). Gravitropism in higher plants. *Plant Physiol.*, **120**, 343–50.

Childress, S., Levandowsky, M. & Spiegel, E. A. (1975). Pattern formation in a suspension of swimming microorganisms: equations and stability theory. *J. Fluid Mech.*, **63**, 591–613.

Christie, J. M., Reymond, P., Powell, G. K., Bernasconi, P., Raibekas, A. A., Liscum, E. & Briggs, W. R. (1998). *Arabidopsis* NPH1: a flavoprotein with the properties of a photoreceptor for phototropism. *Science*, **282**, 1698–701.

Ciferri, O., Tiboni, O., Orlandoni, A. M. & Marchesi, M. L. (1986). Effects of microgravity on genetic recombination in *Escherichia coli*. *Naturwissenschaften*, **73**, 418–21.

Claes, H. (1980). Calcium ionophore-induced stimulation of secretory activity in *Chlamydomonas reinhardii*. *Arch. Microbiol.*, **124**, 81–6.

Clark, K. D. & Nelson, D. L. (1991). An automated assay for quantifying the swimming behavior of *Paramecium* and its use to study cation responses. *Cell Motil. Cytoskel.*, **19**, 91–8.

Clayton, R. (1959). Phototaxis of purple bacteria, in *Hdbch. d. Pfl. Physiol*, **17**, 371–87.

Clifford, J. B. (1897). Notes on some physiological properties of a Myxomycete plasmodium. *Ann. Bot.*, **11**, 179–86.

Coates, T. D., Harman, J. T. & McGuire, W. A. (1985). A microcomputer-based program for video analysis of chemotaxis under agarose. *Comput. Meth. Prog. Biomed.*, **21**, 195–202.

Cogoli, A. (1996). Biology under microgravity conditions in Space Lab International Microgravity Laboratory 2 (IML-2). *J. Biotechnol.*, **47**, 67–70.

Cogoli, A. & Friedrich, U. (1997). Instruments for biological investigations on sounding rockets. Life sciences experiments performed on sounding rockets (1985–1994). *ESA-SP*, **1206**, 109–17.

Cogoli, A., Friedrich, U., Mesland, D. & Demets, R. (1997). Life sciences experiments performed on sounding rockets (1985–1994). *ESA-SP*, **1206**, 117ff.

Cogoli, A. & Gmünder, F. K. (1991). Gravity effects on single cells: techniques, findings and theory. *Adv. Space Biol. Med.*, **1**, 183–248.

Cogoli, A., Tschopp, A. & Fuchs-Bislin, P. (1984). Cell sensitivity to gravity. *Science*, **225**, 228–30.

Cogoli, A., Valluchi-Morf, M., Müller, M. & Briegleb, W. (1980). The effects of hypogravity on human lymphocyte activation. *Aviat. Space Environ. Med.*, **51**, 29–34.

Cogoli, M. (1992). The fast-rotating clinostat: a history of its use in gravitational biology and a comparison of ground-based and flight experiment results. *ASGSB Bull.*, **5**, 59–68.

Cohn, S. A. (1993). Light dependent effects on diatom motility. *Mol. Biol. Cell*, **4**, 168a.

Colombetti, G. (1990). Photomotile responses in ciliated protozoa. *J. Photochem. Photobiol. B Biol.*, **4**, 243–59.

Colombetti, G. & Diehn, B. (1978). Chemosensory responses toward oxygen in *Euglena gracilis*. *J. Protozool.*, **25**, 211–7.

Colombetti, G., Häder, D.-P., Lenci, F. & Quaglia, M. (1982). Phototaxis in *Euglena gracilis*: effect of sodium azide and triphenylmethyl phosphonium ion on the photosensory transduction chain. *Curr. Microbiol.*, **7**, 281–4.

Cooper, M. S. & Schliwa, M. (1986). Transmembrane Ca^{2+} fluxes in the forward and reversed galvanotaxis of fish epidermal cells, in *Ionic Currents in Development.* Alan R. Liss, Inc., New York, pp. 311–8.

Cosson, J., Huitorel, P., Barsanti, L., Walne, P. L. & Gualtieri, P. (2001). Flagellar movements and controlling apparatus in flagellates. *Crit. Rev. Plant Sci.*, **20**, 297–308.

Creutz, C. & Diehn, B. (1976). Motor responses to polarized light and gravity sensing in *Euglena gracilis*. *J. Protozool.*, **23**, 552–6.

Cubitt, A. B., Firtel, R. A., Fischer, G., Jaffe, L. F. & Miller, A. L. (1995). Patterns of free calcium in multicellular stages of *Dictyostelium* expressing jellyfish apoaequorin. *Development*, **121**, 2291–301.

Czapek, F. (1895). Untersuchungen über Geotropismus. *Jahrb. Wiss. Bot.*, **27**, 243–7, 333–9.

Davenport, C. B. (1908). *Experimental Morphology* (3rd edition). Macmillan Publishing Co., New York.

Davison, E. A. & Stross, R. G. (1986). A blue light-reversible reaction in an animal system (*Daphnia pulex*). *Experientia*, **42**, 620–2.

de Araujo, F. F. T., Pires, M. A., Frankel, R. B. & Bicudo, C. E. M. (1986). Magnetite and magnetotaxis in algae. *Biophys. J.*, **50**, 375–8.

De Groot, R. P., Rijken, P. J., Boonstra, J., Verkleij, A. J., De Laat, S. W. & Kruijer, W. (1991). Epidermal growth factor-induced expression of c-fos is influenced by altered gravity conditions. *Aviat. Space Environ. Med.*, **62**, 37–40.

De Groot, R. P., Rijken, P. J., Den Hertog, J., Boonstra, J., Verkleij, A. J., De Laat, S. W. & Kruijer, W. (1990). Microgravity decreases c-fos induction and serum response element activity. *J. Cell Sci.*, **97**, 33–8.

Deininger, W., Kräger, P., Hegemann, U., Lottspeich, F. & Hegemann, P. (1995). Chlamyrhodopsin represents a new type of sensory photoreceptor. *EMBO J.*, **14**, 5849–58.

Delong, E. F., Frankel. R. B. & Bazylinski, D. A. (1993). Multiple evolutionary origins of magnetotaxis in bacteria. *Science*, **259**, 803–6.

del Portillo, H. A. & Dimock, R. V., Jr. (1982). Specificity of the host-induced negative phototaxis of the symbiotic water mite, *Unionicola formosa. Biol. Bull.*, **162**, 163–70.

Dembowski, J. (1929a). Die Vertikalbewegungen von *Paramecium caudatum.* Lage des Gleichgewichtszentrums im Körper des Infusors. *Arch. Protistenk.*, **66**, 104–32.

Dembowski, J. (1929b). Vertikalbewegungen von *Paramecium caudatum.* Einfluß einiger Außenfaktoren. *Arch. Protistenk.*, **68**, 215–60.

Denk, W., Webb, W. W. & Hudspeth, A. J. (1989). Mechanical properties of sensory hair bundles are reflected in their Brownian motion measured with a laser differential interferometer. *Proc. Natl. Acad. Sci. U.S.A.*, **16**, 5371–5.

Derguini, F., Mazur, P., Nakanishi, K., Starace, D. M., Saranak, J. & Foster, K. W. (1991). All-*trans*-retinal is the chromophore bound to the photoreceptor of the green alga *Chlamydomonas reinhardtii. Photochem. Photobiol.*, **54**, 1017–21.

Desroche, P. (1912). *Réactions de Chlamydomonas aux Agents Physiques.* Schultz, Paris.

Deutschlander, M. E., Borland, S. C. & Phillips, J. B. (1999). Extraocular magnetic compass in newts. *Nature*, **400**, 324–5.

DiBella, L. M. & King, S. M. (2001). Dynein motors of the *Chlamydomonas* flagellum. *Int. Rev. Cytol.*, **210**, 227–68.

Diebel, C. E., Proksch, R., Green, C. R., Neilson, P. & Walker, M. M. (2000). Magnetite defines a vertebrate magnetoreceptor. *Nature*, **406**, 299–302.

Diehn, B. (1969a). Action spectra of the phototactic responses in *Euglena. Biochim. Biophys. Acta*, **177**, 136–43.

Diehn, B. (1969b). Phototactic responses of *Euglena* to single and repetitive pulses of actinic light: orientation time and mechanism. *Exp. Cell Res.*, **56**, 375–81.

Diehn, B. (1973). Phototaxis and sensory transduction in *Euglena. Science*, **181**, 1009–15.

Diehn, B., Feinleib, M., Haupt, W., Hildebrand, E., Lenci, F. & Nultsch, W. (1977). Terminology of behavioral responses of motile microorganisms. *Photochem. Photobiol.*, **26**, 559–60.

Diehn, B., Fonseca, J. R. & Jahn, T. R. (1975). High speed cinemicrography of the direct photophobic response of *Euglena* and the mechanism of negative phototaxis. *J. Protozool.*, **22**, 492–4.

Dinallo, M. C., Wohlford, M. & van Houten, J. (1982). Mutants of *Paramecium* defective in chemokinesis to folate. *Genetics*, **102**, 149–58.

Ding, J. P., Badot, P.-M. & Pickard, B. G. (1993). Aluminium and hydrogen ions inhibit a mechanosensory calcium-selective cation channel. *Aust. J. Plant Physiol.*, **20**, 771–8.

Dolle, R. & Nultsch, W. (1987). Effects of calcium ions and of calcium channel blockers on galvanotaxis of *Chlamydomonas reinhardtii. Bot. Act.*, **101**, 18–23.

Donath, R. (1999). Untersuchungen an der Schwerereiz-Reaktionskette von *Loxodes striatus.* Dissertation, Rheinische-Friedrich-Wilhelms-University, Bonn.

Doughty, M. J. (1991). A kinetic analysis of the step-up photophobic response of the flagellated alga *Euglena gracilis* in culture medium. *J. Photochem. Photobiol. B Biol.*, **9**, 75–85.

Doughty, M. J. (1993). Step-up photophobic response of *Euglena gracilis* at different irradiances. *Acta Protozool.*, **32**, 73–7.

Doughty, M. J. & Diehn, B. (1979). Photosensory transduction in the flagellated alga, *Euglena gracilis.* I. Action of divalent cations Ca^{2+} antagonists and Ca^{2+} ionophore on motility and photobehavior. *Biochim. Biophys. Acta*, **588**, 148–68.

Doughty, M. J. & Diehn, B. (1982). Photosensory transduction in the flagellated alga, *Euglena gracilis.* III. Induction of Ca^{2+}-dependent responses by monovalent cation ionophores. *Biochim. Biophys. Acta*, **682**, 32–43.

Doughty, M. J. & Diehn, B. (1983). Photosensory transduction in the flagellated alga, *Euglena gracilis*. IV. Long term effects of ions and pH on the expression of step-down photobehaviour. *Arch. Microbiol.*, **134**, 204–7.

Doughty, M. J. & Diehn, B. (1984). Anion sensitivity of motility and step-down photophobic responses of *Euglena gracilis*. *Arch. Microbiol.*, **138**, 329–32.

Doughty, M. J. & Dodd, G. H. (1976). Fluorimetric determination of the resting potential changes associated with the chemotactic response in *Paramecium*. *Biochim. Biophys. Acta*, **451**, 592–603.

Doughty, M. J., Grieser, R. & Diehn, B. (1980). Photosensory transduction in the flagellated alga *Euglena gracilis*. *Biochim. Biophys. Acta*, **602**, 10–23.

Dryl, S. (1974). Behavior and motor response of *Paramecium*, in *Paramecium: A Current Survey*. Ed., W. J. van Wagtendonk. Elsevier Scientific, Amsterdam, pp. 165–218.

Dubreuil, R. R. & Bouck, G. B. (1985). The membrane skeleton of a universal organism consists of bridged, articulating strips. *J. Cell. Biol.*, **101**, 1884–96.

Dubreuil, R. R., Rosiere, T. K., Rosner, M. C. & Bouck, G. B. (1988). Properties and topography of the major integral membrane protein of a unicellular organism. *J. Cell. Biol.*, **107**, 191–200.

Dulhanty, A. F. & Franzini-Armstrong, C. (1975). The relative contributions of the folds and caveolae to the surface membrane of frog skeletal muscles fibers at different sarcomere length. *J. Physiol. (Lond.)*, **250**, 513–39.

Dusenbery, D. B. (1985). Using a microcomputer and videocamera to simultaneously track 25 animals. *Comput. Biol. Med.*, **15**, 169–75.

Duwe, H.-P., Engelhardt, H., Zilker, A. & Sackmann, E. (1987). Curvature elasticity of smectic A lipid bilayers and cell plasma membranes. *Mol. Cryst. Liq. Cryst.*, **152**,1–7.

Eberhard, M. & Erne, P. (1991). Calcium binding to fluorescent calcium indicators: calcium green, calcium orange and calcium crimson. *Biochem. Biophys. Res. Comm.*, **180**, 209–15.

Eckert, R. (1972). Bioelectric control of ciliary activity. *Science*, **176**, 473–81.

Edelmann, H. G., Gudi, G. & Kühnemann, F. (2002). The gravitropic setpoint angle of dark-grown rye seedlings and the role of ethylene. *J. Exp. Bot.*, **53**, 1627–2634.

Edmunds, L. N., Jr. (1984). Physiology of circadian rhythms in microorganisms. *Adv. Microbiol. Physiol.*, **25**, 61–148.

Edmunds, L. N., Jr., Carre, I. A., Tamponnet, C. & Tong, J. (1992). The role of ions and second messengers in circadian clock function. *Chronobiol. Internat.*, **9**, 180–200.

Edwards, E. S. & Roux, S. J. (1998). Gravity and light control of the developmental polarity of regenerating protoplasts isolated from prothallial cells of the fern *Ceratopteris richardii*. *Plant Cell Rep.*, **17**, 711–6.

Eggersdorfer, B. & Häder, D.-P. (1991a). Phototaxis, gravitaxis and vertical migrations in the marine dinoflagellate *Prorocentrum micans*. *FEMS Microbiol. Ecol.*, **85**, 319–26.

Eggersdorfer, B. & Häder, D.-P. (1991b). Phototaxis, gravitaxis and vertical migrations in the marine dinoflagellates, *Peridinium faeroense* and *Amphidinium caterea*. *Acta Protozool.*, **30**, 63–71.

Ekberg, D. R., Silver, E. C., Bushay, J. L. & Daniels, E. W. (1971). Nuclear and cellular divison in *Pelomyxa carolinesis* during weightlessness, in *The Experiments of Biosatellite II NASA SP*. Ed., J. F. Saunder. NASA, Washington, pp. 273–90.

El Fouly, M. H., Trosko, J. E. & Chang, C. C. (1987). Scrape-loading and dye transfer. A rapid and simple technique to study gap junctional intercellular communication. *Exp. Cell. Res.*, **168**, 422–8.

Emura, R., Ashida, N., Higashi, T. & Takeuchi, T. (2001). Orientation of bull sperms in static magnetic fields. *Bioelectromagnetics*, **22**, 60–5.

Engelmann, T. W. (1862). Zur Naturgeschichte der Infusionsthiere. *Z. Wiss. Zool.*, **11**, 347–93.

Engelmann, T. W. (1882). Über Licht- und Farbenperception niedrigster Organismen. *Pflügers Arch.*, **29**, 387–400.

Engelmann, T. W. (1883). *Bacterium photometricum.* Ein Beitrag zur vergleichenden Physiologie des Licht- und Farbensinnes. *Pflügers Arch.*, **30**, 95–124.

Engelmann, U., Krassnigg, F. & Schill, W. B. (1992). Sperm motility under conditions of weightlessness. *J. Androl.*, **13**, 433–6.

Ensminger, P. A., Chen, X. & Lipson, E. D. (1990). Action spectra for photogravitropism of *Phycomyces* wild type and three behavioral mutants (L150, L152, and L154). *Photochem. Photobiol.*, **51**, 681–7.

Estrada, M., Alcataz, M. & Marrasé (1987). Effect of reversed light gradients on the phytoplankton composition in marine microcosms. *Inv. Pesq.*, **51**, 443–58.

Etter, E. F., Kuhn, M. A. & Fay, F. S. (1994). Detection of changes in near-membrane Ca^{2+}concentration using a novel membrane-associated Ca^{2+}indicator. *J. Biol. Chem.*, **269**, 10141.

Etter, E. F., Minta, A., Poenie, M. & Fay, F. S. (1996). Near-membrane [Ca^{2+}] transients resolved using the Ca^{2+} indicator FFP18. *Proc. Natl. Acad. Sci. U.S.A.*, **93**, 5368.

Evans, E. (1992). Composite membranes and structured interfaces from simple to complex designs in biology, in *Biomembranes Structure and Function – The State of the Art.* Eds., B. P. Gaber & K. R. K. Easwaran. Adenine, New York, pp. 81–101.

Evans, E. & Hochmuth, R. M. (1978). Mechanical properties of membranes, in *Topics in Membrane and Transport*. Eds., A. Kleinzeller & F. Bronner. Academic Press, New York, pp. 1–64.

Evans, E. & Skalar, R. (1980). Mechanics and thermodynamics of membranes. *CRC Crit. Rev. Bioeng.*, **3**, 181–418.

Evans, M. L. & Ishikawa, H. (1997). Cellular specificity of the gravitropic motor response in roots. *Planta*, **203**, 115–22.

Evans, M. L., Ishikawa, H. & Estelle, M. A. (1994). Responses of *Arabidopsis* roots to auxin studied with high temporal resolution: comparison of wild type and auxin-response mutants. *Planta*, **194**, 215–22.

Evans, T. C. & Nelson, D. L. (1989). The cilia of *Paramecium tetraurelia* contain both Ca^{2+}-dependent and Ca^{2+}-inhibitable calmodulin-binding proteins. *Biochem. J.*, **259**, 385–96.

Fabczak, S., Fabczak, H. & Song, P.-S. (1993c). Photosensory transduction in ciliates. III. The temporal relation between membrane potentials and photomotile responses in *Blepharisma japonicum. Photochem. Photobiol.*, **57**, 872–6.

Fabczak, S., Fabczak, H., Tao, N. & Song, P.-S. (1993d). Photosensory transduction in ciliates. I. An analysis of light-induced electrical and motile responses in *Stentor coeruleus. Photochem. Photobiol.*, **57**, 696–701.

Fabczak, H., Tao, N., Fabczak, S. & Song, P.-S. (1993b). Photosensory transduction in ciliates. IV. Modulation of the photomovement response of *Blepharisma japonicum* by cGMP. *Photochem. Photobiol.*, **57**, 889–92.

Fabczak, H., Park, P. B., Fabczak, S. & Song, P.-S. (1993a). Photosensory transduction in ciliates. II. Possible role of G-protein and cGMP in *Stentor coeruleus. Photochem. Photobiol.*, **57**, 702–6.

Farina, M., Kachar, B., Lins, U., Broderick, R. & Lins de Barros, H. (1994). The observation of large magnetite (Fe_3O_4) crystals from magnetotactic bacteria by electron and atomic force microscopy. *J. Microsc.*, **173**, 1–8.

Fazio, M. J., da Silva, A. C., Rosiere, T. K. & Bouck, G. B. (1995). Membrane skeletal proteins and their integral membrane protein anchors are targets for tyrosine and threonine kinases in *Euglena*. *J. Euk. Microbiol.*, **42**, 570–80.

Feinleib, M. E. (1975). Phototactic response of *Chlamydomonas* to flashes of light. I. Response of cell population. *Photochem. Photobiol.*, **21**, 351–4.

Feinleib, M. E. & Curry, G. M. (1971). The nature of the photoreceptor in phototaxis (Chap. 13), in *Handbook of Sensory Physiology. I. Receptor Mechanisms*. Springer-Verlag, Heidelberg, pp. 366–95.

Felle, H. H. & Hepler, P. K. (1997). The cytosolic Ca^{2+} concentration gradient of *Sinapis alba* root hairs as revealed by Ca^{2+}-selective microelectrode tests and fura-dextran ratio imaging. *Plant Physiol.*, **114**, 39.

Fenchel, T. & Finlay, B. J. (1984). Geotaxis in the ciliated protozoon *Loxodes*. *J. Exp. Biol.*, **110**, 17–33.

Fenchel, T. & Finlay, B. J. (1986a). The structure and function of Müller vesicles in loxodid ciliates. *J. Protozool.*, **33**, 69–76.

Fenchel, T. & Finlay, B. J. (1986b). Photobehavior of the ciliated protozoon *Loxodes*: taxic, transient, and kinetic response in the presence and absence of oxygen. *J. Protozool.*, **33**, 139–45.

Fenchel, T. & Finlay, B. J. (1990). Oxygen toxicity, respiration and behavioural responses to oxygen in free-living anaerobic ciliates. *J. Gen. Microbiol.*, **136**, 1953–9.

Ferrara, R., Grassi, S. & del Carratore, G. (1975). An automatic homocontinuous culture apparatus. *Biotechnol. Bioeng.*, **17**, 985–95.

Fetter, D. (1926). Determination of the protoplasmic viscosity of *Paramecium* by the centrifuge method. *J. Exp. Biol.*, **44**, 279–83.

Finlay, B. J. & Fenchel, T. (1986). Photosensitivity in the ciliated protozoon *Loxodes*: pigment granules, absorption and action spectra, blue light perception, and ecological significance. *J. Protozool.*, **33**, 534–42.

Finlay, B. J., Fenchel, T. & Gardener, S. (1986). Oxygen perception and O_2 toxicity in the freshwater ciliated protozoan *Loxodes*. *J. Protozool.*, **33**, 157–65.

Firn, R. D., Wagstaff, C. & Digby, J. (2000). The use of mutants to probe models of gravitropism. *J. Exp. Bot.*, **51**, 1323–40.

Fisher, P. R., Dohrmann, U. & Williams, K. L. (1984). Signal processing in *Dictyostelium discoideum* slugs. *Mod. Cell Biol.*, **3**, 197–248.

Fitton, B. & Moore, D. (1996). National and international life sciences research programmes 1980 to 1993 – and beyond, in *Biological and Medical Research in Space. An Overview of Life Sciences Research in Space*. Eds., D. Moore, P. Bie & H. Oser. Springer, Berlin, pp. 432–541.

Fong, F. & Schiff, J. A. (1978). Blue-light absorbance changes and phototaxis in *Euglena*. *Plant Physiol.*, **61** (Suppl.), 74.

Fong, F. & Schiff, J. A. (1979). Blue-light-inducted absorbance changes associated with carotenoids in *Euglena*. *Planta*, **146**, 119–27.

Fontana, D. R. & Devreotes, P. N. (1984). cAMP-stimulated adenylate cyclase activation in *Dictyostelium discoideum* is inhibited by agents acting at the cell surface. *Devel. Biol.*, **106**, 76–82.

Fornshell, J. A. (1980). Positive geotaxis in *Blepharisma persicinum*. *J. Protozool.*, **27**, 25A–6A.

Fortin, M.-C. & Poff, K. L. (1991). Characterization of thermotropism in primary roots of maize: dependence on temperature and temperature gradient, and interaction with gravitropism. *Planta*, **184**, 410–4.

Foster, K. W. (2001). Action spectroscopy of photomovement, in *Photomovement*. Eds., D.-P. Häder & M. Lebert. Elsevier, Amsterdam, pp. 51–115.

Foster, K. W. & Smyth, R. D . (1980). Light antennas in phototactic algae. *Microbiol. Rev.*, **44**, 572–630.

Fox, H. M. (1925). The effect of light on the vertical movement of aquatic organisms. *Biol. Rev. Cambridge Phil. Soc.*, **1**, 219–24.

Fraenkel, G. S. & Gunn, D. L. (1961). *The Orientation of Animals (Kineses, Taxes and Compass Reactions)*. Dover Publication, Inc., New York, pp. 1–91.

France, R. H. (1908). Experimentelle Untersuchungen über Reizbewegungen und Lichtsinnesorgane der Algen. *Zeitschrift für den Ausbau der Entwicklungs lehre*, **2**, 29–43.

France, R. H. (1909). Untersuchungen über die Sinnesorganfunktion der Augenflecke bei Algen. *Arch. Hydrobiol.*, **4**, 37–48.

Franco, A. Jr., Winegar, B. D. & Lansman, J. B. (1991). Open channel block by gadolinium ion of the stretch-inactivated ion channel in mdx myotubes. *Biochem. J.*, **59**, 1164–70.

Franze, R. (1893). Zur Morphologie und Physiologie der Stigmata der Mastigophoren. *Z. Wiss. Zool.*, **56**, 138–64.

Freed, L. E. & Vunajak-Novakovic, G. (2002). Spaceflight bioreactor studies of cells and tissues, in *Cell Biology and Biotechnology in Space*. Ed., A. Cogoli. Elsevier, Amsterdam, pp. 177–96.

Freeman, H. (1980). Analysis and manipulation of lineal map data, in *Map Data Processing*. Eds., H. Freeman & G. G. Pieroni. Academic Press, New York, pp. 151–68.

Friedrich, U. L. D., Joop, O., Pütz, C. & Willich, G. (1996). The slow rotating centrifuge microscope NIZEMI – a versatile instrument for terrestrial hypergravity and space microgravity research in biology and material science. *J. Biotech.*, **47**, 225–38.

Froehlich, O. & Diehn, B. (1974). Photoeffects in a flavin-containing lipid bilayer membrane and implications for algal phototaxis. *Nature*, **248**, 802–4.

Fuhrmann, M. (1996). Aufbau und Sequenz des Chlamyopsingens. Diploma thesis, University of Regensburg.

Fuhrmann, M., Oertel, W. & Hegemann, P. (1999). A synthetic gene coding for the green fluorescent protein (GFP) is a versatile reporter in *Chlamydomonas reinhardtii*. *Plant J.*, **19**, 353–61.

Fukui, K. & Asai, H. (1985). Negative geotactic behavior of *Paramecium caudatum* is completely described by the mechanism of buoyancy-oriented upward swimming. *Biophys. J.*, **47**, 479–82.

Furness, D. N., Zetes, D. E., Hackney, C. M. & Steele, C. R. (1997). Kinematic analysis of shear displacement as a means of operating mechano-transduction channels in the contact region between adjacent stereocilia of mammalian cochlear hair cells. *Proc. R. Soc. Lond. Ser. B Biol. Sci.*, **264**, 45–51.

Fushimi, K. & Verkman, A. S. (1991). Low viscosity in the aqueous domain of cell cytoplasm measured by picosecond polarization microfluorimetry. *J. Cell. Biol.*, **112**, 719–25.

Galland, P. (2001). Phototropism in *Phycomyces*, in *Photomovement*. Eds., D.-P. Häder & M. Lebert. Elsevier, Amsterdam, pp. 621–57.

Galland, P. & Lipson, E. D. (1984). Photophysiology of *Phycomyces blakesleeanus*. *Photochem. Photobiol.*, **40**, 795–800.

Galland, P., Wallacher, Y., Finger, H., Hannappel, M., Tröster, S., Bold, E. & Grolig, F. (2002). Tropisms in *Phycomyces*: sine law for gravitropism, exponential law for photogravitropic equilibrium. *Planta*, **214**, 931–8.

Gallin, J. I. & Snyderman, R. (1983). Leukocyte chemotaxis. *Fed. Proc.*, **42**, 2851–62.

Gangola, P. & Rosen, B. P. (1987). Maintenance of intracellular calcium in *Escherichia coli. J. Biol. Chem.*, **262**, 12570–4.

Garbers, D. L. (1988). Signal/transduction mechanisms of sea urchin spermatozoa. *ISI Atlas Sci. Biochem.*, **1**, 120–6.

Garcia-Pichel, F., Mechling, M. & Castenholz, R. W. (1994). Diel migrations of microorganisms within a benthic, hypersaline mat community. *Appl. Environ. Microbiol.*, **60**, 1500–11.

Gebauer, H. (1930). Zur Kenntnis der Galvanotaxis von *Polytoma uvella* und einigen anderen Volvocineen. *Beitr. Biol. Pfl.*, **18**, 30–501.

Gebauer, M., Watzke, D. & Machemer, H. (1999). The gravikinetic response of *Paramecium* is based on orientation-dependent mechanotransduction. *Naturwissenschaften*, **86**, 352–6.

Gebeshuber, I. C. & Rattay, F. (2001). Coding efficiency of inner hair cells at the threshold of hearing, in *Computational Models of Auditory Functions*. Eds., S. Greenberg & M. Slaney. IOS Press, Amsterdam, pp. 5–16.

Geleoc, G. S., Forge, A., Casalotti, S. & Ashmore, J. F. (1999). A sugar transporter as a candidate for the outer hair cell motor. *Nature Neurosci.*, **2**, 713–9.

Gerhardt, K. (1913). Beitrag zur Physiologie von *Closterium*. Dissertation, Uni Jena. 7–37.

Ghorai, S. & Hill, N. A. (2000a). Periodic arrays of gyrotactic plumes in bioconvection. *Phys. Fluids*, **12**, 5–22.

Ghorai, S. & Hill, N. A. (2000b). Wavelengths of gyrotactic plumes in bioconvection. *Bull. Math. Biol.*, **62**, 429–50.

Gilroy, S., Read, N. D. & Trewavas, A. J. (1990). Elevation of cytoplasmatic calcium by caged calcium or caged inositol triphosphate initiates stomatal closure. *Nature*, **346**, 769.

Gitelman, S. E. & Witman, G. B. (1980). Purification of calmodulin from *Chlamydomonas*: calmodulin occurs in cell bodies and flagella. *J. Cell Biol.*, **98**, 764–70.

Glazzard, A. N., Hirons, M. R., Mellor, J. S. & Holwill, M. E. J. (1983). The computer assisted analysis of television images as applied to the study of cell motility. *J. Submicrosc. Cytol.*, **15**, 305–8.

Gliwicz, M. Z. (1986). Predation and the evolution of vertical migration in zooplankton. *Nature*, **320**, 746–8.

Gluch, M. F., Typke, D. & Baumeister, W. (1995). Motility and thermotactic responses of *Thermotoga maritima. J. Bacteriol.*, **177**, 5473–9.

Gmünder, F. K. & Cogoli, A. (1988). Cultivation of single cells in space. *Appl. Micrograv. Technol.*, **1**, 115–22.

Godziemba-Czyz, J. (1973). Certain aspects of the chemotaxis reaction of chloroplast in *Funaria hygrometrica. Acta Soc. Bot. Polon.*, **42**, 453–9.

Golovian, V. A. & Blaustein, M. P. (1997). Spatially and functionally distinct Ca^{2+} stores in sarcoplasmic and endoplasmic reticulum. *Science*, **275**, 1643.

Goodenough, U. W., Detmers, P. A. & Hwang, C. (1982). Activation for cell fusion in *Chlamydomonas*: analysis of wild-type gametes and non-fusing mutants. *J. Cell. Biol.*, **92**, 378–86.

Gorby, Y. A., Beveridge, T. J. & Blakemore, R. P. (1988). Characterization of the bacterial magnetosome membrane. *J. Bacteriol.*, **170**, 834–41.

Gordon, D. C., MacDonald, I. R., Hart, J. W. & Berg, A. (1984). Image analysis of geo-induced inhibition, compression, and promotion of growth in an inverted *Helianthus annuus* L. seedling. *Plant Physiol.*, **76**, 589–94.

Görtz, H.-D. (1988). *Paramecium*. Springer-Verlag, Berlin. pp. 1–444.

Gössel, I. (1957). Über das Aktionsspektrum der Phototaxis chlorophyllfreier *Euglenen* und über die Absorption des Augenflecks. *Arch. Microbiol.*, **27**, 288–305.

Gould, J. L. (1984). Magnetic field sensitivity in animals. *Ann. Rev. Physiol.*, **46**, 585–98.

Govorunova, E. G., Sineshchekov, O. A., Gärtner, W., Chunaev, A. S. & Hegemann, P. (2001). Photoreceptor current and photoorientation in *Chlamydomonas* mediated by 9-demethylchlamyrhodopsin. *Biophys. J.*, **81**, 2897–907.

Govorunova, E. G., Sineshchekov, O. A. & Hegemann, P. (1997). Desensitization and dark recovery of the photoreceptor current in *Chlamydomonas reinhardtii*. *Plant Physiol.*, **115**, 633–42.

Graham, C. R. & Kaplan, J. H. (1993). Nitrophenyl-EGTA, a photolabile chelator that selectively binds Ca^{2+} with high affinity and releases it rapidly upon photolysis. *Proc. Natl. Acad. Sci. U.S.A.*, **91**, 187.

Grain, J. (1986). The cytoskeleton in protists: nature, structure and functions. *Int. Rev. Cytol.*, **104**, 153–249.

Grant, G. & Reid, A. F. (1981). An efficient algorithm for boundary tracing and feature extraction. *Comput. Graph. Imag. Process*, **17**, 225–37.

Gressel, J. & Horwitz, B. (1981). Gravitropism and phototropism, in *Molecular Biology of Plant Development*. Eds., H. Smith & D. Grierson, Blackwell, Oxford, pp. 1–42.

Greulich, K. O. & Pilarczyk, G. (1998). Laser tweezers and optical microsurgery in cellular and molecular biology. Working principles and selected applications. *Cell. Mol. Biol.*, **44**, 701.

Grigoriev, A. I., Kalinin, Y. T., Buravkova, L. B. & Mitichkin, O. V. (2002). Space cell physiology and space biotechnology in Russia, in *Cell Biology and Biotechnology in Space*. Ed., A. Cogoli. Elsevier, Amsterdam, pp. 215–36.

Grolig, F., Eibel, P., Schimek, C., Schapat, T., Dennison, D. S. & Galland, P. A. (2000). Interaction between gravitropism and phototropism in sporangiophores of *Phycomyces blakesleeanus*. *Plant Physiol.*, **123**, 765–76.

Gruener, R., Roberts, R. & Reitstetter, R. (1993). Exposure to microgravity alters properties of cultured muscle cells. *ASGSB Bull.*, **7**, 65.

Gruler, H. & Nuccitelli, R. (1986). New insights into galvanotaxis and other directed cell movements: an analysis of the translocation distribution function, in *Ionic Currents in Development*. Alan R. Liss Inc., New York, pp. 337–47.

Gruler, H. & Nuccitelli, R. (1991). Neural crest cell galvanotaxis: new data and a novel approach to the analysis of both galvanotaxis and chemotaxis. *Cell Mot. Cytoskel.*, **19**, 121–33.

Gualtieri, P. (1993). *Euglena gracilis*: is the photoreception enigma solved? *J. Photochem. Photobiol.*, **19**, 3–14.

Gualtieri, P., Barsanti, L. & Passarelli, V. (1989). Absorption spectrum of a single isolated paraflagellar swelling of *Euglena gracilis*. *Biochim. Biophys. Acta*, **993**, 293–6.

Guharay, F. & Sachs, F. (1984). Stretch-activated single ion channel currents in tissue cultured embryonic chick skeletal muscle. *J. Physiol. (Lond.)*, **352**, 685–701.

Gulley, R. L. & Reese, T. S. (1977). Regional specialization of the hair cell plasmalemma in the organ of Corti. *Anat. Rec.*, **189**, 109–24.

Gunn, D. L., Kennedy, J. S. & Pielou, D. P. (1934). Classification of taxes and kinesis. *Nature*, **140**, 1064.

Günther, F. (1928). Über den Bau und die Lebensweise der *Euglenen* besonders der Arten *E. terricola, geniculata, proxima, sanguinea* und *luccus nov.* spec. *Arch. Protistenk.*, **60**, 511–90.

Gurney, A. M. & Bates, S. E. (1995). Use of chelators and photoactivatable caged compounds to manipulate cytosolic calcium, in *Methods in Neurosciences. Measurement and Manipulation of Intracellular Ions*. Ed., J. Kraicer. Academic Press, London, pp. 123–52.

Gustin, M. C., Zhou, X.-L., Martinac, B. & Kung, C. (1988). A mechanosensitive ion channel in the yeast plasma membrane. *Science*, **242**, 762–5.

Häder, D.-P. (1979). Photomovement, in *Encyclopedia of Plant Physiology*. Eds., W. Haupt & M. E. Feinleib. Springer, Berlin, pp. 268–309.

Häder, D.-P. (1986). Effects of solar and artificial UV irradiation on motility and phototaxis in the flagellate *Euglena gracilis*. *Photochem. Photobiol.*, **44**, 651–6.

Häder, D.-P. (1987a). Automatic area calculation by microcomputer-controlled video analysis. *EDV Med. Biol.*, **18**, 33–6.

Häder, D.-P. (1987b). Photomovement, in *The Cyanobacteria*. Eds., P. Fay & C. van Baalen. Elsevier, Amsterdam, pp. 325–45.

Häder, D.-P. (1987c). Photosensory behavior in procaryotes. *Microbiol. Rev.*, **51**, 1–21.

Häder, D.-P. (1987d). Polarotaxis, gravitaxis and vertical phototaxis in the green flagellate, *Euglena gracilis*. *Arch. Microbiol.*, **147**, 179–83.

Häder, D.-P. (1988). Advanced techniques in photobehavioral studies: computer-aided studies, in *Light in Biology and Medicine 1*. Eds., R. H. Douglas, J. Moan & F. Dall'Acqua. Plenum Press, New York, pp. 385–92.

Häder, D.-P. (1990). Tracking of flagellates by image analysis, in *Biological Motion, Proceedings Königswinter 1989*. Eds., W. Alt & G. Hoffmann. Springer-Verlag, Berlin, pp. 343–60.

Häder, D.-P. (1991a). Advanced methods in photobiology of protists. *Acta Protozool.*, **30**, 135–56.

Häder, D.-P. (1991b). Phototaxis and gravitaxis in *Euglena gracilis*, in *Biophysics of Photoreceptors and Photomovements in Microorganisms*. Eds., F. Lenci, F. Ghetti, G. Colombetti, D.-P. Häder & P.-S. Song. Plenum Press, New York, pp. 203–21.

Häder, D.-P. (1991c). Use of image analysis in photobiology, in *Photobiology. The Science and Its Applications*. Ed., E. Riklis. Plenum Press, New York, pp. 329–43.

Häder, D.-P. (1992). Real-time tracking of microorganisms, in *Image Analysis in Biology*. Ed., D.-P. Häder. CRC Press, Boca Raton, pp. 289–313.

Häder, D.-P. (1993). Simulation of phototaxis in the flagellate *Euglena gracilis*. *J. Biol. Phys.*, **19**, 95–108.

Häder, D.-P. (1994a). Gravitaxis in the flagellate *Euglena gracilis* – results from Nizemi, clinostat and sounding rocket flights. *J. Grav. Physiol.*, **1**, P82–P84.

Häder, D.-P. (1994b). Real-time tracking of microorganisms. *Binary*, **6**, 81–6.

Häder, D.-P. (1995). Novel method to determine vertical distributions of phytoplankton in marine water columns. *Environ. Exp. Bot.*, **35**, 547–55.

Häder, D.-P. (1996a). Echtzeit-Bildverarbeitung für Bahnverfolgung und Zellerkennung. *Fachtagung Informationstechnik und Biotechnologie, Bonn*, 81–94.

Häder, D.-P. (1996b). NIZEMI – Experiments on the slow rotating centrifuge microscope during the IML-2 mission. *J. Biotechnol.*, **47**, 223–4.

Häder, D.-P. (1997a). Gravitaxis and phototaxis in the flagellate *Euglena* studied on TEXUS missions, in *Life Science Experiments Performed on Sounding Rockets (1985–1994)*. Eds., A. Cogoli, U. Friedrich, D. Mesland & R. Demets. ESA Publications Division, ESTEC, Noordwijk, The Netherlands, pp. 77–9.

Häder, D.-P. (1997b). Gravitaxis in flagellates. *Biol. Bull.*, **192**, 131–3.

Häder, D.-P. (1997c). Oben oder unten – Schwerkraftperzeption bei dem einzelligen Flagellaten *Euglena gracilis*. *Mikrokosmos*, **86**, 351–6.

Häder, D.-P. (1998). Orientierung im Licht: Phototaxis bei *Euglena gracilis*. *Mikrokosmos*, **87**, 3–11.

Häder, D.-P. (1999). Gravitaxis in unicellular microorganisms. *Adv. Space Res.*, **24**, 843–50.

Häder, D.-P. (2000). Graviperzeption in Pflanzen und Mikroorganismen, in *Bilanzsymposium Forschung unter Weltraumbedingungen. 21–23 Sept. 1998 Wissenschaftliche Projektführung*. Eds., M. H. Keller & P. R. Sahm, Norderney, pp. 633–41.

Häder, D.-P., Colombetti, G., Lenci, F. & Quaglia, M. (1981). Phototaxis in the flagellates, *Euglena gracilis* and *Ochromonas danica*. *Arch. Microbiol.*, **130**, 78–82.

Häder, D.-P. & Griebenow, K. (1987). Versatile digital image analysis by microcomputer to count microorganisms. *EDV Med. Biol.*, **18**, 37–42.

Häder, D.-P. & Griebenow, K. (1988). Orientation of the green flagellate, *Euglena gracilis*, in a vertical column of water. *FEMS Microbiol. Ecol.*, **53**, 159–67.

Häder, D.-P. & Häder, M. A. (1988a). Inhibition of motility and phototaxis in the green flagellate, *Euglena gracilis*, by UV-B radiation. *Arch. Microbiol.*, **150**, 20–5.

Häder, D.-P. & Häder, M. A. (1988b). Ultraviolet-B inhibition of motility in green and dark bleached *Euglena gracilis*. *Curr. Microbiol.*, **17**, 215–20.

Häder, D.-P. & Häder, M. (1989a). Effects of solar radiation on photoorientation, motility and pigmentation in a freshwater *Cryptomonas*. *Bot. Acta*, **102**, 236–40.

Häder, D.-P. & Häder, M. A. (1989b). Effects of solar and artificial radiation on motility and pigmentation in *Cyanophora paradoxa*. *Arch. Microbiol.*, **152**, 453–7.

Häder, D.-P. & Häder, M. A. (1989c). Effects of solar UV-B irradiation on photomovement and motility in photosynthetic and colorless flagellates. *Environ. Exp. Bot.*, **29**, 273–82.

Häder, D.-P. & Häder, M. (1990). Effects of UV radiation on motility, photo-orientation and pigmentation in a freshwater *Cryptomonas*. *J. Photochem. Photobiol. B Biol.*, **5**, 105–14.

Häder, D.-P. & Häder, M. (1991a). Effects of solar and artificial U.V. radiation on motility and pigmentation in the marine *Cryptomonas maculata*. *Environ. Exp. Bot.*, **31**, 33–41.

Häder, D.-P. & Häder, M. A. (1991b). Effects of solar radiation on motility in *Stentor coeruleus*. *Photochem. Photobiol.*, **54**, 423–8.

Häder, D.-P., Häder, M., Liu, S.-M. & Ullrich, W. (1990a). Effects of solar radiation on photoorientation, motility and pigmentation in a freshwater *Peridinium*. *Biosystems*, **23**, 335–43.

Häder, D.-P. & Hansel, A. (1991). Response of *Dictyostelium discoideum* to multiple environmental stimuli. *Bot. Acta.*, **104**, 200–5.

Häder, D.-P. & Hemmersbach, R. (1997). Graviperception and graviorientation in flagellates. *Planta*, **203**, 7–10.

Häder, D.-P. & Hoiczyk, E. (1992). Gliding motility, in *Algal Cell Motility*. Ed., M. Melkonian. Chapman and Hall, New York, pp. 1–38.

Häder, D.-P. & Lebert, M. (1985). Real time computer-controlled tracking of motile microorganisms. *Photochem. Photobiol.*, **42**, 509–14.

Häder, D.-P. & Lebert, M. (1998). The photoreceptor for phototaxis in the photosynthetic flagellate *Euglena gracilis*. *Photochem. Photobiol.*, **68**, 260–5.

Häder, D.-P. & Lebert, M. (2000). Real-time tracking of microorganisms, in *Image Analysis: Methods and Applications*. Ed., D.-P. Häder. CRC Press, Boca Raton, pp. 393–422.

Häder, D.-P. & Lebert, M. (2001). Einzeller als Modelle für das Verständnis der Wirkung der Gravitation auf biologische Systeme, in *Mensch – Leben – Schwerkraft – Kosmos*. Eds., H. Rahmann & K. A. Hirsch. Verlag Günter Heimbach, Stuttgart, pp. 150–61.

Häder, D.-P., Lebert, M. & DiLena, M. R. (1987). Effects of culture age and drugs on phototaxis in the green flagellate, *Euglena gracilis*. *Plant Physiol.*, **6**, 169–74.

Häder, D.-P., Lebert, M. & Richter, P. (1998). Gravitaxis and graviperception in *Euglena gracilis*. *Adv. Space Res.*, **21**, 1277–84.

Häder, D.-P., Lebert, M. & Richter, P. (1999). Gravitaxis and graviperception in flagellates and ciliates., in *Proceedings 14th ESA Symposium on European Rocket and Balloon Programmes and Related Research (ESA SP-437)*. Potsdam, Germany, pp. 479–86.

Häder, D.-P. & Liu, S.-M. (1990a). Effects of artificial and solar UV-B radiation on the gravitactic orientation of the dinoflagellate, *Peridinium gatunense*. *FEMS Microbiol. Ecol.*, **73**, 331–8.

Häder, D.-P. & Liu, S.-M. (1990b). Motility and gravitactic orientation of the flagellate, *Euglena gracilis*, impaired by artificial and solar UV-B radiation. *Curr. Microbiol.*, **21**, 161–8.

Häder, D.-P., Liu, S.-M. & Kreuzberg, K. (1991a). Orientation of the photosynthetic flagellate, *Peridinium gatunense*, in hypergravity. *Curr. Microbiol.*, **22**, 165–72.

Häder, D.-P., Porst, M., Tahedl, H., Richter, P. & Lebert, M. (1997). Gravitactic orientation in the flagellate *Euglena gracilis*. *Micrograv. Sci. Technol.*, **10**, 53–7.

Häder, D.-P. & Reinecke, E. (1991). Phototactic and polarotactic responses of the photosynthetic flagellate, *Euglena gracilis*. *Acta Protozool.*, **30**, 13–8.

Häder, D.-P., Reinecke, E., Vogel, K. & Kreuzberg, K. (1991b). Responses of the photosynthetic flagellate, *Euglena gracilis*, to hypergravity. *Europ. Biophys. J.*, **20**, 101–7.

Häder, D.-P., Rhiel, E. & Wehrmeyer, W. (1988). Ecological consequences of photomovement and photobleaching in the marine flagellate *Cryptomonas maculata*. *FEMS Microbiol. Ecol.*, **53**, 9–18.

Häder, D.-P., Rosum, A., Schäfer, J. & Hemmersbach, R. (1995). Gravitaxis in the flagellate *Euglena gracilis* is controlled by an active gravireceptor. *J. Plant Physiol.*, **146**, 474–80.

Häder, D.-P., Rosum, A., Schäfer, J. & Hemmersbach, R. (1996). Graviperception in the flagellate *Euglena gracilis* during a shuttle space flight. *J. Biotechnol.*, **47**, 261–9.

Häder, D.-P. & Vogel, K. (1991). Simultaneous tracking of flagellates in real time by image analysis. *J. Math. Biol.*, **30**, 63–72.

Häder, D.-P., Vogel, K. & Schäfer, J. (1990b). Responses of the photosynthetic flagellate, *Euglena gracilis*, to microgravity. *Appl. Micrograv. Technol.*, **3**, 110–6.

Hamill, O. P., Holubec, K. V. & Gao, F. (2000). Calcium-activated apoptosis in frog (*Xenopus laevis*) red blood cells. *J. Physiol. (Lond.)*, **527**, 46.

Hamill, O. P. & Martinac, B. (2001). Molecular basis of mechanotransduction in living cells. *Physiol. Rev.*, **81**, 685–740.

Hamill, O. P. & McBride, D. W., Jr. (1994). Molecular mechanisms of mechanoreceptor adaptation. *NIPS*, **9**, 53–9.

Hamill, O. P. & McBride, D. W., Jr. (1996). The pharmacology of mechanogated membrane ion channels. *Pharmacol. Rev.*, **48**, 231–52.

Hammacher, H., Fitton, B. & Kingdom, J. (1987). The environment of Earth-orbiting systems, in *Fluid Sciences and Materials Science in Space. A European Perspective.* Ed. H. U. Walter, Springer-Verlag, Berlin, pp. 1–50.

Hara, R. & Asai, H. (1980). Electrophysiological responses of *Didinium nasutum* to *Paramecium* capture and mechanical stimulation. *Nature*, **283**, 869–70.

Harz, H. & Hegemann, P. (1991). Rhodopsin-regulated calcium currents in *Chlamydomonas*. *Nature*, **351**, 489–91.

Harz, H., Nonnengässer, C. & Hegemann, P. (1992). The photoreceptor current of the green alga *Chlamydomonas*. *Phil. Trans. R. Soc. Lond. B*, **338**, 39–52.

Hasenstein, K. H. (1999). Gravisensing in plants and fungi. *Adv. Space. Res.*, **24**, 677–85.

Hatton, J. P., Gaubert, F., Lewis, M. L., Darsel, Y., Ohlmann, P., Cazenave, J.-P. & Schmitt, D. (1999). The kinetics of translocation and cellular quantity of protein kinase C in human leukocytes are modified during spaceflight. *FASEB J.*, **13**, S23–S33.

Haugland, R. P. (1997). *Handbook of Fluorescent Probes and Research Chemicals. Molecular Probes*. Eugene, Oregon.

Haugland, R. P. (2002). *Handbook of Fluorescent Probes and Research Chemicals. Molecular Probes.* Eugene, Oregon.

Haupt, W. (1959). Die Phototaxis der Algen, in *Handbuch der Pflanzenphysiologie*. Vol. **XVII**, part 2, Ed., W. Ruhland. Springer-Verlag, Berlin, pp. 318–70.

Haupt, W. (1962a). Bewegungen. *Fortschr. Botanik*, **24**, 377–92.

Haupt, W. (1962b). Geotaxis, in *Handbuch der Pflanzenphysiologie*. Vol. **XVII**, part 2, Ed., W. Ruhland. Springer-Verlag, Berlin, pp. 390–5.

Haupt, W. (1962c). Thermotaxis, in *Handbuch der Pflanzenphysiologie*. Vol. **XVII**, part 2, Ed., W. Ruhland. Springer-Verlag, Berlin, pp. 29–33.

Haupt, W. (1966). Phototaxis in plants. *Int. Rev. Cytol.*, **19**, 267–99.

Haupt, W. (1975). Locomotion. *Fortschr. Botanik*, **37**, 177–85.

Hausmann, K. & Hülsmann, N. (1996). *Protozoology*. Thieme Verlag, Stuttgart.

Hegemann, P. & Deininger, W. (2001). Algal eyes and their rhodopsin photoreceptors, in *Photomovement*. Eds., D.-P. Häder & M. Lebert. Elsevier, Amsterdam, pp. 229–43.

Hegemann, P., Gärtner, W. & Uhl, R. (1991). All-trans retinal constitutes the functional chromophore in *Chlamydomonas rhodopsin*. *Biophys. J.*, **60**, 1477–89.

Hegemann, P. & Harz, H. (1998). How microalgae see the light, in *Microbial Responses to Light and Time*. Eds., M. X. Caddick, S. Baumberg, D. A. Hodgson & M. K. Phillip-Jones. Soc. Gen. Microbiol. Symp. Cambridge University Press, pp. 95–105.

Hegemann, P., Hegemann, U. & Foster, K. W. (1988). Reversible bleaching of *Chlamydomonas reinhardtii* rhodopsin in vivo. *Photochem. Photobiol.*, **48**, 123–8.

Helfrich, W. (1973). Elastic properties of lipid bilayers: theory and possible experiments. *Z. Naturforsch.*, **28**, 327–72.

Hellingwerf, K. J., Hoff, W. D. & Crielaard, W. (1996). Photobiology of microorganisms: how photosensors catch a photon to initialize signalling. *Mol. Microbiol.*, **21**, 683–93.

Helmchen, F., Gerard, J., Borst, G. & Sakmann, B. (1997). Calcium dynamics associated with a single action potential in a CNS presynaptic terminal. *Biophys. J.*, **72**, 1458.

Hemmersbach, R., Becker, E. & Stockem, W. (1997). Influence of extremely low frequency electro-magnetic fields on the swimming behavior of ciliates. *Bioelectromagnetics*, **18**, 491–8.

Hemmersbach, R. & Bräucker, R. (2002). Gravity-related behaviour in ciliates and flagellates, in *Cell Biology and Biotechnology in Space*. Ed., A. Cogoli. Elsevier, Amsterdam, pp. 59–75.

Hemmersbach, R. & Briegleb, W. (1987). Experiments concerning gravisensitivity of protozoa. *ESA-SP* **271**, 291–4.

Hemmersbach, R., Bromeis, B., Bräucker, R., Krause, M., Freiberger, N., Stieber, C. & Wilczek, M. (2001). *Paramecium* – a model system for studying cellular graviperception. *Adv. Space Res.*, **27**, 893–8.

Hemmersbach, R. & Häder, D.-P. (1999). Graviresponses of certain ciliates and flagellates. *FASEB J.*, **13**, S69–S75.

Hemmersbach, R., Tairbekov, M., Gawrilowa, O., Rieder, N., Send, W., Bromeis, B., Wilczek, M., Neubert, J. & Mulisch, M. (1999a). Morphology and physiology of *Loxodes* after cultivation in space. *ESA-SP*, **1222**, 119–26.

Hemmersbach, R., Volkmann, D. & Häder, D.-P. (1999b). Graviorientation in protists and plants. *J. Plant Physiol.*, **154**, 1–15.

Hemmersbach, R., Voormanns, R., Briegleb, W., Rieder, N. & Häder, D.-P. (1996a). Influence of accelerations on the spatial orientation of *Loxodes* and *Paramecium*. *J. Biotechnol.*, **47**, 271–8.

Hemmersbach, R., Voormanns, R., Bromeis, B., Schmidt, N., Rabien, H. & Ivanova, K. (1998). Comparative studies of the graviresponses of *Paramecium* and *Loxodes*. *Adv. Space Res.*, **21**, 1285–9.

Hemmersbach, R., Voormanns, R. & Häder, D.-P. (1996b). Graviresponses in *Paramecium biaurelia* under different accelerations: studies on the ground and in space. *J. Exp. Biol.*, **199**, 2199–205.

Hemmersbach, R., Voormanns, R., Rabien, R., Bromeis, B. & Ivanova, K. (1996c). Graviperception and signal transduction in unicellular organisms. *ESA-SP (Special Publication)*, **390**, 193–7.

Hemmersbach, R., Wilczek, M., Stieber, C., Bräucker, R. & Ivanova, K. (2002). Variable acceleration influences cAMP in *Paramecium biaurelia*. *ESA-SP*, **501**, 343–4.

Hemmersbach-Krause, R., Briegleb, W. & Häder, D.-P. (1991a). Dependence of gravitaxis in *Paramecium* on oxygen. *Europ. J. Protistol.*, **27**, 278–82.

Hemmersbach-Krause, R., Briegleb, W., Häder, D.-P. & Plattner, H. (1991b). Gravity effects on *Paramecium* cells: an analysis of a possible sensory function of trichocysts and of simulated weightlessness of trichocyst exocytosis. *Europ. J. Protistol.*, **27**, 85–92.

Hemmersbach-Krause, R., Briegleb, W. & Häder, D.-P. (1992). Swimming behavior of *Paramecium* – first results with the low-speed centrifuge microscope (NIZEMI). *Adv. Space Res.*, **12**, 113–6.

Hemmersbach-Krause, R., Briegleb, W., Häder, D.-P., Vogel, K., Grothe, D. & Meyer, I. (1993a). Orientation of *Paramecium* under the conditions of weightlessness. *J. Euk. Microbiol.*, **40**, 439–46.

Hemmersbach-Krause, R., Briegleb, W., Häder, D.-P., Vogel, K., Klein, S. & Mulisch, M. (1994). Protozoa as model systems for the study of cellular responses to altered gravity conditions. *Adv. Space Res.*, **14**, 49–60.

Hemmersbach-Krause, R., Briegleb, W., Vogel, K. & Häder, D.-P. (1993b). Swimming velocity of *Paramecium* under the conditions of weightlessness. *Acta Protozool.*, **32**, 229–36.

Hemmersbach-Krause, R. & Häder, D.-P. (1990). Negative gravitaxis (geotaxis) of *Paramecium* – demonstrated by image analysis. *Appl. Micrograv. Technol.*, **4**, 221–3.

Hendriks, J., Gensch, T., Hviid, L., van der Horst, M. A., Hellingwerf, K. J. & van Thor, J. J. (2002). Transient exposure of hydrophobic surface in the photoactive yellow protein monitored with Nile Red. *Biophys. J.*, **82**, 1632–43.

Hennessey, T., Machemer, H. & Nelson, D. L. (1985). Injected cyclic AMP increases ciliary beat frequency in conjunction with membrane hyperpolarization. *Europ. J. Cell Biol.*, **36**, 153–6.

Hénon, S., Lenormand, G., Richert, A. & Gallet, F. (1999). A new determination of the shear modulus of the human erythrocyte membrane using optical tweezers. *Biophys. J.*, **76**, 1145–51.

Hensel, W. (1985). Cytochalasin B affects the structural polarity of statocytes from cress roots (*Lepidium sativum* L.). *Protoplasma*, **129**, 178–87.

Hensel, W. & Iversen, T. H. (1980). Ethylene production during clinostat rotation and effect on root geotropism. *Z. Pflanzenphys.*, **97**, 343–52.

Hensel, W. & Sievers, A. (1980). Effects of prolonged omnilateral gravistimulation on the ultrastructure of statocytes and on the graviresponse of roots. *Planta*, **150**, 338–46.

Hildebrandt, A. (1974). Effects of some inhibitors in *Euglena* photomotion, in *Progress in Photobiology, Proc. 6th Int. Congr. Photobiol.* Ed., G. Schenk. Dtsch. Ges. Lichtforsch. E. V., Frankfurt, p. 358.

Hill, N. A. & Häder, D.-P. (1997). A biased random walk for the trajectories of swimming micro-organisms. *J. Theor. Biol.*, **186**, 503–26.

Hinrichsen, R. & Schultz, J. E. (1988). *Paramecium*: a model system for the study of excitable cells. *Trends Neurosci.*, **11**, 27–32.

Hinrichsen, R. D., Fraga, D. & Reed, M. W. (1992). 3′-Modified antisense oligodeoxyribonucleotides complementary to calmodulin mRNA alter behavioral responses in *Paramecium*. *Proc. Natl. Acad. Sci. U.S.A.*, **89**, 8601–5.

Hiramoto, Y. & Kamitsubo, E. (1995). Centrifuge microscope as a tool in the study of cell motility. *Int. Rev. Cytol.*, **157**, 99–128.

Hochmuth, R. M. & Waugh, R. E. (1987). Erythrocyte membrane elasticity and viscosity. *Annu. Rev. Physiol.*, **49**, 209–19.

Holland, E.-M., Braun, F.-J., Nonnengässer, C., Harz, H. & Hegemann, P. (1996). The nature of rhodopsin-triggered photocurrents in *Chlamydomonas*. I. Kinetics and influence of divalent ions. *Biophys. J.*, **70**, 924–31.

Holland, E.-M., Harz, H., Uhl, A. & Hegemann, F. P. (1997). Control of phobic behavioral responses by rhodopsin-induced photocurrents in *Chlamydomonas*. *Biophys. J.*, **73**, 1395–401.

Holmes, R. W., Williams, P. M. & Eppley, R. W. (1967). Red water in La Jolla Bay, 1964–1966. *Limnol. Oceanogr.*, **12**, 503–12.

Holton, T. & Hudspeth, A. J. (1986). The transduction channel of hair cells from the bull-frog characterized by noise analysis. *J. Physiol.*, **375**, 195–227.

Horie, T., Schimek, C. & Ootaki, T. (1998). Gravitropic responses of *Phycomyces* and *Pilobolus* sporangiophores and possible mode of action (Abstract). Jerusalem, Israel, p. 103.

Horneck, G., Baumstark-Khan, C. & Reitz, G. (2002a). Space microbiology, effects of ionizing radiation on microorganisms in space, in *The Encyclopedia on Environmental Microbiology*. Ed., G. Bitton. Wiley, New York, pp. 2985–96.

Horneck, G., Mileikowsky, Melosh, H. J., Wilson, J. W., Cucinotta, F. A. & Gladman, B. (2002b). Viable transfer of microorganisms in the solar system and beyond, in *Astrobiology: The Quest for the Conditions of Life*. Eds., G. Horneck & C. Baumstark-Khan. Springer-Verlag, Berlin, pp. 57–76.

Hoson, T., Kamisaka, S., Masuda, Y., Yamashita, M. & Buchen, B. (1997). Evaluation of the three-dimensional clinostat as a simulator of weightlessness. *Planta*, **203**, 187–97.

Hoson, T., Kamisaka, S., Yamamoto, R., Yamashita, M. & Masuda, Y. (1995). Automorphosis of maize shoots under simulated microgravity on a three-dimensional clinostat. *Physiol. Plant.*, **93**, 346–51.

Howard, J. E., Roberts, W. M. & Hudspeth, A. J. (1988). Mechanoelectrical transduction by hair cells. *Annu. Rev. Biophys. Biophys. Chem.*, **17**, 99–124.

Hsu, C.-Y. & Li, C.-W. (1994). Magnetoreception in honeybees. *Science*, **265**, 95–7.

Hubert, G., Rieder, N., Schmitt, G. & Send, W. (1975). Bariumanreicherung in den Müllerschen Körperchen der Loxodidae (Ciliata, Holotricha). *Z. Naturforsch.*, **30**, 422–3.

Hudspeth, A. J. (1989). How the ear's works work. *Nature*, **341**, 397–404.

Hudspeth, A. J. (1997). Mechanical amplification of stimuli by hair cells. *Curr. Opin. Neurobiol.*, **7**, 480–6.

Hudspeth, A. J., Choe, Y., Mehta, A. D. & Martin, P. (2000). Putting ion channels to work: mechanoelectrical transduction, adaptation, and amplification by hair cells. *Proc. Natl. Acad. Sci. U.S.A.*, **97**, 11765–72.

Hughes-Fulford, M. & Lewis, M. L. (1996). Effects of microgravity on osteoblast growth activation. *Exp. Cell Res.*, **224**, 103–9.

Huijser, R. H., Bergh, L. C. v. d., Postema, R. J. & Schelling, R. (1995). Cell biology on sounding rockets – report on CIS-3 and CIS4 flights, in *Proceedings 12th Symp. Europ. Rocket and Balloon Programmes and Related Results*, Lillehammer, Norway. ESA-SP, pp. 161–74.

Hunte, C., Schulz, M. & Schnabl, H. (1993). Influence of clinostat rotation on plant proteins. 2. Effects on membrane bound enzyme activities and ubiquitin-protein-conjugates in leaves of *Vicia faba* L. *J. Plant Physiol.*, **142**, 31–6.

Huttenlauch, I., Geisler, N., Plessmann, U., Peck, R. K., Weber, K. & Stick, R. (1995). Major epiplasmic proteins of ciliates are articulins: cloning, recombinant expression and structural characterization. *J. Cell. Biol.*, **130**, 1401–12.

Huttenlauch, I., Peck, R. K., Plessmann, U., Weber, K. & Stick, R. (1998a). Characterization of two articulins, the major epiplasmic proteins comprising the membrane skeleton of the ciliate *Pseudomicrothorax*. *J. Cell Sci.*, **111**, 1909–19.

Huttenlauch, I., Peck, R. K. & Stick, R. (1998b). Articulins and epiplasmins: two distinct classes of cytoskeletal proteins of the membrane skeleton of protists. *J. Cell Sci.*, **111**, 3367–78.

Ijiri, K. (1995). *The First Vertebrate Mating in Space – A Fish Story*. RICUT, Tokyo, Japan.

Ikemura, T. (1985). Codon usage and t-RNA content in unicellular and multicellular organisms. *Mol. Biol. Evol.*, **2**, 13–34.

Imae, Y. & Atsumi, T. (1989). Na^+-driven bacterial flagellar motors. *J. Bioenerg. Biomembr.*, **21**, 705–16.

Iseki, M., Matsunaga, S., Murakami, A., Ohno, K., Shiga, K., Yoshida, C., Sugai, M., Takahashi, T., Hori, T. & Watanabe, M. (2002). A blue-light-activated adenylyl cyclase mediates photoavoidance in *Euglena gracilis*. *Nature*, **415**, 1047–51.

Iwatsuki, K. (1992). *Stentor coeruleus* shows positive photokinesis. *Photochem. Photobiol.*, **55**, 469–71.

Iwatsuki, K., Hirano, T., Kawase, M., Chiba, H., Michibayashi, N., Yamada, C., Sumiyoshi, N., Yagi, K. & Mizoguchi, T. (1996). Thigmotaxis in *Paramecium caudatum* is induced by hydrophobic or polyaniline-coated glass surface to which liver cells from rat adhere with forming multicellular spheroids. *Europ. J. Protistol.*, **32**, 58–61.

Jaffe, M. J., Wakefield, A. H., Telewski, F., Gulley, E. & Biro, R. (1985). Computer-assisted image analysis of plant growth, thigmomorphogenesis, and gravitropism. *Plant Physiol.*, **77**, 722–30.

Jahn, T. L. (1946). The Euglenoid flagellates. *Anat. Rev. Biol.*, **21**, 246–74.

Jahn, T. L. & Votta, J. J. (1972). Locomotion of protozoa. *Ann. Rev. Fluid Mech.*, **4**, 93–116.

Jähne, B. (1989). *Digitale Bildverarbeitung*, Springer-Verlag, Berlin.

Jenkins, G. I., Courtice, G. R. M. & Cove, D. J. (1986). Gravitropic responses of wild-type and mutant strains of the moss *Physcomitrella patens*. *Plant Cell Environ.*, **9**, 637–44.

Jennings, H. S. (1906). *Behavior of the Lower Organisms*. Columbia University Press, New York.

Jennings, H. S. (1910). *Das Verhalten der niederen Organismen*. Translated into German by E. Mangold, Leipzig, Berlin.

Jensen, P. (1893). Über den Geotropismus niederer Organismen. *Pflüger's Arch. Ges. Physiol.*, **53**, 428–80.

Jimenez, C., Figueroa, F. L., Aguilera, J., Lebert, M. & Häder, D.-P. (1996). Phototaxis and gravitaxis in *Dunaliella bardawil*: influence of UV radiation. *Acta Protozool.*, **35**, 287–95.

Jobbagy, A. & Furnee, E. H. (1994). Marker centre estimation algorithms in CCD camera-based motion analysis. *Med. Biol. Eng. Comput.*, **32**, 85–91.

Johnson, C. H., Knight, M. R., Kondo, T., Masson, P., Sedbrook, J., Haley, A. & Trewavas, A. (1995). Circadian oscillations of cytosolic and chloroplastic free calcium in plants. *Science*, **269**, 1863.

Joop, O., Kreuzberg, K. & Treichel, R. (1989). The slow-rotating centrifuge microscope (NIZEMI). *ASGSB Bull*, **3**, 55.

Julez, B. & Harmon, L. D. (1984). Noise and recognizability of coarse quantized images. *Nature*, **308**, 211–2.

Juniper, B. E., Gorves, S., Landau-Schachar, B. & Audus, L. J. (1966). Root cap and the perception of gravity. *Nature*, **209**, 93–4.

Kalinec, F., Holley, M. C., Iwasa, K. H., Lim, D. J. & Kachar, B. (1992). A membrane-based force generation mechanism in auditory sensory cells. *Proc. Natl. Acad. Sci. U.S.A.*, **89**, 8671–5.

Kamphuis, A. (1999). Digitale Pfadanalyse am Beispiel der Schwerkraftausrichtung von *Euglena gracilis* in Flachküvetten. *Dissertation*, Rheinische-Friedrich-Wilhelms-Universität, Bonn.

Kamykowski, D. & Zentara, S.-J. (1977). The diurnal vertical migration of motile phytoplankton through temperature gradients. *Limnol. Oceanogr.*, **22**, 148–51.

Kanda, S. (1914). On the geotropism of *Paramecium* and *Spirostomum*. *Biol. Bull.*, **26**, 1–24.

Kanda, S. (1918). Further studies on the geotropism of *Paramecium caudatum*. *Biol. Bull.*, **34**, 108–19.

Kang, B. G. (1979). Epinasty, in *Physiology of Movements. Encyclopedia of Plant Physiology. N.S.* Eds., W. Haupt & M. E. Feinleib. Springer-Verlag, Berlin, pp. 647–67.

Kaplan, J. H., Graham, C. R. & Ellis-Davies, C. R. (1988). Photolabile chelators for the rapid photorelease of divalent cations. *Proc. Natl. Acad. Sci. U.S.A.*, **85**, 6571.

Katnik, C. & Waugh, R. E. (1990a). Alterations of the apparent expansivity modulus of red blood cell membrane by electric fields. *Biophys. J.*, **57**, 877–82.

Katnik, C. & Waugh, R. E. (1990b). Electric fields induce reversible changes in surface to volume ratio of micropipette-aspirated erythrocytes. *Biophys. J.*, **57**, 865–75.

Kato, R. (1990). Effects of a very low magnetic field on the gravitropic curvature of *Zea* roots. *Plant Cell Physiol.*, **31**, 565–8.

Keim, C. N., Lins, U. & Farina, M. (2001). Elemental analysis of uncultured magnetotactic bacteria exposed to heavy metals. *Can. J. Microbiol.*, **47**, 1132–6.

Kern, V. D. & Hock, B. (1993). *Gravitropism of fungi – experiments in Space. ESA-SP*, 366, pp. 49–60.

Kern, V. D., Mendgen, K. & Hock, B. (1997). *Flammulina* as a model system for fungal graviresponses. *Planta*, **203**, 23–32.

Kern, V. D., Smith, J. D., Schwuchow, J. M. & Sack, F. D. (2001). Amyloplasts that sediment in protonemata of the moss *Ceratodon purpureus* are nonrandomly distributed in microgravity. *Plant Physiol.*, **125**, 2085–94.

Kessler, J. O. (1985a). Cooperative and concentrative phenomena of swimming microorganisms. *Contemp. Phys.*, **26**, 147–66.

Kessler, J. O. (1985b). Hydrodynamic focusing of motile algal cells. *Nature*, **313**, 218–20.

Kessler, J. O. (1986). Individual and collective fluid dynamics of swimming cells. *J. Fluid Mech.*, **173**, 191–205.

Kessler, J. O. (1989). Path and pattern – the mutual dynamics of swimming cells and their environment. *Comments Theor. Biol.*, **1**, 85–108.

Kessler, J. O. (1992). Theory and experimental results on gravitational effects on monocellular algae. *Adv. Space Res.*, **12**, 33–42.

Kessler, J. O., Hill, N. A. & Häder, D.-P. (1992). Orientation of swimming flagellates by simultaneously acting external factors. *J. Phycol.*, **28**, 816–22.

Kincaid, R. L. (1991). Signaling mechanisms in microorganisms: common themes in the evolution of signal transduction pathways, in *Advances in Second Messenger and Phosphoprotein Res.* Eds., P. Greengard & G. A. Robison. Raven Press, Ltd., New York, pp. 165–84.

Kinsman, R., Ibelings, B. W. & Walsby, A. E. (1991). Gas vesicle collapse by turgor pressure and its role in buoyancy regulation by *Anabaena flos-aquae. J. Gen. Microbiol.*, **137**, 1171–8.

Kirschvink, J. L. (1980). South-seeking magnetic bacteria. *J. Exp. Biol.*, **86**, 345–7.

Kirschvink, J. L. (1989). Magnetite biomineralization and geomagnetic sensitivity in higher animals: an update and recommendations for future study. *Bioelectromagnetics*, **10**, 239–59.

Kirschvink, J. L. (1997). Homing in on vertebrates. *Nature*, **390**, 339–40.

Kiss, J. Z. (1994). The response to gravity is correlated with the number of statoliths in *Chara* rhizoids. *Plant Physiol.*, **105**, 937–40.

Kiss, J. Z. (2000). Mechanisms of the early phases of plant gravitropism. *Crit. Rev. Plant Sci.*, **19**, 551–73.

Kiss, J. Z., Guisinger, M. M., Miller, A. J. & Stackhouse, K. S. (1997). Reduced gravitropism in hypocotyls of starch-deficient mutants of *Arabidopsis. Plant Cell Physiol.*, **38**, 518–25.

Kiss, J. Z., Hertel, R. & Sack, F. D. (1989). Amyloplasts are necessary for full gravitropic sensitivity in roots of *Arabidopsis thaliana. Planta*, **177**, 198–206.

Kiss, J. Z., Katembe, W. J. & Edelmann, R. E. (1998). Gravitropism and development of wild-type and starch-deficient mutants of *Arabidopsis* during spaceflight. *Physiol. Plant.*, **102**, 493–502.

Kiss, J. Z., Wright, J. B. & Caspar, T. (1996). Gravitropism in roots of intermediate-starch mutants of *Arabidopsis. Physiol. Plant.*, **97**, 237–44.

Klaus, D. (2002). Space microbiology: microgravity and microorganisms, in *Encyclopedia of Environmental Microbiology*. Ed., D. Britton. J. Wiley and Sons, New York, pp. 2996–3004.

Klaus, D., Schatz, A., Neubert, J., Höfer, M. & Todd, P. (1997a). *Escherichia coli* growth kinetics: a definition of "functional weightlessness" and a comparison of clinostat and space flight results. *Naturwissenschaften*, **143**, 449–55.

Klaus, D., Simske, S., Todd, P. & Stodieck, L. (1997b). Investigation of space flight effects on *Escherichia coli* and a proposed model of underlying physical mechanisms. *Microbiology*, **143**, 449–55.

Klaus, D. M., Todd, P. & Schatz, A. (1998). Functional weightlessness during clinorotation of cell suspensions. *Adv. Space Res.*, **21**, 1315–8.

Kloda, A. & Martinac, B. (2001a). Common evolutionary origins of mechanosensitive ion channels in archaea, bacteria and cell-walled eukaryotic organisms. *Archaea*, **1**, 1–10.

Kloda, A. & Martinac, B. (2001b). Molecular identification of a mechanosensitive channel in Archaea. *Biophys. J.*, **80**, 229–40.

Kloda, A. & Martinac, B. (2002). Mechanosensitive channels of bacteria and archaea share a common ancestral origin. *Europ. Biophys. J.*, **31**, 14–25.

Klopocka, W. (1983). The question of geotaxis in *Amoeba proteus. Acta Protozool.*, **22**, 211–7.

Knecht, D. A. (1993). Preparation of figures for publication using a digital color printer. *Biotechnology*, **14**, 1006–8.

Knight, M. R., Campell, A. K., Smith, S. M. & Trewavas, A. J. (1991). Transgenic plant aequorin reports the effects of touch and cold-shock and elictors on cytoplasmatic calcium. *Nature*, **352**, 524.

Knight, T. A. (1806). On the direction of radicle and germen during the vegetation of seeds. *Phil. Trans. Roy. Soc. (Lond.) B*, **96**, 99–108.

Kogan, A. B. & Tikhonova, N. A. (1965). Effect of a constant magnetic field on the movement of *Paramecia*. *Biophysics*, **10**, 322–8.

Köhler, O. (1921). Über die Geotaxis von *Paramecium. Verh. Dtsch. Zool. Ges.*, **26**, 69–71.

Köhler, O. (1922). Über die Geotaxis von *Paramecium.* I. *Arch. Protistenk.*, **45**, 1–94.

Köhler, O. (1930). Über die Geotaxis von *Paramecium.* II. *Arch. Protistenk.*, **70**, 297–306.

Köhler, O. (1939). Ein Filmprotokoll zum Reizverhalten querzertrennter *Paramecien. Zool. Anz. Suppl.*, **12**, 132–42.

Kokubo, Y. & Hardy, W. H. (1982). Digital image processing: a path to better pictures. *Ultramicroscopy*, **8**, 277–86.

Koller, D. (2001). Solar navigation by plants, in *Photomovement.* Eds., D.-P. Häder & M. Lebert. Elsevier Science B.V., Amsterdam, pp. 833–95.

Kondepudi, D. & Prigogine, I. (1981). Sensitivity of non-equilibrium systems. *Physica*, **107A**, 1–24.

Kordyum, V. A., Polidova, L. S. & Mashinsky, A. L. (1978). Spaceflight effects on developing organisms. *Naukova Dumka Publishers*, 64–8.

Kordyum, E. L. (1997). Biology of plant cells in microgravity and under clinostating. *Int. Rev. Cytol.*, **171**, 1–78.

Korohoda, W., Mycielska, M., Janda, E. & Madeja, Z. (2000). Immediate and long-term galvanotactic responses of *Amoeba proteus* to dc electric fields. *Cell Motil. Cytoskel.*, **45**, 10–26.

Kowalewski, U., Bräucker, R. & Machemer, H. (1998). Responses of *Tetrahymena pyriformis* to the natural gravity vector. *Micrograv. Sci. Technol.*, **11**, 167–72.

Krah, M., Marwan, W., Vermeglio, A. & Oesterhelt, D. (1994). Phototaxis of *Halobacterium salinarium* requires a signalling complex of sensory rhodopsin I and its methyl-accepting transducer Htrl. *EMBO J.*, **13**, 2150–5.

Krause, M. (1999). Elektrophysiologie, Mechanosensitivität und Schwerkraftbeantwortung von *Bursaria truncatella. Diploma thesis*. Ruhr-Universität Bochum.

Krause, M. (2003). Schwerkraftwahrnehmung des Ciliaten *Stylonychia mytilus*: elektro- und verhaltensphysiologische Untersuchungen. *Dissertation*, Rheinische-Friedrich-Wilhelms-Universität, Bonn.

Kreimer, G. (1994). Cell biology of phototaxis in flagellate algae. *Int. Rev. Cytol.*, **148**, 229–309.

Kreimer, G. (2001). Light perception and signal modulation during photoorientation of flagellate green algae, in *Photomovement.* Eds., D.-P. Häder & M. Lebert. Elsevier, Amsterdam, pp. 193–227.

Kreimer, G., Marner, F.-J., Brohsonn, U. & Melkonian, M. (1991). Identification of 11-*cis* and all-*trans* retinal in the photoreceptive organelle of a flagellate green alga. *FEBS Lett.*, **293**, 49–52.

Kreimer, G. & Melkonian, M. (1990). Reflection confocal laser scanning microscopy of eyespots in flagellated green algae. *Europ. J. Cell Biol.*, **53**, 101–11.

Kreuels, T., Joerres, R., Martin, W. & Brinkmann, K. (1984). System analysis of the circadian rhythm of *Euglena gracilis*. II. Masking effects and mutual interactions of light and temperature responses. *Z. Naturf.*, **39c**, 801–11.

Kuhlmann, H.-W. (1994). Escape response of *Euplotes octocarinatus* to turbellarian predators. *Arch. Protistenk.*, **144**, 163–71.

Kuhlmann, H.-W. (1998). Photomovements in ciliated protozoa. *Naturwissenschaften*, **85**, 143–54.

Kuhlmann, H.-W., Bräucker, R. & Schepers, A. G. (1997a). Phototaxis in *Porpostoma notatum*, a marine scuticociliate with a composed crystalline organelle. *Europ. J. Protistol.*, **33**, 295–304.

Kuhlmann, H.-W., Brünen-Nieweler, C. & Heckmann, K. (1997b). Pheromones of the ciliate *Euplotes octocarinatus* not only induce conjugation but also function as chemoattractants. *J. Exp. Zool.*, **277**, 38–48.

Kuhlmann, H. W. & Hemmersbach-Krause, R. (1993a). Phototaxis in the "stigma"-forming ciliate *Nassula citrea*. *J. Photochem. Photobiol.*, **21**, 191–5.

Kuhlmann, H.-W. & Hemmersbach-Krause, R. (1993b). Phototaxis in an "eyespot"-exposing ciliate. *Naturwissenschaften*, **80**, 139–41.

Kuhlmann, H.-W., Kusch, J. & Heckmann, K. (1998). Predator-induced defenses in ciliated protozoa, in *The Evolution of Inducible Defenses*. Eds., R. Tollrian & C. D. Harvell. Princeton University Press, Princeton, New Jersey, pp. 142–61.

Kühnel-Kratz, C. & Häder, D.-P. (1993). Real time three-dimensional tracking of ciliates. *J. Photochem. Photobiol.*, **19**, 193–200.

Kühnel-Kratz, C. & Häder, D.-P. (1994). Light reactions of the ciliate *Stentor coeruleus* – a three-dimensional analysis. *Photochem. Photobiol.*, **59**, 257–62.

Kühnel-Kratz, C., Schäfer, J. & Häder, D.-P. (1993). Phototaxis in the flagellate, *Euglena gracilis*, under the effect of microgravity. *Micrograv. Sci. Technol.*, **6**, 188–93.

Kung, C., Chang, S.-Y., Satow, Y., van Houten, J. & Hansma, H. (1975). Genetic dissection of behavior in *Paramecium*. *Science*, **188**, 898–905.

Kung, C. & Eckert, R. (1972). Genetic modification of electric properties in an excitable membrane. *Proc. Natl. Acad. Sci. U.S.A.*, **69**, 93–7.

Kuroda, K. & Kamiya, N. (1989). Propulsive force of *Paramecium* as revealed by the video centrifuge microscope. *Exp. Cell Res.*, **184**, 268–72.

Kuroda, K., Kamiya, N. M. J. A., Yoshimoto, Y. & Hiramoto, Y. (1986). *Paramecium* behavior during video centrifuge-microscopy. *Proc. Jap. Acad. Ser. B*, **62**, 117–21.

Kuwada, Y. (1916). Some peculiarities observed in the culture of *Chlamydomonas*. *Bot. Mag.*, **30**, 347–58.

Kuznetsov, O. A. & Hasenstein, K. H. (1995). Intracellular magnetophoresis of amyloplasts and induction of root curvature. *Planta*, **198**, 87–94.

Kuznetsov, O. A., Schwuchow, J., Sack, F. D. & Hasenstein, K. H. (1999). Curvature induced by amyloplast magnetophoresis in protonemata of the moss *Ceratodon purpureus*. *Plant Physiol.*, **119**, 645–50.

Kuznicki, L. (1968). Behavior of *Paramecium* in gravity fields. I. Sinking of immobilized specimens. *Acta Protozool.*, **5**, 109–17.

Lacampagne, A., Gannier, F., Argibay, J., Garnier, D. & Le Guennec, J.-Y. (1994). The stretch-activated ion channel blocker gadolinium also blocks L-type calcium channels in isolated ventricular myocytes of the guinea-pig. *Biochim. Biophys. Acta*, **1191**, 205–8.

Lachney, C. L. & Lonergan, T. A. (1985). Regulation of cell shape in *Euglena gracilis*. III. Involvement of stable microtubules. *J. Cell Sci.*, **74**, 219–37.

Land, M. (1972). The physics and biology of animal reflectors. *Progr. Biophys. Mol. Biol*, **24**, 75–106.

Lapchine, L., Moatti, N., Richoilley, G., Templier, J., Gasset, G. & Tixador, R. (1987). The Antibo Experiment, in *Biorack on Spacelab D1 ESA-SP-1091*. Eds., N. Longdon & V. David. ESH, Paris, pp. 45–51.

Lebert, M. & Häder, D.-P. (1996). How *Euglena* tells up from down. *Nature*, **379**, 590.

Lebert, M. & Häder, D.-P. (1997a). Behavioral mutants of *Euglena gracilis*: functional and spectroscopic characterization. *J. Plant Physiol.*, **151**, 188–95.

Lebert, M. & Häder, D.-P. (1997b). Effects of hypergravity on the photosynthetic flagellate, *Euglena gracilis*. *J. Plant Physiol.*, **150**, 153–9.

Lebert, M. & Häder, D.-P. (1999a). Aquarack: long-term growth facility for 'professional' gravisensing cells. Proc. of the 2nd European Symposium on the Utilisation of the International Space Station, ESTEC, Noordwijk, The Netherlands. 16–18 Nov. 1998. *ESA-SP* 433, 533–7.

Lebert, M. & Häder, D.-P. (1999b). Image analysis: a versatile tool for numerous applications. *G. I. T. Special Ed. Imag. Microsc.*, **1**, 5–6.

Lebert, M. & Häder, D.-P. (1999c). Negative gravitactic behavior of *Euglena gracilis* can not be described by the mechanism of buoyancy-oriented upward swimming. *Adv. Space Res.*, **24**, 843–50.

Lebert, M., Porst, M. & Häder, D.-P. (1995). Long-term culture of *Euglena gracilis*: an AQUARACK progress report. *Proceedings of the C.E.B.A.S. Workshops. Annual Issue 1995, 11th C.E.B.A.S. Workshop*, Ruhr-University of Bochum, pp. 223–31.

Lebert, M., Porst, M. & Häder, D.-P. (1999a). Circadian rhythm of gravitaxis in *Euglena gracilis*. *J. Plant Physiol.*, **155**, 344–9.

Lebert, M., Porst, M., Richter, P. & Häder, D.-P. (1999b). Physical characterization of gravitaxis in *Euglena gracilis*. *J. Plant Physiol.*, **155**, 338–43.

Lebert, M., Richter, P. & Häder, D.-P. (1997). Signal perception and transduction of gravitaxis in the flagellate *Euglena gracilis*. *J. Plant Physiol.*, **150**, 685–90.

Lebert, M., Richter, P., Porst, M. & Häder, D.-P. (1996). Mechanism of gravitaxis in the flagellate *Euglena gracilis*. *Proceedings of the C.E.B.A.S. Workshops. Annual Issue*, 225–34.

Leech, D. M. & Williamson, C. E. (2001). In situ exposure to ultraviolet radiation alters the depth distribution of *Daphnia*. *Limnol. Oceanogr.*, **46**, 416–20.

Legue, V., Blancaflor, E., Wymer, C., Perbal, G., Fantin, D. & Gilroy, S. (1997). Cytoplasmic free Ca^{2+} in *Arabidopsis* roots changes in response to touch but not gravity. *Plant Physiol.*, **114**, 789–800.

Lenci, F., Ghetti, F. & Song, P.-S. (2001). Photomovement in ciliates, in *Photomovement*. Eds., D.-P. Häder & M. Lebert. Elsevier, Amsterdam, pp. 475–503.

Levandowsky, M., Childress, W. S., Spiegel, E. A. & Hunter, S. H. (1975). A mathematical model of pattern formation by swimming microorganisms. *J. Protozool.*, **22**, 296–306.

Lewis, M. L. (2002). The cytoskeleton, apoptosis, and gene expression in T lymphocytes and other mammalian cells exposed to altered gravity, in *Cell Biology and Biotechnology in Space*. Ed., A. Cogoli. Elsevier, Amsterdam, pp. 77–128.

Lewis, M. L., Reynolds, J. L., Cubano, L. A., Hatton, J. P., Lawless, B. D. & Piepmeier, E. H. (1998). Spaceflight alters microtubules and increases apoptosis in human lymphocytes (Jurkat). *FASEB J.*, **12**, 1007–18.

Linden, L. & Kreimer, G. (1995). Calcium modulates rapid protein phosphorylation/dephosphorylation in isolated eyespot apparatuses of the green alga *Spermatozopsis similis*. *Planta*, **197**, 343–51.

Lindes, D. A., Diehn, B. & Tollin, G. (1965). Phototaxigraph: recording instrument for determination of rate of response of phototactic microorganisms to light of controlled intensity and wavelength. *Rev. Sci. Inst.*, **36**, 1721–5.

Lipp, P., Lüscher, C. & Niggli, E. (1996). Photolysis of caged compounds characterized by ratiometric confocal microscopy: a new approach to homogeneously control and measure the calcium concentration in cardiac myocytes. *Cell Calcium*, **19**, 255.

Litvin, F. F., Sineshchekov, O. A. & Sineshchekov, V. A. (1978). Photoreceptor electric potential in the phototaxis of the alga *Haematococcus pluvialis*. *Nature*, **271**, 476–8.

Loeb, J. (1897). Zur Theorie der physiologischen Licht- und Schwerkraftwirkungen. *Pflüger's Arch. Ges. Physiol.*, **66**, 439–66.

Loeb, J. & Wasteneys, H. (1916). The relative efficiency of various parts of the spectrum for the heliotropic reactions of animals and plants. *J. Exp. Zool.*, **20**, 217–36.

Lohmann, K. J., Pentcheff, N. D., Nevitt, G. A., Stetten, G. D., Zimmer-Faust, R. K., Jarrard, H. E. & Boles, L. C. (1995). Magnetic orientation of spiny lobsters in the ocean: experiments with undersea coil systems. *J. Exp. Bot.*, **198**, 2041–8.

Lonergan, T. A. (1984a). A diurnal rhythm in the rate of light-induced electron flow in maize bundle sheath strips. *Photochem. Photobiol.*, **39**, 89–93.

Lonergan, T. A. (1984b). Regulation of cell shape in *Euglena gracilis*. *J. Cell Sci.*, **71**, 37–50.

Lonergan, T. A. (1984c). Regulation of cell shape in *Euglena gracilis*. II. The effect of altered extra- and intracellular Ca^{2+} concentrations and the effect of calmodulin antagonists. *J. Cell Sci.*, **71**, 37–50.

Lonergan, T. A. (1985). Regulation of cell shape in *Euglena gracilis*. IV. Localization of actin, myosin and calmodulin. *J. Cell. Sci.*, **77**, 197–208.

Lonergan, T. A. (1986). The photosynthesis and cell shape rhythms can be naturally uncoupled from the biological clock in *Euglena gracilis*. *J. Exp. Bot.*, **37**, 1334–40.

Lonergan, T. A. (1990). Role of actin, myosin, microtubules, and calmodulin in regulating the cellular shape of *Euglena gracilis*, in *Calcium as an Intracellular Messenger in Eukaryotic Microbes*. Ed., D. H. O'Day. American Society for Microbiology, Washington, D.C., pp. 258–77.

Lonergan, T. A. & Williamson, L. C. (1988). Regulation of cell shape in *Euglena gracilis*. V. Time-dependent responses to Ca^{2+} agonists and antagonists. *J. Cell Sci.*, **89**, 365–71.

Lorenz, M., Bisikirska, B., Hanus-Lorenz, B., Strzalka, K. & Sikorski, A. F. (1995). Proteins reacting with anti-spectrin antibodies are present in *Chlamydomonas* cells. *Cell Biol. Internat.*, **19**, 625–32.

Lüttge, U., Kluge, M. & Bauer, G. (1994). Botanik. 2nd edition. VCH Verlagsgesellschaft, Weinheim.

Lynch, T. M., Lintilhac, P. M. & Domozych, D. (1998). Mechanotransduction molecules in the plant gravisensory response: amyloplast/statolith membranes contain an integrin-like protein. *Protoplasma*, **201**, 92–100.

Lyon, E. P. (1905). On the theory of geotropism in *Paramecium*. *Am. J. Physiol.*, **14**, 421–32.

MacLennan, D. H., Rice, W. J. & Green, N. M. (1997). The mechanism of Ca^{2+} transport by sarcoplasmic reticulum Ca^{2+}-ATPases. *J. Biol. Chem.*, **272**, 28815.

Machemer, H. (1974). Frequency and directional responses of cilia to membrane potential changes in *Paramecium*. *J. Comp. Physiol.*, **92**, 293–316.

Machemer, H. (1988a). Electrophysiology, in *Paramecium*. Ed., H.-D. Görtz. Springer-Verlag, Berlin, pp. 185–215.

Machemer, H. (1988b). Motor control of cilia, in *Paramecium*. Ed., H.-D. Görtz. Springer-Verlag, Berlin, pp. 216–35.

Machemer, H. (1998). Unicellular responses to gravity transitions. *Space Forum*, **3**, 3–44.

Machemer, H. & Bräucker, R. (1992). Gravireception and graviresponses in ciliates. *Acta Protozool.*, **31**, 185–214.

Machemer, H. & Bräucker, R. (1996). Gravitaxis screened for physical mechanism using *g*-modulated cellular orientational behaviour. *Micrograv. Sci. Technol.*, **9**, 2–9.

Machemer, H., Bräucker, R., Takahashi, K. & Murakami, A. (1992). Short-term microgravity to isolate graviperception in cells. *Micrograv. Sci. Technol.*, **5**, 119–23.

Machemer, H. & de Peyer, J. E. (1977). Swimming sensory cells: electrical membrane parameters, receptor properties and motor control in ciliated Protozoa. *Verh. Dtsch. Zool. Ges.*, 86–110.

Machemer, H. & Deitmer, J. W. (1985). Mechanoreception in ciliates, in *Progress in Sensory Physiology*. Eds., H. Autrum, D. Ottoson, E. R. Perl, R. F. Schmidt, H. Schimazu & W. D. Willis. Springer-Verlag, Berlin, pp. 81–118.

Machemer, H. & Deitmer, J. W. (1987). From structure to behaviour: *Stylonychia* as a model system for cellular physiology, in *Progress in Protistology*. Eds., J. O. Corliss & D. J. Patterson. Biopress, Bristol, pp. 213–330.

Machemer, H. & Machemer-Röhnisch, S. (1996). Is gravikinesis in *Paramecium* affected by swimming velocity? *Europ. J. Protistol.*, **32**, 90–3.

Machemer, H., Machemer-Röhnisch, S. & Bräucker, R. (1993). Velocity and graviresponses in *Paramecium* during adaptation and varied oxygen concentrations. *Arch. Protistenk.*, **143**, 285–96.

Machemer, H., Machemer-Röhnisch, S., Bräucker, R. & Takahashi, K. (1991). Gravikinesis in *Paramecium*: theory and isolation of a physiological response to the natural gravity vector. *J. Comp. Physiol. A.*, **168**, 1–12.

Machemer, H., Nagel, U. & Bräucker, R. (1997). Assessment of *g*-dependent cellular gravitaxis: determination of cell orientation from locomotion track. *J. Theor. Biol.*, **185**, 201–11.

Machemer, H. & Teunis, P. F. M. (1996). *Sensory Coupling and Motor Responses.* Gustav Fischer Verlag, Stuttgart.

Machemer-Röhnisch, S., Bräucker, R. & Machemer, H. (1993). Neutral gravitaxis of gliding *Loxodes* exposed to normal and raised gravity. *J. Comp. Physiol. A.*, **171**, 779–90.

Machemer-Röhnisch, S., Machemer, H. & Bräucker, R. (1996). Electric-field effects on gravikinesis in *Paramecium*. *J. Comp. Physiol. A*, **179**, 213–26.

Machemer-Röhnisch, S., Nagel, U. & Machemer, H. (1999). A gravity-induced regulation of swimming speed in *Euglena gracilis*. *J. Comp. Phys.*, **185**, 517–27.

MacNab, R. M. (1977). Bacterial flagella rotating in bundles: a study in helical geometry. *Proc. Natl. Acad. Sci. U.S.A.*, **74**, 221–5.

MacNab, R. M. (1990). Genetics, structure, and assembly of the bacterial flagellum. *Microbiology*, **46**, 77–106.

MacNab, R. M. & Aizawa, S.-I. (1984). Bacterial motility and the bacterial flagellar motor. *Ann. Rev. Biophys. Bioeng.*, **13**, 51–83.

Maihle, N. J., Dedman, J. R., Means, A. R., Chafouleas, J. G. & Satir, B. H. (1981). Presence and indirect immunofluorescent localization of calmodulin in *Paramecium tetraurelia*. *J. Cell Biol.*, **89**, 695–9.

Majima, T., Hamasaki, T. & Arai, T. (1986). Increase in cellular cyclic GMP level by potassium stimulation and its relation to ciliary orientation in *Paramecium*. *Experienta*, **42**, 62–4.

Malik, N. R. (1980). Microcomputer realisations of Lynn's fast digital-filtering designs. *Med. Biol. Eng. Comput.*, **18**, 638–42.

Malik, N. R. & Huang, G. (1988). Integer filters for image processing. *Med. Biol. Eng. Comput.*, **26**, 62–7.

Mann, S., Sparks, N. H. C. & Blakemore, R. P. (1987). Ultrastructure and characterization of anisotropic magnetic inclusions in magnetotactic bacteria. *Proc. R. Soc. Lond. B*, **231**, 469–76.

Manson, M. D., Armitage, J. P., Hoch, J. A. & MacNab, R. M. (1998). Bacterial locomotion and signal transduction. *J. Bacteriol.*, **180**, 1009–22.

Marangoni, R., Colombetti, G. & Gualtieri, P. (2000). Digital filters in image analysis, in *Image Analysis: Methods and Applications*. Ed., D.-P. Häder. CRC Press, Boca Raton, pp. 93–106.

Marchese-Ragona, S. P., Mellor, J. S. & Holwill, M. E. J. (1983). Calmodulin inhibitors cause flagellar wave reversal. *J. Submicrosc. Cytol.*, **15**, 43–7.

Mardia, K. V. (1972). *Statistics of Directional Data*. Academic Press, London.

Markin, V. S. & Hudspeth, A. J. (1995). Gating-spring models of mechanoelectrical transduction by hair cells of the internal ear. *Annu. Rev. Biophys. Biomol. Struct.*, **24**, 59–83.

Marrs, J. A. & Bouck, G. B. (1992). The major membrane skeletal proteins (articulins) of *Euglena gracilis* define a novel class of cytoskeletal proteins. *J. Cell. Biol.*, **118**, 1465–75.

Martens, H., Novotny, J., Oberstrass, J., Steck, T. L., Postlethwait, P. & Nellen, W. (2002). RNAi in *Dictyostelium*: the role of RNA-directed RNA polymerases and double-stranded RNase. *Mol. Biol. Cell*, **13**, 445–53.

Marthy, H.-J. (2002). Developmental biology of animal models under varied gravity conditions: a review. *Vie Milieu*, **52**, 149–66.

Mason, W. T., Dempster, J., Zorec, R., Hoyland, J. & Lledo, P. M. (1995). Fluorescence measurements of cytosolic calcium: combined photometry with electrophysiology, in *Methods in Neurosciences. Measurement and Manipulation of Intracellular Ions*. Ed., J. Kraicer. Academic Press, Inc., London, pp. 81–122.

Mason, W. T., Hoyland, J., McCann, T. J., Somasundaram, B. & O'Brien, W. (1999). Strategies for quantitative digital imaging of biological activity in living cells with ion-sensitive fluorescence probes, in *Imaging Living Cells*. Eds., R. Rizzuto & C. Fasolato. Springer, Berlin, p. 3.

Massart, J. (1891). Recherches sur les organismes inférieurs (1). *Acad. R. Sci. Lett. Beaux-arts Belg.*, **22**, 148–67.

Mast, S. O. (1911). *Light and Behavior of Organisms*. John Wiley & Sons, New York.

Mast, S. O. (1914). Orientation in *Euglena* with some remarks on tropisms. *Biol. Zentralbl.*, **34**, 641–64.

Matsunaga, T. & Sakaguchi, T. (2000). Molecular mechanism of magnet formation in bacteria. *J. Biosci. Bioeng.*, **90**, 1–13.

Matsuoka, K. & Nakaoka, Y. (1988). Photoreceptor potential causing phototaxis of *Paramecium bursaria*. *J. Exp. Biol.*, **137**, 477–85.

Matsuoka, T., Sato, M. & Matsuoka, S. (1999). Photoreceptor pigment mediating swimming acceleration of *Blepharisma*, a unicellular organism. *Microbios*, **99**, 89–94.

Mattoni, R. H., Ebersold, W. T., Eiserling, F. A., Keller, E. D. & Romig, W. R. (1971). Induction of lysogenic bacteria in the space environment, in *The Experiments of Biosatellite 2 NASA-SP-204*. Ed., J. F. Saunders. NASA, Washington, pp. 304–24.

McNeil, P. L., Murphy, R. F. & Taylor, D. L. (1984). A method for incorporating macromolecules into adhaerent cells. *J. Cell Biol.*, **98**, 1556.

McBride, D. W., Jr. & Hamill, O. P. (1992). Pressure-clamp: a method for rapid step perturbation of mechanosensitive channels. *Pflüger's Arch. Ges. Physiol.*, **421**, 606–12.

Meldrum, F. C., Mann, S., Heywood, B. R., Frankel, R. B. & Bazylinski, D. A. (1993). Electron microscopy study of magnetosomes in two cultured vibrioid magnetotactic bacteria. *Phil. Trans. R. Soc. Lond. B*, **251**, 237–42.

Meleard, P., Gerbeaud, C., Bardusco, P., Jeandaine, N., Mitov, M. D. & Fernandez-Puente, L. (1998). Mechanical properties of model membranes studied from shape transformations of giant vesicles. *Biochimie*, **80**, 401–13.

Melkonian, M. (1992). *Algal Cell Motility*. Chapman and Hall, New York.

Melkonian, M. & Robenek, H. (1984). The eyespot apparatus of flagellated green algae: a critical review. in *Progress in Phycological Research*. Eds., F. E. Round & D. J. Chapman. Biopress Ltd., pp. 193–268.

Mendelssohn, M. (1902). Travaux originaux. I. Recherches sur la thermotaxie des organismes unicellulaires. *J. Physiol. Pathol. Gener.*, **4**, 393–409.

Mennigmann, H. D. & Lange, M. (1986). Growth and differentiation of *Bacillus subtilis* under microgravity. *Naturwissenschaften*, **73**, 415–7.

Mesland, D. A. M., Anton, A. H., Willemsen, H. & van den Ende, H. (1996). The free fall machine – a ground-based facility for microgravity research in life sciences. *Micrograv. Sci. Technol.*, **9**, 10–4.

Metzner, P. (1920). Die Bewegung und Reizbeantwortung der bipolar begeißelten Spirillen. *Jahrb. Wiss. Bot.*, **59**, 325–410.

Metzner, P. (1929). Bewegungsstudien an *Peridineen*. *Z. Bot.*, **22**, 225–65.

Mikolajczyk, E. (1972). Patterns of body movements of *Euglena gracilis*. *Acta Protozool.*, **11**, 317–31.

Mikolajczyk, E. (1973). Effect of some chemical factors on the euglenoid movement in *Euglena gracilis*. *Acta Protozool.*, **12**, 133–43.

Mikolajczyk, E. & Walne, P. L. (1990). Photomotile responses and ultrastructure of the euglenoid flagellate *Astasia fritschii*. *J. Photochem. Photobiol. B Biol.*, **6**, 275–82.

Miller, S. & Diehn, B. (1978). Cytochrome c oxidase as the receptor molecule for chemoaccumulation (chemotaxis) of *Euglena* toward oxygen. *Science*, **200**, 548–9.

Millet, B. & Pickard, B. G. (1988). Gadolinium ion is the inhibitor suitable for testing the putative role of stretch-activated ion channels in geotropism and thigmotropism. *Biophys. J.*, **53**, 155a.

Mogami, Y., Ishii, J. & Baba, S. A. (2001). Theoretical and experimental dissection of gravity-dependent mechanical orientation in gravitactic microorganisms. *Biol. Bull.*, **201**, 26–33.

Moir, J. (1996). Aerotactic response. *Microbiology UK*, **142**, 718–9.

Montgomery, P. O. B., Cook, J. E. & Franz, R. (1965). The effects of prolonged centrifugation on *Amoeba proteus*. *Exp. Cell Res.*, **40**, 140–2.

Moore, A. (1903). Some facts concerning geotropic gatherings of paramecia. *Am. J. Physiol.*, **9**, 238–44.

Morris, B. M., Reid, B. & Gow, N. A. R. (1992). Electrotaxis of zoospores of *Phytophthora palmivora* at physiologically relevant field strengths. *Plant Cell Environ.*, **15**, 645–53.

Morris, C. E. (1990). Mechanosensitive ion channels. *J. Membr. Biol.*, **113**, 93–107.

Müller, F. & Kaupp, U. B. (1998). Signaltransduktion in Sehzellen. *Naturwissenschaften*, **85**, 49–61.

Muller, H. J. (1959). Approximation to a gravity-free situation for the human organism achievable at moderate expense. *Science*, **128**, 772.

Nagel, U. (1993). Elektrische Eigenschaften und elektromotorische Kopplung im Verhalten von *Loxodes striatus. Dissertation*, Faculty of Biology, Ruhr-University.

Nagel, U. & Machemer, H. (2000a). Effects of gadolinium on electrical membrane properties and behavior in *Paramecium tetraurelia*. *Europ. J. Protistol.*, **36**, 161–8.

Nagel, U. & Machemer, H. (2000b). Physical and physiological components of the graviresponses of wild-type and mutant *Paramecium tetraurelia*. *J. Exp. Biol.*, **203**, 1059–70.

Nagy, K. & Stieve, H. (1995). Light-transduction in photoreceptors, in *Bioelectrochemistry of Cells and Tissues*. Eds., D. Walz, H. Berg & G. Milazzo. Birkhäuser, Basel, pp. 57–133.

Naitoh, Y. (1984). Mechanosensory transduction in protozoa, in *Membranes and Sensory Transduction*. Eds., G. Colombetti & F. Lenci. Plenum Press, New York, pp. 113–35.

Naitoh, Y. & Eckert, R. (1968). Electrical properties of *Paramecium caudatum*: modification by bound and free cations. *Zeits. Vergl. Physiol.*, **61**, 427–52.

Naitoh, Y. & Eckert, R. (1969). Ionic mechanisms controlling behavioral responses of *Paramecium* to mechanical stimulation. *Science*, **164**, 963–5.

Nakanishi, K. & Crouch, R. (1995). Application of artificial pigments to structure determination and study of photoinduced transformants of retinal proteins. *Israel J. Chem.*, **35**, 253–72.

Nakaoka, Y., Kurotani, T. & Itoh, H. (1987). Ion mechanism of thermoreception in *Paramecium. J. Exp. Biol.*, **127**, 95–103.

Nakaoka, Y. & Machemer, H. (1990). Effects of cyclic nucleotides and intracellular Ca^{2+} on voltage-activated ciliary beating in *Paramecium. J. Comp. Physiol. A*, **166**, 401–6.

Nakaoka, Y., Tanaka, T., Kuriu, T. & Murata, T. (1997). Possible mediation of G-proteins in cold-sensory transduction in *Paramecium multimicronucleatum. J. Exp. Biol.*, **200**, 1025–30.

Needham, D. & Hochmuth, R. M. (1989). Electro-mechanical permeabilization of lipid vesicles. *Biophys. J.*, **55**, 1001–9.

Needham, D. & Nunn, R. S. (1990). Elastic deformation and failure of lipid bilayer membranes containing cholesterol. *Biophys. J.*, **58**, 997–1009.

Nemec, P., Altmann, J., Marhold, S., Burda, H. & Oelschläger, H. H. A. (2001). Neuroanatomy of magnetoreception: the superior colliculus involved in magnetic orientation in a mammal. *Science*, **294**, 366–8.

Neubert, J., Rahmann, H., Briegleb, W., Slenzka, K., Schatz, A. & Bromeis, B. (1991). Statex II on Spacelab mission D-2 – an overview of the joint project "graviperception and neuronal plasticity" and preliminary pre-flight results. *Microgravity*, **1**, 173–82.

Neugebauer, D. C. & Machemer, H. (1997). Is there an orientation-dependent excursion of the Müller body in the "statocystoid" of *Loxodes*? *Cell Tissue Res.*, **287**, 577–82.

Neugebauer, D. C., Machemer-Röhnisch, S., Nagel, U., Bräucker, R. & Machemer, H. (1998). Evidence of central and peripheral gravireception in the ciliate *Loxodes striatus. J. Comp. Phys. A*, **183**, 303–11.

Neumann, R. & Iino, M. (1997). Phototropism of rice (*Oryza sativa L.*) coleoptiles: fluence–response relationships, kinetics and photogravitropic equilibrium. *Planta*, **201**, 288–92.

Nichol, J. A. & Hutter, O. F. (1996). Tensile strength and dilatational elasticity of giant sarcolemmal vesicles shed from rabbit muscles. *J. Physiol. (Lond.)*, **493**, 187–98.

Nick, P. & Schäfer, E. (1988). Interaction of gravi- and phototropic stimulation in the response of maize (*Zea mays* L.) coleoptiles. *Planta*, **173**, 213–20.

Nitschke, R., Wilhelm, S., Borlinghaus, R., Leipziger, J., Bindels, R. & Greger, R. (1997). A modified confocal laser scanning microscope allows fast ultraviolet ratio imaging of intracellular Ca^{2+} activity using Fura-2. *Pflügers Arch. Europ. J. Physiol.*, **433**, 653.

Noble, P. B. & Levine, M. D. (1986). *Computer Assisted Analysis of Cell Locomotion and Chemotaxis* CRC Press, Inc., Boca Raton, Florida.

Noever, D. A., Cronise, R. & Matsos, H. C. (1994). Preferred negative geotactic orientation in mobile cells: *Tetrahymena* results. *Biophys. J.*, **67**, 2090–5.

Noguchi, M., Nakamura, Y. & Okamoto, K.-I. (1991). Control of ciliary orientation in ciliated sheets from *Paramecium* – differential distribution of sensitivity to cyclic nucleotides. *Cell Motil. Cytoskel.*, **20**, 38–46.

Noguchi, M., Ogawa, T. & Taneyama, T. (2000). Control of ciliary orientation through cAMP-dependent phosphorylation of axonemal proteins in *Paramecium caudatum. Cell Motil. Cytoskel.*, **45**, 263–71.

Noguchi, M., Sawada, T. & Akazawa, T. (2001). ATP-regenerating system in the cilia of *Paramecium caudatum. J. Exp. Biol.*, **204**, 1063–71.

Nonnengässer, C., Holland, E.-M., Harz, H. & Hegemann, P. (1996). The nature of rhodopsin-triggered photocurrents in *Chlamydomonas*. II. Influence of monovalent ions. *Biophys. J.*, **70**, 932–8.

Norris, V., Grant, S., Freestone, P., Canvin, J., Sheikh, F. N., Toth, I., Trinei, M., Modha, K. & Norman, R. I. (1996). Calcium signalling in bacteria. *J. Bacteriol.*, **178**, 3677–82.

Nowakowska, G. & Grebecki, A. (1977). On the mechanism of orientation of *Paramecium caudatum* in the gravity field. II. Contributions to hydrodynamic model of geotaxis. *Acta Protozool.*, **16**, 359–76.

Nowycky, M. C. & Thomas, A. P. (2002). Intracellular calcium signaling. *J. Cell Sci.*, **115**, 3715–6.

Nultsch, W. (1973). Phototaxis and photokinesis in bacteria and blue-green algae, in *Behaviour of Microorganisms*. Ed., A. Perez-Miravete. Plenum Press, New York, pp. 70–82.

Nultsch, W. (1975). Phototaxis and photokinesis, in *Primitive Sensory and Communication Systems*. Ed., M. J. Carlile. Academic Press, New York, pp. 29–90.

Nultsch, W. (2001). *Allgemeine Botanik,* 11th edition. Thieme Verlag, Stuttgart.

Nultsch, W. & Häder, D.-P. (1979). Photomovement of motile microorganisms. *Photochem. Photobiol.*, **29**, 423–37.

Nultsch, W. & Häder, D.-P. (1988). Photomovement in motile microorganisms. II. *Photochem. Photobiol.*, **47**, 837–69.

Ohata, K., Murakami, T. & Miwa, I. (1997). Circadian rhythmicity of negative gravitaxis in *Paramecium bursaria. Zool. Sci. Tokyo*, **14**(Suppl.), 58.

Okamoto, K.-I. & Nakaoka, Y. (1994a). Reconstitution of metachronal waves in ciliated cortical sheets of *Paramecium*. I. Wave stabilities. *J. Exp. Biol.*, **192**, 61–72.

Okamoto, K.-I. & Nakaoka, Y. (1994b). Reconstitution of metachronal waves in ciliated cortical sheets of *Paramecium*. II. Asymmetry of the ciliary movements. *J. Exp. Biol.*, **192**, 73–81.

Olcese, J., Reuss, S. & Semm, P. (1988). Geomagnetic field detection in rodents. *Life Sci.*, **42**, 605–13.

Olson, J. E. & Li, G. Z. (1997). Increased potassium, chloride and taurine conductance in astrocytes during osmotic swelling. *Glia*, **20**, 254–61.

Oltmanns, F. (1917). Über Phototaxis. *Z. Bot.*, **9**, 257–338.

Omasa, K. & Aiga, I. (1987). Environmental measurement: image instrumentation for evaluating pollution effects on plants, in *System & Control Encyclopedia*. Ed., M. G. Singh. Pergamon Press, Oxford, pp. 1516–22.

Omasa, K. & Onoe, M. (1984). Measurement of stomatal aperture by digital image processing. *Plant Cell Physiol.*, **25**, 1379–88.

Omodeo, P. (1975). Morphology of the phototactic apparatus in eucariotic flagellated cells, in *Biophysics of Photoreceptors and Photobehaviour of Microorganisms*. Ed., G. Colombetti. Lito Felici, Pisa, pp. 24–48.

Omoto, C. K. & Brokaw, C. J. (1985). Bending patterns of *Chlamydomonas* flagella. II. Calcium effects on reactivated *Chlamydomonas* flagella. *Cell Motil.*, **5**, 53–60.

Ootaki, T., Ito, K., Abe, M., Lazarova, G., Miyazaki, A. & Tsuru, T. (1995). Parameters governing gravitropic response of sporangiophores in *Phycomyces blakesleeanus*. *Mycoscience*, **36**, 263–70.

Ooya, M., Mogami, Y., Izumi-Kurotani, A. & Baba, S. A. (1992). Gravity-induced changes in propulsion of *Paramecium caudatum*: a possible role of gravireception in protozoan behaviour. *J. Exp. Biol.*, **163**, 153–67.

Overmann, J. & Pfennig, N. (1992). Buoyancy regulation and aggregate formation in *Amoebobacter purpureus* from Mahoney Lake. *FEMS Microbiol. Ecol.*, **101**, 67–79.

Pasquale, S. M. & Goodenough, U. W. (1988). Calmodulin sensitivity of the flagellar membrane adenylate cyclase and signaling of motile responses by cAMP in gametes of *Chlamydomonas reinhardtii*. *Bot. Acta*, **101**, 118–22.

Pavlidis, T. (1982). *Graphics and Image Processing*. Springer-Verlag, Berlin.

Pech, L. L. (1995). Regulation of ciliary motility in *Paramecium* by cAMP and cGMP. *Comp. Biochem. Physiol.*, **111**, 31–7.

Pedley, T. J., Hill, N. A. & Kessler, J. O. (1988). The growth of bioconvection patterns in a uniform suspension of gyrotactic micro-organisms. *J. Fluid Mech.*, **195**, 223–37.

Pedley, T. J. & Kessler, J. O. (1987). The orientation of spheroidal microorganisms swimming in a flow field. *Proc. R. Soc. Lond. B*, **231**, 47–70.

Pedley, T. J. & Kessler, J. O. (1990). A new continuum model for suspensions of gyrotactic micro-organisms. *J. Fluid Mech.*, **212**, 155–82.

Penard, E. (1917). Le genre *Loxodes*. *Rev. Suisse de Zool.*, **25**, 453–89.

Perbal, G., Jeune, B., Lefranc, A., Carnero-Diaz, E. & Driss-Ecole, D. (2002). The dose-response curve of the gravitropic reaction: a re-analysis. *Physiol. Plant.*, **114**, 336–42.

Petersen-Mahrt, S. K., Ekelund, N. G. A. & Widell, S. (1994). Influence of UV-B radiation and nitrogen starvation on daily rhythms in phototaxis and cell shape of *Euglena gracilis*. *Physiol. Plant.*, **92**, 501–5.

Petrov, A. G. & Bivas, I. (1984). Elastic and flexoelectric aspects of out-of-plane fluctuations in biological and model membranes. *Progr. Surface Sci.*, **16**, 386–512.

Petrov, A. G. & Usherwood, P. N. R. (1994). Mechanosensitivity of cell membranes, ion channels, lipid matrix and cytoskeleton. *Europ. Biophys. J.*, **23**, 1–19.

Pfeffer, W. (1881). *Pflanzenphysiologie*. Verlag Wilhelm Engelmann, Leipzig.

Pfeffer, W. (1897). Locomotorische Richtungsbewegungen durch chemische Reize. *Ber. Dtsch. Bot. Ges.*, **15**, 524–33.

Piccinni, E., Albergoni, V. & Coppellotti, O. (1975). ATPase activity in flagella from *Euglena gracilis*. Localization of the enzyme and effects of detergents. *J. Protozool.*, **22**, 331–5.

Piccinni, E. & Mammi, M. (1978). Motor apparatus of *Euglena gracilis*: ultrastructure of the basal portion of the flagellum and the paraflagellar body. *Boll. Zool.*, **45**, 405–14.

Piccinni, E. & Omodeo, P. (1975). Photoreceptors and phototactic programs in protista. *Boll. Zool.*, **42**, 57–79.

Pickard, B. G. & Thimann, K. V. (1966). Geotropic response of wheat coleoptiles in absence of amyloplast starch. *J. Gen. Physiol.*, **49**, 1065–86.

Planel, H., Richoilley, G., Tixador, R., Templier, J., Bes, J. C. & Gasset, G. (1981). Space flight effects on *Paramecium tetraurelia* flown aboard Salyut 6 in the Cytos 1 and Cytos M experiment. *Adv. Space Res.*, **1**, 95.

Planel, H., Richoilley, G., Tixador, R., Templier, J., Bes, J. C. & Gasset, G. (1987). The "*Paramecium* experiment" – demonstration of a role of microgravity on the cell, in *Proceedings Norderney Symp. Scientific Res. of the German Spacelab Mission D1, DLR Köln*. Eds., P. R. Sahm, R. Jansen & M. H. Keller. WPF c/o DLR, Köln, pp. 376–82.

Planel, H., Tixador, R., Nefedov, Y., Grtechko, G. & Richoilley, G. (1982). Effect of space flight factors at the cellular level: results of the Cytos experiment. *Aviation, Space Environ. Med.*, **53**, 370–4.

Platt, J. B. (1899). On the specific gravity of *Spirostomum, Paramecium* and the tadpole in relation to the problem of geotaxis. *Am. Nat.*, **33**, 31.

Plattner, H. & Klauke, N. (2001). Calcium in ciliated protozoa: sources, regulation, and calcium-regulated cell functions. *Int. Rev. Cytol.*, **201**, 115–208.

Pohl, R. (1948). Tagesrhythmus im phototaktischen Verhalten in der *Euglena gracilis*. *Z. Naturf.*, **3b**, 367–74.

Pollard, E. C. (1965). Theoretical studies on living systems in the absence of mechanical stress. *J. Theor. Biol.*, **8**, 113–23.

Popescu, T., Zängler, F., Sturm, B. & Fukshansky, L. (1989). Image analyzer used for data acquisition in phototropism studies. *Photochem. Photobiol.*, **50**, 701–5.

Porst, M. (1998). *Euglena gracilis*: Langzeitversuche in artifiziellen Ökosystemen und Untersuchungen zur Gravitaxis. *Dissertation*, Friedrich-Alexander-Universität, Erlangen-Nürnberg, pp. 1–100.

Porst, M., Lebert, M. & Häder, D.-P. (1996). *Long-term culture of Euglena gracilis and Aquarack progress report. Proc. 12th C.E.B.A.S. Workshop*, 217–23.

Porterfield, D. M. (1997). Orientation of motile unicellular algae to oxygen: oxytaxis in *Euglena*. *Biol. Bull.*, **193**, 229–30.

Posudin, Y. I., Massjuk, N. P., Radchenko, M. I. & Lilitskaya, G. G. (1988). Photokinetic reactions of two *Dunaliella* TEOD species. *Mikrobiologiya*, **57**, 1001–6.

Potma, E. O., de Boeij, W. P., Bosgraaf, L., Roelofs, J., van Haastert, P. J. M. & Wiersma, D. A. (2001). Reduced protein diffusion rate by cytoskeleton in vegetative and polarized *Dictyostelium* cells. *Biophys. J.*, **81**, 2010–9.

Preston, R. R. & Saimi, Y. (1990). Calcium ions and the regulation of motility in *Paramecium*, in *Ciliary and Flagellar Membranes*. Ed., R. A. Bloodgood. Plenum Press, New York, pp. 173–94.

Preston, R. R., Wallen-Friedman, M. A., Saimi, Y. & Kung, C. (1990). Calmodulin defects cause the loss of Ca^{2+}-dependent K^+ currents in two pantophobiac mutants of *Paramecium tetraurelia*. *J. Membr. Biol.*, **115**, 51–60.

Pringsheim, E. G. (1922). Zur Physiologie saprophytischer Flagellaten. *Beitr. Allg. Bot.*, **2**, 88–137.

Pringsheim, E. G. (1937). Über das Stigma bei farblosen Flagellaten. *Cytologia*, **1**, 234–55.

Pringsheim, E. G. (1948). The loss of chromatophores in *Euglena gracilis*. *N. Phytol.*, **47**, 52–87.

Pringsheim, E. G. (1912). Das Zustandekommen der taktischen Reaktionen. *Biol. Zentralbl.*, **32**, 337–65.

Prowazek, S. (1910). *Einführung in die Physiologie der Einzelligen (Protozoen)*. Verlag Teubner, Leipzig, pp. 129–70.

Purcell, E. M. (1977). Life at low Reynolds numbers. *Am. J. Phys.*, **45**, 3–11.

Rai, S., Singh, U. P., Awasthi, M. & Pandey, S. (1998). Physiological response of the cyanobacterium *Anacystis nidulans* to a magnetic field. *Electro- and Magnetobiol.*, **17**, 145–60.

Raikov, I. B. (1985). Primitive never-dividing macronuclei of some lower ciliates. *Int. Rev. Cytol.*, **95**, 267–325.

Ramanathan, R., Saimi, Y., Hinrichsen, R., Burgess-Cassler, A. & Kung, C. (1988). A genetic dissection of ion-channel function, in *Paramecium*. Ed., H.-D. Görtz. Springer-Verlag, Berlin, pp. 236–53.

Ranjeva, R., Graziana, A. & Mazars, C. (1999). Plant graviperception and gravitropism: a newcomer's view. *FASEB J.*, **13**, S135–S141.

Raper, K. B. (1940). Pseudoplasmodium formation and organisation in *Dictyostelium discoideum*. *J. Elisha Mitchell Sci. Soc.*, **36**, 241–82.

Reifarth, F. W., Clauss, W. & Weber, W.-M. (1999). Stretch-independent activation of the mechanosensitive cation channel in oocytes of *Xenopus laevis*. *Biochim. Biophys. Acta*, **1417**, 63–76.

Reimers, H. (1928). Über die Thermotaxis niederer Organismen. *Jahrb. Wiss. Bot.*, **67**, 242–90.

Reisser, W. & Häder, D.-P. (1984). Role of endosymbiotic algae in photokinesis and photophobic responses of ciliates. *Photochem. Photobiol.*, **39**, 673–8.

Relkin, E. M. & Doucet, J. R. (1991). Recovery from prior stimulation. I. Relationship to spontaneous firing rates of primary auditory neurons. *Hearing Res.*, **55**, 215–22.

Reynolds, C. S., Oliver, R. L. & Walsby, A. E. (1987). Cyanobacterial dominance: the role of buoyancy regulation in dynamic lake environments. *N. Z. J. Mar. Freshwater Res.*, **21**, 379.

Rhode, S. C., Pawlowski, M. & Tollrian, R. (2001). The impact of ultraviolet radiation on the vertical distribution of zooplankton of the genus *Daphnia*. *Nature*, **412**, 69–72.

Richoilley, G., Tixador, R., Templier, J., Bes, J. C., Gasset, G. & Planel, H. (1988). The *Paramecium* experiment, in *Biorack on Spacelab D1 ESA-SP-1091*. Eds., N. Longdon & V. David, ESH, Paris, pp. 69–73.

Richter, P., Lebert, M., Korn, R. & Häder, D.-P. (2001a). Possible involvement of the membrane potential in the gravitactic orientation of *Euglena gracilis*. *J. Plant Physiol.*, **158**, 35–9.

Richter, P., Lebert, M., Tahedl, H. & Häder, D.-P. (2001b). Calcium is involved in the gravitactic orientation in colorless flagellates. *J. Plant Physiol.*, **158**, 689–97.

Richter, P. R., Schuster, M., Wagner, H., Lebert, M. & Häder, D.-P. (2002). Physiological parameters of gravitaxis in the flagellate *Euglena gracilis* obtained during a parabolic flight campaign. *J. Plant Physiol.*, **159**, 181–90.

Rieder, N. (1977). Die Müllerschen Körperchen von *Loxodes magnus* (Ciliata, Holotricha): Ihr Bau und ihre mögliche Funktion als Schwererezeptor. *Verh. Dt. zool. Ges.*, **70**, 254.

Rieder, N., Ott, H. A., Pfundstein, P. & Schoch, R. (1982). X-ray microanalysis of the mineral contents of some protozoa. *J. Protozool.*, **29**, 15–8.

Rikmenspoel, R. & Isles, C. A. (1985). Digitized precision measurements of the movement of sea urchin sperm flagella. *Biophys. J.*, **47**, 395–410.

Ringelberg, J. (1999). The photobehaviour of *Daphnia* spp. as a model to explain diel vertical migration in zooplankton. *Biol. Rev.*, **74**, 397–423.

Ritz, T., Adem, S. & Schulten, K. (2000). A model for photoreceptor-based magnetoreception in birds. *Biophys. J.*, **78**, 707–18.

Roberts, A. M. (1970). Geotaxis in motile microorganisms. *J. Exp. Biol.*, **53**, 687–99.

Roberts, A. M. (1981). Hydrodynamics of protozoan swimming, in *Biochemistry and Physiology of Protozoa*. Eds., M. Levandowsky & S. M. Hunter, Acad. Press, New York, pp. 5–66.

Roberts, A. M. & Deacon, F. M. (2002). Gravitaxis in motile micro-organisms: the role of fore-aft body asymmetry. *J. Fluid. Mech.*, **452**, 405–23.

Rosiere, T. K., Marrs, J. A. & Bouck, G. B. (1990). A 39-kD plasma membrane protein (IP39) is an anchor for the unusual membrane skeleton of *Euglena gracilis*. *J. Cell. Biol.*, **110**, 1077–88.

Ross, P. E., Garber, S. S. & Cahalan, M. D. (1994). Membrane chloride conductance and capacitance in Jurkat T lymphocytes during osmotic swelling. *Biophys. J.*, **66**, 169–78.

Rüsch, A. & Thurm, U. (1989). Cupula displacement, hair bundle deflection, and physiological responses in the transparent semicircular canal of young eel. *Pflügers Arch.*, **413**, 533–45.

Sachs, F. (1991). Mechanical transduction by membrane ion channels: a mini review. *Mol. Cell Biochem.*, **104**, 57–60.

Sachs, F. & Morris, C. E. (1998). Mechanosensitive ion channels in nonspecialized cells, in *Reviews of Physiology and Biochemistry and Pharmacology*. Eds., M. P. Blaustein, R. Greger, H. Grunicke, R. Jahn, L. M. Mendell, A. Miyajima, D. Pette, G. Schultz & M. Schweiger. Springer-Verlag, Berlin, pp. 1–78.

Sack, F. D. (1997). Plastids and gravitropic sensing. *Planta*, **203**, 63–8.

Sack, F. D. (2002). Tip growing cells of the moss *Ceratonon purpureus* are gravitropic in high-density media. *Plant Physiol.*, **130**, 2095–100.

Sanderson, M. J. & Dirksen, E. R. (1985). A versatile and quantitative computer-assisted photoelectronic technique used for the analysis of ciliary beat cycles. *Cell Motil.*, **5**, 267–92.

Satir, P., Wais-Steider, J., Lebduska, S., Nasr, A. & Avolio, J. (1981). The mechanochemical cycle of the dynein arm. *Cell Motil.*, **1**, 303–27.

Sauer, H. W. (1982). *Developmental Biology of Physarum.* Cambridge University Press, Cambridge.

Saxton, M. J. (1994). Single-particle tracking: models of directed transport. *Biophys. J.*, **67**, 2110–9.

Schaap, P., Nebl, T. & Fisher, P. R. (1996). A slow sustained increase in cytosolic Ca^{2+} levels mediates stalk gene induction by differentiation inducing factor in *Dictyostelium. EMBO J.*, **15**, 5177.

Schaefer, G. (1922). Studien über den Geotropismus von *Paramecium aurelia. Pflüger's Arch. Ges. Physiol.*, **195**, 227–44.

Schäfer, J., Sebastian, C. & Häder, D.-P. (1993). Effects of solar radiation on motility, orientation, pigmentation and photosynthesis in a green dinoflagellate *Gymnodinium. Acta Protozool.*, **33**, 59–65.

Schatz, A., Briegleb, W. & Neubert, J. (1973). 0-*g* conditions in laboratory experiments – the important parameters and limitations. München, XXI International Congress of Aviation and Space Medicine, Preprints of Scientific Programme, pp. 111–2.

Schatz, A., Linke-Hommes, A. & Neubert, J. (1996). Gravity dependency of gramicidin A channel conductivity. A model for gravity perception on the cellular level. *Europ. Biophys. J.*, **25**, 37–41.

Schimek, C., Eibel, P., Grolig, F., Horie, T., Ootaki, T. & Galland, P. (1999). Gravitropism in *Phycomyces*: a role for sedimenting protein crystals and floating lipid globules. *Planta*, **210**, 132–42.

Schlatterer, C., Knoll, G. & Malchow, D. (1992). Intracellular calcium during chemotaxis of *Dictyostelium discoideum*: a new fura-2 derivative avoids sequestration of the indicator and allows long-term calcium measurements. *Europ. J. Cell Biol.*, **58**, 172–81.

Schleicher, M., Lukas, T. J. & Watterson, D. M. (1984). Isolation and characterization of calmodulin from the motile green alga *Chlamydomonas reinhardtii. Arch. Biochem. Biophys.*, **229**, 33–42.

Schletz, K. (1976). Phototaxis bei *Volvox* – Pigmentsysteme der Lichtrichtungsperzeption. *Z. Pflanzenphys.*, **77**, 189–211.

Schlicher, U., Linden, L., Calenberg, M. & Kreimer, G. (1995). G-Proteins and Ca^{2+}-modulated protein kinases of a plasma membrane enriched fraction and isolated eyespot apparatusses of *Spermatozopsis similis* (Chlorophyceae). *Europ. J. Phycol.*, **30**, 319–30.

Schmidt, W. & Galland, P. (2000). Gravity-induced absorbance changes in *Phycomyces*: a novel method for detecting primary responses of gravitropism. *Planta*, **210**, 848–52.

Schuh, K., Uldrijan, S., Telkamp, M., Röthlein, N. & Neyses, L. (2001). The plasmamembrane calmodulin-dependent calcium pump: a major regulator of nitric oxide synthase I. *J. Cell. Biol.*, **155**, 201–5.

Schultz, J. E., Guo, Y.-L., Kleefeld, G. & Völkel, H. (1997). Hyperpolarization- and depolarisation-activated Ca^{2+} currents in *Paramecium* trigger behavioral changes and cGMP formation independently. *J. Membr. Biol.*, **156**, 251–9.

Schultz, J. E. & Klumpp, S. (1980). Guanylate cyclase in the excitable ciliary membrane of *Paramecium*. *FEBS Lett.*, **122**, 64–6.

Schultz, J. E. & Klumpp, S. (1988). Biochemistry of cilia, in *Paramecium*. Ed., H.-D. Görtz. Springer-Verlag, Berlin, pp. 254–70.

Schultz, J. E., Pohl, T. & Klumpp, S. (1986). Voltage-gated Ca^{2+} entry into *Paramecium* linked to intraciliary increase in cyclic GMP. *Nature*, **322**, 271–3.

Schultz, J. E. & Schönborn, C. (1994). Cyclic AMP formation in *Tetrahymena pyriformis* is controlled by K^+-conductances. *FEBS Lett.*, **356**, 322–6.

Schwarz, F. (1884). Der Einfluß der Schwerkraft auf die Bewegungsrichtung von *Chlamydomonas* und *Euglena*. *Ber. Dtsch. Bot. Ges.*, **2**, 51–72.

Schwarzacher, J. C. & Audus, L. J. (1973). Further studies in magnetotropism. *J. Exp. Bot.*, **24**, 459–74.

Schwarzenberg, M., Pippia, P., Meloni, M. A., Cossu, G., Cogoli-Greuter, M. & Cogoli, A. (1998). Microgravity simulations with human lymphocytes in the free fall machine and in the random positioning machine. *J. Grav. Physiol.*, **5**, P23–P26.

Schwarzenberg, M., Pippia, P., Meloni, M. A., Cossu, G., Cogoli-Greuter, M. & Cogoli, A. (1999). Signal transduction in T lymphocytes – a comparison of the data from space, the free fall machine and the random positioning machine. *Adv. Space Res.*, **24**, 793–800.

Sciola, L., Cogoli-Greuter, M., Spano, A. & Pipia, P. (1999). Influence of microgravity on mitogen binding and cytoskeleton in Jurkat cells. *Adv. Space Res.*, **24**, 801–5.

Scott, A. C. & Allen, N. S. (1999). Changes in cytosolic pH within *Arabidopsis* root columella cells play a key role in the early signaling pathway for root gravitropism. *Plant Physiol.*, **121**, 1291–8.

Sebastian, C., Scheuerlein, R. & Häder, D.-P. (1994). Graviperception and motility of three *Prorocentrum* strains impaired by solar and artificial ultraviolet radiation. *Mar. Biol.*, **120**, 1–7.

Sellick, P. M., Patuzzi, R. & Johnstone, B. M. (1982). Measurement of the basilar membrane motion in guinea pig using Mössbauer technique. *J. Acoust. Soc. Am.*, **72**, 131–41.

Serra, J. (1980). Digitalization. *Mikroskopie*, **37**(Suppl.), 109–18.

Sharma, V. K., Engelmann, W. & Johnsson, A. (2000). Effects of static magnetic field on the ultradian lateral leaflet movement rhythm in *Desmodium gyrans*. *Z. Naturforsch.*, **55**, 638–42.

Shi, W., Stocker, B. A. D. & Adler, J. (1996). Effect of the surface composition of motile *Escherichia coli* and motile *Salmonella* species on the direction of galvanotaxis. *J. Bacteriol.*, **178**, 1113–9.

Shimmen, F. (1997). Studies of mechano-perception in characeae: effects of external Ca^{2+} and Cl^-. *Plant Cell Physiol.*, **38**, 691.

Shvirst, E. M., Krinskii, V. J. & Ivanitskii, V. J. (1984). Biophysics of complex systems. Role of oxytaxis in the origin of dissipative structures in a culture of *Tetrahymena*. *Biophysics*, **4**, 710–5.

Sievers, A. (1999). Gravitational biology in Bonn. *ASGSB Newslett.*, **15**, 15–22.

Sievers, A., Buchen, B., Volkmann, D. & Hejnowicz, Z. (1991). Role of the cytoskeleton in gravity perception, in *The Cytoskeletal Basis of Plant Growth and Form*. Ed., C. W. Lloyd. Academic Press, London, pp. 169–82.

Sievers, A. & Hejnowicz, Z. (1992). How well does the clinostat mimic the effect of microgravity on plant cells and organs. *ASGSB Bull.*, **5**, 69–75.

Sievers, A. & Volkmann, D. (1977). Ultrastructure of gravity-perceiving cells in plant roots. *Proc. R. Soc. Lond. B*, **199**, 525–36.

Sievers, A. & Volkmann, D. (1979). Gravitropism in single cells, in *Physiology of Movements. Encyclopedia of Plant Physiology. N.S.* Eds., W. Haupt & M. E. Feinleib. Springer-Verlag, Berlin, pp. 567–72.

Sikora, J., Baranowski, Z. & Zajaczkowska, M. (1992). Two-state model of *Paramecium bursaria* thigmotaxis. *Experientia*, **48**, 789–92.

Silverman, M. & Simon, M. (1974). Flagellar rotation and the mechanism of bacterial motility. *Nature*, **249**, 73–4.

Simons, P. J. (1981). The role of electricity in plant movements. *New Phytol.*, **87**, 11–37.

Sineshchekov, O. A. (1991a). Electrophysiology of photomovements in flagellated algae, in *Biophysics of Photoreceptors and Photomovements in Microorganisms*. Ed., F. Lenci. Plenum Press, New York, pp. 191–202.

Sineshchekov, O. A. (1991b). Photoreception in unicellular flagellates: bioelectric phenomena in phototaxis, in *Light in Biology and Medicine*. Ed., R. H. Douglas. Plenum Press, New York, pp. 523–32.

Sineshchekov, O. A. & Govorunova, E. G. (1991). Rhythmic motion activity of unicellular flagellated algae and its role in phototaxis. *Biofisika*, **36**, 603–8.

Sineshchekov, O. A. & Govorunova, E. G. (1999). Rhodopsin-mediated photosensing in green flagellated algae. *Trends Plant Sci.*, **4**, 58–63.

Sineshchekov, O. A., Govorunova, E. G., Der, A., Keszthelyi, L. & Nultsch, W. (1992). Photoelectric responses in phototactic flagellated algae measured in cell suspension. *J. Photochem. Photobiol. B Biol.*, **13**, 119–34.

Sineshchekov, O., Lebert, M. & Häder, D.-P. (2000). Effects of light on gravitaxis and velocity in *Chlamydomonas reinhardtii*. *J. Plant Physiol.*, **157**, 247–54.

Sineshchekov, V. A. & Sineshchekov, A. V. (1988). Wheat roots orientation under the effect of geomagnetic field. *Biofizika SSSR*, **32**, 110–3.

Sitte, P., Ziegler, H., Ehrendorfer, F. & Bresinsky, A. (1998). *Strasburger, Lehrbuch der Botanik.* 34 ed. Gustav Fischer, Stuttgart.

Slenzka, K. (2002). Life support for aquatic species – past, present, future. *Adv. Space Res.*, **30**, 789–95.

Smith, A. H. (1992). Centrifuges: their development and use in gravitational biology. *ASGSB Bull.*, **5**, 33–41.

Sobick, V. & Briegleb, W. (1983). Influence of zero gravity simulation on time course of mitosis in microplasmodia of *Physarum polycephalum*. *Adv. Space Res.*, **3**, 259–62.

Sobick, V., Briegleb, W. & Block, J. (1983). Is there an orientation of the nuclei in microplasmodia of *Physarum polycephalum*? *The Physiologist*, **26**, S129–S130.

Sokabe, M., Sachs, F. & Jing, Z. (1991). Quantitative video microscopy of patch clamped membranes stress, strain, capacitance, and stretch channel activation. *Biophys. J.*, **59**, 722–8.

Solsona, C., Innocenti, B. & Fernandez, J. M. (1998). Regulation of exocytotic fusion by cell inflation. *Biophys. J.*, **74**, 1061–73.

Solter, K. M. & Gibor, A. (1977). Evidence for role of flagella as sensory transducers in mating of *Chlamydomonas reinhardi*. *Nature*, **265**, 444–5.

Song, P.-S. (1985). Primary molecular events in aneural cell photoreceptors, in *Sensory Perception and Transduction in Aneural Organisms*. Eds., G. Colombetti, F. Lenci & P.-S. Song. Plenum Press, New York, pp. 47–59.

Song, P.-S., Häder, D.-P. & Poff, K. L. (1980). Step-up photophobic response in the ciliate, *Stentor coeruleus*. *Arch. Microbiol.*, **126**, 181–6.

Song, P.-S. & Poff, K. L. (1989). Photomovement, in *The Science of Photobiology*, 2nd ed. Ed. K. C. Smith, Plenum, New York. pp. 305–46.

Spencer, C. N. & King, D. L. (1985). Interactions between light, NH_4^+, and CO_2 in buoyancy regulation of *Anabaena* flos-aquae (Cyanophyceae). *J. Phycol.*, **21**, 194–9.

Sperber, D., Darnsfeld, K., Maret, G. & Weisenseel, H. M. (1981). Oriented growth of pollen tubes in strong magnetic fields. *Naturwissenschaften*, **68**, 40–1.

Spring, S., Lins, U., Amann, R., Schleifer, K.-H., Ferreira, L. C. S., Esquivel, D. M. S. & Farina, M. (1998). Phylogenetic affiliation and ultrastructure of uncultured magnetic bacteria with unusually large magnetosomes. *Arch. Microbiol.*, **169**, 136–47.

Spudich, J. L., Zacks, D. N. & Bogomolni, R. A. (1995). Microbial sensory rhodopsins: photochemistry and function. *Israel J. Chem.*, **35**, 495–513.

Spurny, M. (1974). Interactions of photo- and geotropism with periodical oscillations of growing pea root (*Pisum sativum* L.). *Biol. Plantarum (Praha)*, **16**, 43–9.

Srinivasan, K. R., Kay, R. L. & Nagle, J. F. (1974). The pressure dependence of the lipid bilayer phase transition. *Biochemistry*, **13**, 3494–6.

Stahl, E. (1880). Über den Einfluß von Richtung und Stärke der Beleuchtung auf einige Bewegungserscheinungen im Pflanzenreich. *Bot. Ztg.*, **38**, 297–413.

Stahl, E. (1884). Zur Biologie der Myxomyceten. *Bot. Ztg.*, **42,** 145–75, 187–92.

Stallwitz, E. & Häder, D.-P. (1993). Motility and phototactic orientation of the flagellate *Euglena gracilis* impaired by heavy metal ions. *J. Photochem. Photobiol.*, **18**, 67–74.

Stallwitz, E. & Häder, D.-P. (1994). Effects of heavy metals on motility and gravitactic orientation of the flagellate, *Euglena gracilis*. *Europ. J. Protistol.*, **30**, 18–24.

Stalmans, P. & Himpens, B. (1997). Confocal imaging of Ca^{2+} signaling in cultured rat retinal pigment epithelial cells during mechanical and pharmacologic stimulation. *Invest. Ophthalmol. Vis. Sci.*, **38**, 176.

Staves, M. P. (1997). Cytoplasmic streaming and gravity sensing in *Chara* internodal cells. *Planta*, **203**, 79–84.

Staves, M. P., Wayne, R. & Leopold, A. C. (1992). Hydrostatic pressure mimics gravitational pressure in characean cells. *Protoplasma*, **168**, 141–52.

Stavis, R. L. (1974a). Phototaxis in *Chlamydomonas*: a sensory receptor system. *Doctoral dissertation*, Albert Einstein College of Medicine, New York. University Microfilms No. 75-27, 884.

Stavis, R. L. (1974b). The effect of azide on phototaxis in *Chlamydomonas reinhardi*. *Proc. Natl. Acad. Sci. U.S.A.*, **71**, 1824–7.

Steinitz, B., Hagiladi, A. & Anav, D. (1992). Thigmomorphogenesis and its interaction with gravity in climbing plants of *Epipremum aureum*. *J. Plant Physiol.*, **140**, 571–4.

Streb, C., Richter, P., Lebert, M. & Häder, D.-P. (2001). Gravi-sensing microorganisms as model systems for gravity sensing in eukaryotes. Proceeding of the First European Workshop on Exo-/Astro-Biology, Frascati. 251–4.

Stricker, S. A. (1996). Repetitive calcium waves induced by fertilization in the nemertean worm *Cerebratus lacteus*. *Devel. Biol.*, **176**, 496.

Stricker, S. A. (1997). Intracellular injections of a soluble sperm factor trigger calcium oscillations and meiotic maturation in unfertilized oocytes of a marine worm. *Devel. Biol.*, **186**, 185.

Strong, D. R. & Ray, T. S. (1975). Host tree location behavior of a tropical vine (*Monstera gigantea*) by skototropism. *Science*, **190**, 804–6.

Sugimori, M., Lang, E. J., Silver, R. B. & Llinás, R. (1994). High-resolution measurement of the time-course of calcium concentration microdomains at squid presynaptic terminals. *Biol. Bull.*, **187**, 300–3.

Sukharev, S. I., Sigurdson, W. J., Kung, C. & Sachs, F. (1999). Energetics and spatial parameters for gating of the bacterial large conductance mechanosensitive channel, MscL. *J. Gen. Physiol.*, **113**, 525–39.

Suzaki, T. & Williamson, R. E. (1986a). Cell surface displacement during euglenoid movement and its computer simulation. *Cell Motil. Cytoskel.*, **6**, 186–92.

Suzaki, T. & Williamson, R. E. (1986b). Pellicular ultrastructure and euglenoid movement in *Euglena ehrenbergii* Klebs and *Euglena oxyuris* Schmarda. *J. Protozool.*, **33**, 165–71.

Suzaki, T. & Williamson, R. E. (1986c). Ultrastructure and sliding of pellicular structures during euglenoid movement in *Astasia longa* Pringsheim (Sarcomastigophora, Euglenida). *J. Protozool.*, **33**, 179–84.

Svrcek-Seiler, W. A., Gebeshuber, I. C., Rattay, F., Biro, T. S. & Markum, H. (1998). Micromechanical models for the Brownian motion of hair cell stereocilia. *J. Theor. Biol.*, **193**, 623–30.

Swaminathan, R., Bicknese, S., Periasamy, N. & Verkman, A. S. (1996). Cytoplasmic viscosity near cell plasma membrane: translational diffusion of a small fluorescent solute measured by total internal reflection-fluorescence photobleaching recovery. *Biophys. J.*, **71**, 1140–51.

Tabony, J., Glade, N., Papseit, C. & Demongeot, J. (2002). Microtubule self-organisation and its gravity dependence, in *Cell Biology and Biotechnology in Space*. Ed., A. Cogoli. Elsevier, Amsterdam, pp. 19–58.

Tahedl, H., Richter, P., Lebert, M. & Häder, D.-P. (1998). cAMP is involved in gravitaxis signal transduction of *Euglena gracilis*. *Micrograv. Sci. Technol.*, **11**, 173–8.

Tairbekov, M. G., Parfyonov, R. W., Platonova, R. W., Abramova, V. M., Golov, V. K., Rostophsina, A. V., Lyubchenko, V. Y. & Chuchkin, V. G. (1981). Biological investigations aboard the biosatellite Cosmos-1129. *Adv. Space Res.*, **1**, 88–94.

Takahashi, H. (1997). Gravimorphogenesis: gravity-regulated formation of the peg in cucumber seedlings. *Planta*, **203**, 164–9.

Takahashi, H. & Scott, T. K. (1991). Hydrotropism and its interaction with gravitropism in maize roots. *Plant Physiol.*, **96**, 558–64.

Takahashi, M., Onimaru, H. & Naitoh, Y. (1980). A mutant of *Tetrahymena* with non-excitable membrane. *Proc. Jap. Acad. Ser. B*, **56**, 585–92.

Taneda, K. (1987). Geotactic behavior in *Paramecium caudatum*. I. Geotaxis assay of individual specimen. *Zool. Sci.*, **4**, 781–8.

Taneda, K. & Miyata, S. (1995). Analysis of motile tracks of *Paramecium* under gravity field. *Comp. Biochem. Physiol.*, **111A**, 673–80.

Taneda, K., Miyata, S. & Shiota, A. (1987). Geotactic behavior in *Paramecium caudatum*. II. Geotaxis assay in a population of the specimens. *Zool. Sci.*, **4**, 789–95.

Tash, J. S. & Bracho, G. E. (1999). Microgravity alters protein phosphorylation changes during initiation of sea urchin sperm motility. *FASEB J.*, **13**, S43–S54.

Tash, J. S., Johnson, D. C. & Enders, G. C. (2002). Long-term (6 wk) hindlimb suspension inhibits spermatogenesis in adult male rats. *J. Appl. Physiol.*, **92**, 1191–8.

Tash, J. S., Kim, S., Schuber, M., Seibt, D. & Kinsey, W. H. (2001). Fertilization of sea urchin eggs and sperm motility are negatively impacted under low hypergravitional forces significant to space flight. *Biol. Reprod.*, **65**, 1224–31.

Taylor, C. J., Cootes, T. F., Lanitis, A., Edwards, G., Smyth, P. & Kotcheff, A. C. W. (1997). Model-based interpretation of complex and variable images. *Phil. Trans. R. Soc. Lond. B*, **352**, 1267–74.

Taylor, W. R., Seliger, H. H., Fastie, W. G. & McElroy, W. D. (1966). Biological and physical observations on a phosphorescent bay in Falmouth harbor, Jamaica. *J. Mar. Res.*, **24**, 28–43.

Tevini, M. & Häder, D.-P. (1985). *Allgemeine Photobiologie*. Georg Thieme Verlag, Stuttgart.

Thiele, R. (1960). Über Lichtadaption und Musterbildung bei *Euglena gracilis*. *Arch. Microbiol.*, **37**, 379–98.

Thomas, G. H., Newbern, E. C., Korte, C. C., Bales, M. A., Muse, S. V., Clark, A. G. & Kiehart, D. P. (1997). Intragenic duplication and divergence in the spectrin superfamily of proteins. *Mol. Biol. Evol.*, **14**, 1285–95.

Thomas, J. R., Schrot, J. & Liboff, A. R. (1986). Low-intensity magnetic fields alter operant behavior in rats. *Bioelectromagnetics*, **7**, 349–57.

Thomas, R. H. & Walsby, A. E. (1986). The effect of temperature on recovery of buoyancy by *Microcystis*. *J. Gen. Microbiol.*, **132**, 1665–72.

Thornton, R. M. (1968). The fine structure of *Phycomyces*. II. Organization of the stage I sporangiophore apex. *Protoplasma*, **66**, 269–85.

Thurm, U. (1983). Mechano-electrical transduction, in *Biophysics*. Eds., W. Hoppe, W. Lohmann, H. Markl & H. Ziegler. Springer, Berlin, pp. 666–71.

Thurm, U., Erler, G., Gödde, J., Kastrup, H., Keil, T. H., Völker, W. & Vohwinkel, B. (1983). Cilia specialized for mechanoreception. *J. Submicrosc. Cytol.*, **15**, 151–5.

Timm, U. & Okubo, A. (1994). Gyrotaxis: a plume model for self-focusing micro-organisms. *Bull. Math. Biol.*, **56**, 187–206.

Tirlapur, U., Scheuerlein, R. & Häder, D.-P. (1993). Motility and orientation of a dinoflagellate, *Gymnodinium*, impaired by solar and ultraviolet radiation. *FEMS Microbiol. Ecol.*, **102**, 167–74.

Tixador, R., Richoilley, G., Gasset, G., Templier, J., Bes, J. C., Moatti, N. & Lapchine, L. (1985). Study of minimal inhibitory concentration of antibiotics on bacteria cultivated in vitro in space (Cytos 2 Experiment). *Aviat. Space Environ. Med.*, **56**, 748–51.

Toda, H., Yazawa, M. & Yagi, K. (1992). Amino acid sequence of calmodulin from *Euglena gracilis*. *Europ. J. Biochem.*, **205**, 653–60.

Tominaga, T. & Naitoh, Y. (1994). Comparison between thermoreceptor and mechanoreceptor currents in *Paramecium caudatum*. *J. Exp. Biol.*, **189**, 117–31.

Tong, I. & Edmunds, L. N., Jr. (1993). Role of cyclic GMP in the mediation of circadian rhythmicity of the adenylate cyclase-cyclic AMP-phosphodiesterase system in *Euglena*. *Biochem. Pharmacol.*, **45**, 2087–91.

Tong, J., Carre, I. A. & Edmunds, L. N. (1991). Circadian rhythmicity in the activities of adenylate cyclase and phosphodiesterase in synchronously dividing and stationary-phase cultures of the achlorophyllous ZC mutant of *Euglena gracilis*. *J. Cell Sci.*, **100**, 365–9.

Travis, S. M. & Nelson, D. L. (1988). Purification and properties of dyneins from *Paramecium* cilia. *Biochim. Biophys. Acta*, **966**, 73–83.

Trollinger, D. R., Isseroff, R. R. & Nuccitelli, R. (2002). Calcium channel blockers inhibit galvanotaxis in human keratinocytes. *J. Cell Physiol.*, **193**, 1–9.

Trzeciak, H. I., Grzesik, J., Bortel, M., Kuska, R., Duda, D., Michnik, J. & Malecki, A. (1993). Behavioral effects of long-term exposure to magnetic fields in rats. *Bioelectromagnetics*, **14**, 287–97.

Tucker, J. B. (1971). Development and deployment of cilia, basal bodies, and other microtubular organelles in the cortex of the ciliate *Nassula*. *J. Cell Sci.*, **9**, 539–67.

Tyler, M. A. & Seliger, H. H. (1978). Annual subsurface transport of a red tide dinoflagellate to its bloom area: water circulation patterns and organism distributions in the Chesapeake Bay. *Limnol. Oceanogr.*, **23**, 227–46.

Tyler, M. A. & Seliger, H. H. (1981). Selection for a red tide organism: physiological responses to the physical environment. *Limnol. Oceanogr.*, **26**, 310–24.

Ueda, M., Sako, Y., Tanaka, T., Devreotes, P. & Yanagida, T. (2001). Single-molecule analysis of chemotactic signaling in *Dictyostelium* cells. *Science*, **294**, 864–7.

Ueda, T., Nakagaki, T. & Yamada, T. (1990). Dynamic organization of ATP and birefringent fibrils during free locomotion and galvanotaxis in the plasmodium of *Physarum polycephalum*. *J. Cell Biol.*, **110**, 1097–102.

Uhl, R. & Hegemann, P. (1990). Probing visual transduction in a plant cell. Optical recording of rhodopsin-induced structural changes from *Chlamydomonas reinhardtii*. *Biophys. J.*, **58**, 1295–302.

Umrath, K. (1959). Galvanotaxis. *Handbuch der Pflanzenphysiologie*. Vol. **XVII**, part 1, Ed., W. Ruhland. Springer-Verlag, Berlin, 164–7.

Urban, J. E. (2000). Adverse effects of microgravity on the magnetotactic bacterium *Magnetospirillum magnetotacticum*. *Acta Astronaut.*, **47**, 775–80.

Vaija, J., Lagaude, A. & Ghommidh, C. (1995). Evaluation of image analysis and laser granulometry for microbial cell sizing. *Antonie van Leeuwenhoek*, **67**, 139–49.

Vainshtein, M., Suzina, N., Kudryashova, E. & Ariskina, E. (2002). New magnet-sensitive structures in bacterial and archaeal cells. *Biol. Cell*, **94**, 29–35.

Vainshtein, M. B., Suzina, N. E., Kudryashova, E. B., Ariskina, E. V. & Sorokin, V. V. (1998). On the diversity of magnetotactic bacteria. *Microbiology*, **67**, 670–6.

van den Hoek, C., Jahns, H. M. & Mann, D. G. (1993). *Algen*. 3rd ed. Georg Thieme Verlag, Stuttgart.

van Eldik, L. J., Piperno, G. & Watterson, D. M. (1980). Similarities and dissimilarities between calmodulin and a *Chlamydomonas* flagellar protein. *Proc. Natl. Acad. Sci. U.S.A.*, **77**, 4779–83.

van Houten, J., Martel, E. & Kasch, T. (1982). Kinetic analysis of chemokinesis of *Paramecium*. *J. Exp. Zool.*, **29**, 226–30.

van Houten, J. L. (1990). Chemosensory transduction in *Paramecium*, in *Biology of the Chemotactic Response*. Eds., J. Armitage & J. Lackie. Cambridge University Press, Cambridge, pp. 297–322.

van Houten, J. L. & Preston, R. R. (1987). Chemoreception in single-celled organisms, in *Neurobiology of Taste and Smell*. Ed., T. E. Finger. J. Wiley and Sons, New York, pp. 11–38.

Vassy, J., Portet, S., Beil, M., Millot, G., Fauvel-Lafeve, F., Karniguian, A., Gasset, G., Irinopoulou, T., Calvo, F., Rigaut, J. P. & Schoevaert, D. (2001). The effect of weightlessness on cytoskeleton architecture and proliferation of human breast cancer cell line MCF-7. *FASEB J.*, **15**, 1104–6.

Verworn, M. (1889a). Die polare Erregung der Protisten durch den galvanischen Strom. *Pflügers Arch.*, **45**, 1–36.

Verworn, M. (1889b). *Psychophysiologische Protistenstudien.* Gustav Fischer Verlag, Jena. pp. 25–130.

Vitha, S., Zhao, L. & Sack, F. D. (2000). Interaction of root gravitropism and phototropism in *Arabidopsis* wild-type and starchless mutants. *Plant Physiol.*, **122**, 453–61.

Vogel, K., Hemmersbach-Krause, R., Kühnel, C. & Häder, D.-P. (1993). Swimming behavior of the unicellular flagellate, *Euglena gracilis*, in simulated and real microgravity. *Micrograv. Sci. Technol.*, **4**, 232–7.

Volkmann, D., Baluska, F., Lichtscheidl, I., Driss-Ecole, D. & Perbal, G. (1999). Statoliths motions in gravity-perceiving plant cells: does actomyosin counteract gravity? *FASEB J.*, **13**, S143–S147.

Volkmann, D. & Sievers, A. (1979). Graviperception in multicellular organs, in *Encyclopedia of Plant Physiology. N.S.* Vol. **7** *Physiology of Movements*. Eds., W. Haupt & M. E. Feinleib. Springer-Verlag, Berlin, pp. 573–600.

Voss, K. (1990). Differentialgeometrie und digitale Bildverarbeitung. *Bild und Ton*, **43**, 165.

Votta, J. J. & Jahn, T. L. (1972a). Galvanotaxis of *Chilomonas paramecium* and *Trachelomonas volvocina*. *J. Protozool.*, **19**(Suppl.), 43.

Votta, J. J. & Jahn, T. L. (1972b). Galvanotaxis of *Euglena gracilis*. *J. Protozool.*, **19** (Suppl.), 43.

Wager, H. (1911). On the effect of gravity upon the movements and aggregation of *Euglena viridis*, Ehrb., and other micro-organisms. *Phil. Trans. R. Soc. Lond. B*, **201**, 333–90.

Wagner, G. (1998). The physiology of tropisms, in *Progress in Botany: Genetics, Cell Biology and Physiology, Ecology and Vegetation Science*, **59**, 396–428.

Wagtendonk, W. J. (1974). *Paramecium. A Current Survey.* Elsevier, Amsterdam.

Walcott, C., Gould, J. L. & Lednor, A. J. (1988). Homing of magnetized and demagnetized pigeons. *J. Exp. Biol.*, **134**, 27–41.

Walker, M. M. & Bitterman, M. E. (1989a). Attached magnets impair magnetic field discrimination by honeybees. *J. Exp. Biol.*, **141**, 447–51.

Walker, M. M. & Bitterman, M. E. (1989b). Conditioning analysis of magnetoreception in honeybees. *Bioelectromagnetics*, **10**, 261–75.

Walker, M. M. & Bitterman, M. E. (1989c). Honeybees can be trained to respond to very small changes in geomagnetic field intensity. *J. Exp. Biol.*, **145**, 489–94.

Walne, P. L. & Arnott, H. J. (1967). The comparative ultrastructure and possible function of eyespots: *Euglena granulata* and *Chlamydomonas eugemetos*. *Planta*, **77**, 325–53.

Walne, P. L., Lenci, F., Mikolajczyk, E. & Colombetti, G. (1984). Effect of pronase treatment on step-down and step-up photophobic responses in *Euglena gracilis*. *Cell Biol. Intern. Rep.*, **8**, 1017–27.

Walsby, A. E. (1987). Mechanisms of buoyancy regulation by planktonic cyanobacteria with gas vesicles, in *The Cyanobacteria*. Eds., P. Fay & C. van Baalen. Elsevier Science Publishers, New York, pp. 385–92.

Walter, I. (2002). Bioreactors and their applications in *Cell Biology and Biotechnology in Space*. Ed., A. Cogoli, Elsevier, Amsterdam, pp. 197–213.

Walter, M. F. & Schultz, J. E. (1981). Calcium receptor protein calmodulin isolated from cilia and cells of *Paramecium tetraurelia*. *Europ. J. Cell Biol.*, **24**, 97–100.

Ward, S. (1978). Nematode chemotaxis and chemoreceptors, in *Taxes and Behavior*. Ed., G. L. Hazelbauer. Chapman, London, pp. 141–68.

Watzke, D., Bräucker, R. & Machemer, H. (1998). Graviresponses of iron-fed *Paramecium* under hypergravity. *Europ. J. Protistol.*, **34**, 82–92.

Wayne, R. & Staves, M. P. (1996). A down to earth model of gravisensing or Newton's law of gravitation from the apple's perspective. *Physiol. Plant*, **98**, 917–21.

Wayne, R., Staves, M. P. & Leopold, A. C. (1990). Gravity-dependent polarity of cytoplasmic streaming in *Nitellopsis*. *Protoplasma*, **155**, 43–57.

Weaver, J. C. (1993). Electroporation: a general phenomenon for manipulating cells and tissues. *J. Cell. Biochem.*, **51**, 426–35.

Webb, A. A. R., McAinsh, M. R., Taylor, J. E. & Hetherington, A. M. (1996). Calcium ions as intracellular second messengers in higher plants. *Adv. Bot. Res.*, **22**, 45–96.

Weise, S. E., Kuznetsov, O. A., Hasenstein, K. H. & Kiss, J. Z. (2000). Curvature in *Arabidopsis* inflorescence stems is limited to the region of amyloplast displacement. *Plant Cell Physiol.*, **41**, 702–9.

Whalen, T. A. (1997). Volume and surface area estimation from microscopic images. *J. Microsc.*, **188**, 93.

Wichtermann, R. (1986). *The Biology of Paramecium*. Plenum, New York.

Wilczek, M. (2001). *Paramecium biaurelia* im niederfrequenten Magnetfeld: Auswirkungen auf das Schwimmverhalten, die cAMP-, cGMP- und 5′-Methoxytryptamin-Konzentration. *Dissertation*, Rheinische-Friedrich-Wilhelms-University Bonn.

Williams, N. E., Honts, J. E. & Stuart, K. R. (1989). Properties of microtuble-free cortical residues isolated from *Paramecium tetraurelia*. *J. Cell. Sci.*, **92**, 427–32.

Wiltschko, W., Munro, U., Ford, H. & Wiltschko, R. (1993). Red light disrupts magnetic orientation of migratory birds. *Nature*, **364**, 525–7.

Wiltschko, W. & Wiltschko, R. (1996). Magnetic orientation in birds. *J. Exp. Biol.*, **199**, 29–38.

Wiltschko, W., Wiltschko, R. & Munro, U. (2000). Light-dependent magnetoreception in birds: does directional information change with light intensity. *Naturwissenschaften*, **87**, 36–40.

Winet, H. (1973). Wall drag on free-moving ciliated micro-organisms. *J. Exp. Biol.*, **59**, 753–66.

Winet, H. & Jahn, T. L. (1974). Geotaxis in protozoa. I. A propulsion-gravity model for *Tetrahymena* (Ciliata). *J. Theor. Biol.*, **46**, 449–65.

Witman, G. B. (1993). *Chlamydomonas* phototaxis. *Trends Cell Biol.*, **3**, 403–8.

Wohlfarth-Bottermann, K. E. (1979). Oscillatory contraction activity in *Physarum*. *J. Exp. Biol.*, **81**, 15–32.

Wolke, A., Niemeyer, F. & Achenbach, F. (1987). Geotactic behavior of the acellular myxomycete *Physarum polycephalum*. *Cell Biol. Inter. Rep.*, **11**, 525–8.

Wolken, J. J. (1969). Microspectrophotometry and the photoreceptor of *Phycomyces* I. *J. Cell Biol.*, **43**, 354–60.

Wolken, J. J. & Shin, E. (1958). Photomotion in *Euglena gracilis*. I. Photokinesis. II. Phototaxis. *J. Protozool.*, **5**, 39–46.

Wood, D. C. (1982). Membrane permeabilities determining resting, action and mechanoreceptor potentials in *Stentor coeruleus*. *J. Comp. Physiol.*, **146**, 537–50.

Wortmann, J. (1887). Einige weitere Versuche über die Reizbewegungen vielzelliger Organe. *Ber. Dtsch. Bot. Ges.*, **5**, 459–68.

Wu, H.-S. & Barba, J. (1995). An efficient semi-automatic algorithm for cell contour extraction. *J. Microsc.*, **179**, 270–6.

Wu, T. Y. (1977). Hydrodynamics of swimming at low Reynolds numbers, in *Physiology of Movement – Biomechanics (Fortschr. Zool. 24)*. Ed., W. Nachtigall. Fischer-Verlag, Stuttgart, pp. 149–69.

Wunsch, C. & Volkmann, D. (1993). Immunocytological detection of myosin in the root tip cells of *Lepidium sativum*. *Europ. J. Cell Biol.*, **61**(Suppl.), 46.

Yamasaki, F. & Hayashi, H. (1982). Comparison of properties of the cellular and extracellular phosphodiesterases induced by cyclic adenosine 3′,5′-monophosphate in *Dictyostelium discoideum*. *J. Biochem.*, **91**, 981–8.

Yang, W. Q., Braun, C., Plattner, H., Purvee, J. & van Houten, J. L. (1997). Cyclic nucleotids in glutamate chemosensory signal transduction of *Paramecium*. *J. Cell Sci.*, **110**, 2567–72.

Yang, X.-C. & Sachs, F. (1989). Block of stretch-activated ion channels in *Xenopus* oocytes by gadolinium and calcium ions. *Science*, **243**, 1068.

Yentsch, C. S., Backus, R. H. & Wing, A. (1964). Factors affecting the vertical distribution of bioluminescence in the euphotic zone. *Limnol. Oceanogr.*, **9**, 519–24.

Yoshimura, K. (1994). Chromophore orientation in the photoreceptor of *Chlamydomonas* as probed by stimulation with polarized light. *Photochem. Photobiol.*, **60**, 594–7.

Yoshimura, K. (1996). A novel type of mechanoreception by the flagella of *Chlamydomonas*. *J. Exp. Biol.*, **199**, 295–302.

Zampighi, G. A., Kreman, M., Boorer, K. J., Loo, D. D. F., Bezanilla, F., Chandy, G., Hall, J. E. & Wright, E. M. (1995). A method for determining the unitary functional capacity of cloned channels and transporters expressed in *Xenopus laevis* oocytes. *J. Membr. Biol.*, **148**, 65–78.

Zhang, Y. & Hamill, O. P. (2000). On the discrepancy between whole cell and membrane patch mechanosensitivity of *Xenopus* oocytes. *J. Physiol. (Lond)*, **523**, 101–15.

Zhelev, D. V., Needham, D. & Hochmuth, R. V. (1994). A novel micropipette method for measuring the bending modulus of vesicle membranes. *Biophys. J.*, **67**, 720–7.

Zhenan, M. & Shouyu, R. (1983). The effect of red light on photokinesis of *Euglena gracilis*, in *Proceedings of the Joint China–U.S. Phycology Symposium*. Ed., C. K. Tseng. Science Press Beijing, China, pp. 311–21.

Zhulin, I. B. & Armitage, J. P. (1993). Motility, chemokinesis, and methylation-independent chemotaxis in *Azospirillum brasilense*. *J. Bacteriol.*, **175**, 952–8.

Zilker, A., Engelhard, H. & Sackmann, E. (1987). Dynamic reflection interference contrast (RIC-) microscopy: a new method to study surface exitations of cells and to measure membrane bending elastic moduli. *J. Physiol.*, **48**, 2139–51.

Zoeger, J., Dunn, J. R. & Fuller, M. (1981). Magnetic material in the head of the common Pacific dolphin. *Science*, **213**, 892–4.

Index

Printed in the United States
by Baker & Taylor Publisher Services